Mobile Health Technologies for Ambient Assisted Living and Healthcare

Mobile Health Technologies for Ambient Assisted Living and Healthcare

Editor

Ivan Miguel Pires

MDPI • Basel • Beijing • Wuhan • Barcelona • Belgrade • Manchester • Tokyo • Cluj • Tianjin

Editor
Ivan Miguel Pires
Instituto de
Telecomunicações,
Universidade da Beira
Interior
Covilhã
Portugal

Editorial Office
MDPI
St. Alban-Anlage 66
4052 Basel, Switzerland

This is a reprint of articles from the Special Issue published online in the open access journal *Sensors* (ISSN 1424-8220) (available at: https://www.mdpi.com/journal/sensors/special_issues/Mobile_Health_Tech).

For citation purposes, cite each article independently as indicated on the article page online and as indicated below:

LastName, A.A.; LastName, B.B.; LastName, C.C. Article Title. *Journal Name* **Year**, *Volume Number*, Page Range.

ISBN 978-3-0365-7026-6 (Hbk)
ISBN 978-3-0365-7027-3 (PDF)

Contents

About the Editor vii

Preface to "Mobile Health Technologies for Ambient Assisted Living and Healthcare" ix

Arielle Verri Lucca, Luís Augusto Silva, Rodrigo Luchtenberg, Leonardo Garcez, Raúl García Ovejero, Ivan Miguel Pires, et al.
A Case Study on the Development of a Data Privacy Management Solution Based on Patient Information
Reprinted from: *Sensors* **2020**, *20*, 6030, doi:10.3390/s20216030 1

Diogo Luís Marques, Henrique Pereira Neiva, Ivan Miguel Pires, Eftim Zdravevski, Martin Mihajlov, Nuno M. Garcia, et al.
An Experimental Study on the Validity and Reliability of a Smartphone Application to Acquire Temporal Variables during the Single Sit-to-Stand Test with Older Adults
Reprinted from: *Sensors* **2021**, *21*, 2050, doi:10.3390/s21062050 25

Tan-Hsu Tan, Jin-Hao Hus, Shing-Hong Liu, Yung-Fa Huang and Munkhjargal Gochoo
Using Direct Acyclic Graphs to Enhance Skeleton-Based Action Recognition with a Linear-Map Convolution Neural Network
Reprinted from: *Sensors* **2021**, *21*, 3112, doi:10.3390/s21093112 41

Rodrigo Pérez-Rodríguez, Elena Villalba-Mora, Myriam Valdés-Aragonés, Xavier Ferre, Cristian Moral-Martos, Marta Mas-Romero, et al.
Usability, User Experience, and Acceptance Evaluation of CAPACITY: A Technological Ecosystem for Remote Follow-Up of Frailty
Reprinted from: *Sensors* **2021**, *21*, 6458, doi:10.3390/s21196458 55

Parisa Fard Moshiri, Reza Shahbazian, Mohammad Nabati and Seyed Ali Ghorashi
A CSI-Based Human Activity Recognition Using Deep Learning
Reprinted from: *Sensors* **2021**, *21*, 7225, doi:10.3390/s21217225 73

Alexandre Neto, José Camara and António Cunha
Evaluations of Deep Learning Approaches for Glaucoma Screening Using Retinal Images from Mobile Device
Reprinted from: *Sensors* **2022**, *22*, 1449, doi:10.3390/s22041449 93

Carlos A. S. Cunha and Rui P. Duarte
Multi-Device Nutrition Control
Reprinted from: *Sensors* **2022**, *22*, 2617, doi:10.3390/s22072617 115

Maria Krizea, John Gialelis, Grigoris Protopsaltis, Christos Mountzouris and Gerasimos Theodorou
Empowering People with a User-Friendly Wearable Platform for Unobtrusive Monitoring of Vital Physiological Parameters
Reprinted from: *Sensors* **2022**, *22*, 5226, doi:10.3390/s22145226 139

Research Dawadi, Teruhiro Mizumoto, Yuki Matsuda and Keiichi Yasumoto
PATROL: Participatory Activity Tracking and Risk Assessment for Anonymous Elderly Monitoring
Reprinted from: *Sensors* **2022**, *22*, 6965, doi:10.3390/s22186965 161

Francisco José Melero-Muñoz, María Victoria Bueno-Delgado, Ramón Martínez-Carreras, Rafael Maestre-Ferriz, Miguel Ángel Beteta-Medina, Tomás Puebla-Martínez, et al.
Design and Development of a Heterogeneous Active Assisted Living Solution for Monitoring and Following Up with Chronic Heart Failure Patients in Spain
Reprinted from: *Sensors* **2022**, *22*, 8961, doi:10.3390/s22228961 . . . . **189**

Luís Pimenta, Nuno M. Garcia, Eftim Zdravevski, Ivan Chorbev, Vladimir Trajkovik, Petre Lameski, et al.
Can the Eight Hop Test Be Measured with Sensors? A Systematic Review
Reprinted from: *Sensors* **2022**, *22*, 3582, doi:10.3390/s22093582 . . . . **213**

About the Editor

Ivan Miguel Pires

Ivan Miguel Pires holds an MSc and a European Ph.D. from Universidade da Beira Interior (UBI, 2012 and 2018). Since April 2020, he has also been an Integrated Researcher at Instituto de Telecomunicações. Between 2012 and 2019, he worked as a developer and consultant in the industry. In 2019, he became a Post-Doctoral Researcher at UBI on a subject related to the monitoring of vehicles. Between October 2019 and September 2021, he was an Invited Adjunct Professor at the Polytechnic Institute of Viseu. Between October 2021 and July 2022, he was an Invited Assistant Professor at Universidade de Trás-os-Montes e Alto Douro. He was General (Co-)Chair of the EAI GOODTECHS 2021, EAI GOODTECHS 2022, and EAI MOBIHEALTH 2022 conferences. He is (co-)editor of four books, and he is (co-)author of seven book chapters, 62 journal papers, 50 conference papers, and 15 datasets. His research during his Ph.D. studies was related to developing a method for automatically identifying activities of daily living and the environment with the sensors available on a mobile device. During his Ph.D. studies, he also participated in different projects related to medical research, with the development of a preliminary technological method for the heel-rise test. He was co-supervisor of one MSc dissertation about developing a method for the timed-up and go test analysis. His research studies mainly use sensors and mobile devices for movement detection based on techniques such as sensor data fusion, sensor data imputation, and others.

Preface to "Mobile Health Technologies for Ambient Assisted Living and Healthcare"

The use of telemedicine and mobile devices is growing, and sensors might aid in creating creative solutions. Developing these solutions is crucial for monitoring senior citizens, lifestyles, and medical procedures.

This Special Issue's goal is to bring together academics and professionals in healthcare and medicine interested in using information and communication technologies (ICT) to serve people with special needs. The development of assistive technology for various users to follow sports and other activities is strongly tied to this study area. Data protection is crucial, and the development of these solutions for medical uses should be verified. The security and privacy of the information may be tied to other recognized research projects for their acceptability. ICT research has considerably improved quality of life and fully assimilated all citizens into society through medical rehabilitation and assistive technology. The technologies and research fields that influence medical informatics include databases, networking, graphical user interfaces, data mining, machine learning, intelligent decision support systems, and specialized programming languages.

Because mobile devices are commonly used for several everyday chores and are equipped with sensors that monitor various physical and physiological indicators, it is crucial to encourage the development of m-Health and e-Health solutions for healthcare practitioners. In this area, several solutions are now being developed. In addition, these can collaborate with emerging technologies for social assistance while enhancing life quality.

This book presents a collection of 11 studies related to the use of technological equipment for an improvement in quality of life, establishing its relation with other subjects. The editor would like to thank the authors for their interest in submission to this Special Issue, the reviewers that collaborated with the different reviews, the different people that helped in the promotion of this Special Issue, and the different institutions involved. This work is also funded by FCT/MEC through national funds and, when applicable, is co-funded by the FEDER-PT2020 partnership agreement under the project UIDB/50008/2020.

Ivan Miguel Pires
Editor

sensors

MDPI

Article

A Case Study on the Development of a Data Privacy Management Solution Based on Patient Information

Arielle Verri Lucca [1], Luís Augusto Silva [1,2], Rodrigo Luchtenberg [1], Leonardo Garcez [1], Xuzeng Mao [2], Raúl García Ovejero [3], Ivan Miguel Pires [4,5,6], Jorge Luis Victória Barbosa [7] and Valderi Reis Quietinho Leithardt [8,9,10,*]

1 Laboratory of Embedded and Distribution Systems, University of Vale do Itajaí, Rua Uruguai 458, C.P. 360, Itajaí 88302-901, Brazil; arielle@edu.univali.br (A.V.L.); luisaugustos@usal.es (L.A.S.); rodrigol@edu.univali.br (R.L.); leonardo.conceicao@edu.univali.br (L.G.)
2 Expert Systems and Applications Lab, Faculty of Science, University of Salamanca, Plaza de los Caídos s/n, 37008 Salamanca, Spain; xuzengmao@usal.es
3 Expert Systems and Applications Lab., E.T.S.I.I of Béjar, University of Salamanca, 37008 Salamanca, Spain; raulovej@usal.es
4 Instituto de Telecomunicações, Universidade da Beira Interior, 6200-001 Covilhã, Portugal; impires@it.ubi.pt
5 Computer Science Department, Polytechnic Institute of Viseu, 3504-510 Viseu, Portugal
6 UICISA:E Research Centre, School of Health, Polytechnic Institute of Viseu, 3504-510 Viseu, Portugal
7 Applied Computing Graduate Program, University of Vale do Rio dos Sinos, Av. Unisinos 950, São Leopoldo, RS 93.022-750, Brazil; jbarbosa@unisinos.br
8 Departamento de Informática da Universidade da Beira Interior, 6200-001 Covilhã, Portugal
9 COPELABS, Universidade Lusófona de Humanidades e Tecnologias, 1749-024 Lisboa, Portugal
10 VALORIZA, Research Center for Endogenous Resources Valorization, Instituto Politécnico de Portalegre, 7300-555 Portalegre, Portugal
* Correspondence: valderi@ipportalegre.pt

Received: 23 September 2020; Accepted: 20 October 2020; Published: 23 October 2020

Abstract: Data on diagnosis of infection in the general population are strategic for different applications in the public and private spheres. Among them, the data related to symptoms and people displacement stand out, mainly considering highly contagious diseases. This data is sensitive and requires data privacy initiatives to enable its large-scale use. The search for population-monitoring strategies aims at social tracking, supporting the surveillance of contagions to respond to the confrontation with Coronavirus 2 (COVID-19). There are several data privacy issues in environments where IoT devices are used for monitoring hospital processes. In this research, we compare works related to the subject of privacy in the health area. To this end, this research proposes a taxonomy to support the requirements necessary to control patient data privacy in a hospital environment. According to the tests and comparisons made between the variables compared, the application obtained results that contribute to the scenarios applied. In this sense, we modeled and implemented an application. By the end, a mobile application was developed to analyze the privacy and security constraints with COVID-19.

Keywords: data privacy; taxonomy; IoT; COVID-19

1. Introduction

Internet of Things (IoT) devices can be applied in various sectors, acting as a facilitating tool [1]. Devices may help monitor health conditions without the presence of healthcare professionals [2]. There are also wireless technologies that monitor older adults and remotely send data such as heart rate and blood pressure to their caregivers [3]. In addition to monitoring, other devices have auxiliary functions, such as automatic insulin injection devices [4]. These are directly linked to sensitive patient data and provide additional control in critical situations by, for example, setting the dose to be injected into the insulin pump. Both privacy settings and control information must have an extreme level of security.

For hospital environments, IoT devices are distributed not only for patient use but also for other functionalities. According to Farahani et al. [5], some of the IoT applications used in hospital settings collect patient data, such as heart rate, blood pressure, or glucose level. As far as the environment is concerned, some sensors detect temperature changes or control the air conditioning; cameras are used to detect intruders and send alerts. In this context, the devices' scope ranges from patient monitoring to evaluate the environment and the equipment used by health professionals. Thus, the data is recorded from the moment that patients are registered at the reception until they are discharged.

When the patient is registered for admission, basic information is collected and complemented after screening. In a first-aid environment, to ensure all patients' safety, many hospitals use a screening technique known as the Manchester Protocol [6]. After screening, the information is added to the patient's record. Next, the person is given a classification according to their condition; this varies from non-urgent cases to emergency intervention cases. Sensitive information is added to the user record, whose preservation and confidentiality level must be treated as critical. There is information that should not be disclosed or related to the patient, as is the case with a patient suspected of having viral and infectious diseases.

The current pandemic of Severe Acute Respiratory Syndrome Coronavirus 2 (COVID-19 SARS-CoV-2) causes the patient to be identified as a possible carrier even during the screening process, based on certain symptoms. According to Rothan and Siddappa [7], those infected usually show symptoms after approximately five days, the most common signs of illness being fever, cough, and fatigue; the patient may also present headaches, phlegm, hemoptysis, diarrhea, shortness of breath, and lymphopenia. These symptoms are identifiable without specific examinations that are directly documented in the patient's medical record. Liang et al. [8] mention that for most patients diagnosed with COVID-19, 85.7% had fever, 42.9% had cough, 33.3% had expectoration, 57.1% had fatigue and 38.1% had headache and dizziness. For this reason, one can see that fever is a common symptom. Thus, this condition must be checked as soon as the patient is admitted to the hospital. Due to COVID-19's high rate of contagion, the patient's referral to medical care and subsequent isolation should be done quickly and strictly in confirmation.

When it is confirmed that the patient has a COVID-19 infection, this information is directly linked to their record, which should remain confidential. Soares and Dall'Agnol [9] comment that privacy is considered an individual right that includes the protection of the intimacy of the subjects, respect for dignity, limitation of access to the body, intimate objects, family and social relationships. In addition, in this same bias, the concern also covers the complete information collected during the patient care process. Even though patients' data must be confident among all parties in general, due to the current pandemic situation and contagion rate, an extra precaution must be taken to join the statistics without having their information revealed. The application of privacy on patient data must be given to all levels with access to any information, be it registration, device, or image.

The main purpose of this work is to apply privacy constraints in patients with suspected COVID-19. The basis for the application of privacy is the same for patients in general, but using as a basis the fact that it is a pandemic situation, and the discretion in handling data of a suspected patient is crucial. Also, as it is a highly contagious virus, the process from admission to the emergency room to the patient's referral must be done quickly. In this way, a taxonomy was proposed that covers four topics and five subtopics regarding the entities/environments participating in the hospital admission process.

The scientific contribution of this paper is a system to support the privacy constraints related to COVID-19. It started with the study of the state-of-the-art in hospital environment. Next, we defined a taxonomy, and a mobile application was implemented to test and validate the use of the mobile application to cover the privacy constraints defined in the taxonomy.

The main results of this study are related to the identification of the users. Cryptography methods were implemented control the users according to the diagnosis of COVID-19. As these data are related to health, it must be secure and anonymous. The data collected included reliable data related to temperature parameters for the detection of the symptoms, such as fever.

For a better understanding of the matter and a clearer overview of the relevant details, this work is organized as follows: Section 2 lists the related works; Section 3 describes the taxonomic definition developed for this project and the attributes of the user parameter, environment, privacy, and device; Section 4 demonstrates the modeling of the project, including the use cases, sequence and context diagrams; in Section 5, we present the prototype with the application developed to be validated. Section 6 presents experiments and results. Finally, in Section 7, we conclude and discuss the future work.

2. Related Work

Studies on the application of privacy in hospital settings cover different aspects. Various studies were selected to identify privacy targeting, including encryption, profile privacy, device privacy, and taxonomic definitions. The focus among the related papers vary from studies on security over mobile application to systems conceived to protect user privacy.

Barket et al. [10] present a broad study on the context of privacy, developing a taxonomy meant to connect privacy and technology based on the following aspects: purpose, visibility, and granularity. According to the authors, the aim is related to why the information is requested; depending on the cause, more or fewer details about the user are passed on. Visibility refers to who is allowed to access the user data. Granularity designates the data transfer required for the type of access and purpose for that particular request.

The work of Asaddok et al. [11] involves mobile devices in the area of health (Mobile Health (mHealth) and the parameters: usability, security, and privacy. The authors propose a taxonomy that involves the three parameters mentioned, and, for each, it branches into taxonomies. One taxonomy is defined by usability, effectiveness, efficiency, satisfaction, and learning. Next, for security, confidentiality, integrity, and availability is restricted to another taxonomy. Finally, for privacy, identity, access, and disclosure, the last taxonomy is defined.

Coen-Porisini et al. [12] describe a conceptual model for defining privacy policies that cover the user, the user's profile, the information, and the action that will be taken by a third party to request the information. The authors revealed the link between the three topics mentioned in a Unified Modeling Language (UML) format. The user is divided into personnel—the person to whom the data is referred; processor—the person who will request the data; controller—the person who controls the actions requested by the processor. Data is divided into: identifiable—in situations when it is clear who the data refers to, such as the name; sensitive—it refers to information, processing, and purpose. We can also observe that there is an interaction between the medical user and the controller, along with the processes of access (processing), treatment (purpose), and communication (obligation). The diagram demonstrates how information is delivered to the medical user through requests, based on their access profile.

Silva et al. [13] use a notification management system focused on user privacy in this context. It contributed to the development of an application that can handle different types of notifications. Moreover, the network made it possible for those involved to ensure that the messages sent and received followed the rules defined earlier. If applied to health notifications or to alert cases of COVID-19, this is a strategic tool, addressing messages with defined priorities while also linking privacy in the traffic sent. Therefore, this work contributes to finding a link between IoT requirements and definitions. In [14], the authors implemented a system for monitoring and profiling based on data privacy in IoT. From the results obtained in the tests, they identified different profiles assigned to random situations. In this case, the health system user's profile priorities would apply and determine which profiles would be authorized to receive data. In this work, it was also possible to address the evolution and reduction of the hierarchy based on factors that identify users' frequency in the environments tested.

Concerning the relationship between data privacy and its use in situations such as the COVID-19 crisis, Zwitter et al. [15] deals with the basic concept of human rights that relates data privacy with the need to use certain information, such as someone's location. The authors mention features of applications developed by China, South Korea, and the United States that use tracking techniques to indicate close contact with virus carriers or identify specific individuals or groups' movements. The study concludes that location data is important in the fight against the spread of the virus, but other relevant information, such as genetic data, should be considered. It is necessary to use this information correctly, as stipulated by the law. It also states that data sensitivity classification is contextual; data protection and privacy are important and must be maintained even in crisis. Information leaks are inevitable, so organizations should always protect themselves; ethics in data manipulation is mandatory for more efficient analysis.

Yesmin et al. [16] deal with the privacy of patients' data in terms of the interoperability of systems and the employees' access to information. Also, they tell us that there is no framework for evaluating privacy audit tools in hospitals yet. The application of a framework would help identify any trend in accessing the data and allow the hospital to improve its performance in detecting possible data leaks. According to the authors, the literature reveals that the most significant leakage of information occurs through employees (nurses, doctors, sellers, and others). An evaluation framework was then developed and tested using the black box concept, which uses usability testing information. The following must be monitored through machine learning or artificial intelligence tools: employee access to information, validation of entry and non-standard behavior, and unexplained access to files.

The work of Islam et al. [17] deal with a survey on the application of IoT devices in the health system. The authors deal with the IoT network's topology for health, which facilitates the transmission and reception of medical data and enables data transmission on demand. They also mention features of wearable devices, which capture and store patient data. These may include blood sugar levels, cardiac monitoring, body temperature, and oxygen saturation. The authors explain that the security requirements applied to healthcare IoT equipment are similar to those of other communication scenarios. Therefore, the following must be considered: confidentiality, integrity, authentication, availability, data update, non-denial, authorization, resilience, fault tolerance, and fault recovery.

Sun et al. [18] designed the HCPP (Healthcare System for Patient Privacy) system to protect privacy and enable patient care in emergency cases. The entities defined for the system are the patient, the doctor, the data server, the family, the personal device, and the authentication server. According to the authors, the system meets the following security criteria: privacy, data preservation by backup, access control, accountability, data integrity, confidentiality, and availability.

Samaila et al. [19] developed a survey in which information was collected regarding work on security and privacy in IoT in general. The study's scope ranges from security, encryption, communication protocols, authentication to privacy, among others. The authors also collected information on applications, reliability, and other technical issues, combining ten related works. Additionally, the authors claim that the work covers a system model, a threat model, protocols and technologies, and security requirements. The work discusses the IoT architecture considering nine application domains: home automation, energy, developed urban areas, transport, health, manufacturing, supply chain, wearables, and agriculture. Security measures and system and threat models were defined for each application domain, including protocols and communications. The security properties covered were confidentiality, integrity, availability, authenticity, authorization, non-repudiation, accountability, reliability, privacy, and physical security. These also describe mechanisms that can be applied to achieve the desired security requirements: authentication, access control, encryption, secure boot, security updates, backup, physical security of the environment, and device tampering detection.

Plachkinova, Andrés and Chatterjee [20] elaborated a taxonomy focused on privacy over mHealth apps. Downloadable apps through the app store do not have a unified way to provide terms of use or privacy policies for the user. Apps mostly communicate between patients and doctors, access to patient medical records, self-diagnosis based on symptoms, etc. The management of user data after the app is installed may not be precise. The authors elaborated a taxonomy that embraces the following three dimensions: mHealth app (patient care and monitoring; health apps for the layperson; communication, education and research; physician or student reference apps), mHealth security (authentication; authorization; accountability; integrity; availability; ease of use; confidentiality; management; physical security) and mHealth privacy (identity threats; access threats; disclosure threats).

Alsubaei, Abuhussein, and Shiva [21] proposed a taxonomy aiming to enhance security among IoT medical devices, as it has life-threatening risks when a device is not secure. According to the authors, since security and privacy are becoming challenging due to the sensitivity of data in healthcare, it is crucial to enhance these measures. The taxonomy is based on the following topics: IoT layer, intruders, compromise level, attack impact, attack method, CIA compromise, attack origin, attack level, and attack difficulty. For each topic, some subsections embrace items from that topic. Since new attacks are always being created, this taxonomy can be updated, according to the authors. The related works we have selected cover the topics that we cited as critical to privacy. Some applied cryptography in the study as a reference of types of attacks, and others used cryptography to prevent data from being accessed from third parties. Most of them applied user profile privacy to prevent any unauthorized access or mitigate when it happens.

Data encryption is necessary so that in the event of an attack, a third party cannot gain access to information [22]. Cryptography is part, directly, from [17,18]. Islam et al. [17] mentioned cryptography among security threats, where cryptographic keys can be stolen to collect user sensitive data. The work of Sun et al. [18] mentioned encryption as a way to protect health information and applied identity-based cryptography for encryption, authentication, and deriving shared keys for their Healthcare system for Patient Privacy (HCPP) protocols. Also, they made use of searchable symmetric encryption to return encrypted documents to the owner.

The application of private profile was mentioned in all works, except by [21]. The user's profile privacy serves to protect any information from being used by third parties [23]. A security layer should be applied at the device level to prevent third parties from accessing information or even gaining control of it [24]. The work of Alsubaei, Shiva, and Abuhussein [21] mentions about attacks that influences on Confidentiality, Integrity and Availability (CIA) triad, which is a basic thread on privacy, but does not explore ways to protect user privacy concerning data access based on authorization. Barker et al. [10] are concerned about private profile through who can access the data and which data can be accessed, based on the purpose of this access request.

Asaddok and Ghazali [11] defined data access based on access to patient identity information, personal health information, and personal health records, moreover defined in their taxonomy as identity, access, and disclosure. Coen-Porisini et al. [12] say that data access must be based on access control based on the users and their roles. Thus, data access must be granted based on a consent given by the patient. Silva et al. [13] defined their privacy requirements based on the user permissions, environment, and hierarchy. Leithardt et al. [14] proposed a middleware in which the user's permission can be changed due to the environment and the frequency in which the user frequent it. This way, the given information will vary based on this environment, and the rules of its context.

Zwitter and Gstrein [15] say that data collection and its use must be done concerning the principle of proportionality and individual's interests. Their work is based on data collected over the individual's location and genetic data. Thus, the authors exposed user data principles as: sensitivity, privacy and protection, breaches precaution, ethics. The study of Yesmin and Carter [16] was concerned about the patient data through authorized and unauthorized access. The authors developed a framework that audits this access, although the study was limited as real patient information could not validate the tool. Instead, they used real data and could evaluate the amount of unauthorized/unexplained accesses to the patient's data.

Islam et al. [17] treated data with CIA triad, so that confidentiality is related to the medical information and its protection against unauthorized users. Their study gathered information on various aspects related to the use of IoT devices in medical care. Thus, they say that policies and security measures must be introduced for data protection when sharing data with users, organizations, and applications. Sun et al. [18] combined cryptography with user privacy and their trust relationship with entities, such as family members, physicians, or his device. Thus, these entities are allowed to access the patient's protected health information. In Plachkinova, Andrés, and Chatterjee [20], the authors studied mHealth apps and the concern about the use of information, terms of use, and privacy policies. The authors mentioned that it is not clear how the data is managed, neither who gets access to it. They developed a taxonomy in which user data is part of the identity threats, access threats, and disclosure threats.

The concern for privacy regarding the device was found in most papers. In Alsubaei, Shiva, and Abuhussein [21], the IoT device is part of the proposed security taxonomy. As their work concerns about mHealth devices, it is part of the proposed taxonomy's wearable devices, which embraces numerous sensors. The authors describe potential attacks for these devices, as side-channel, tag cloning, tampering devices, and sensor tracking. In the work of Asaddok and Ghazali [11], the authors classified mobile devices as part of the application dimension of the taxonomy, present in the topic 'patient care and monitoring', as they are used for observation of the patient.

The work of Silva et al. [13] applies privacy over mobile devices regarding aspects such as the environment. Thus, privacy used on mobile devices is part of their taxonomy and a base point of their study. Leithardt et al. [14] are guided on device privacy. This topic is the central part of their work. Zwitter and Gstrein [15] mention mobile devices, although their concern focuses on apps and location data, not the device itself. Islam et al. [17] treat devices like mobile, connected to the Internet through IoT providers. Thus, they are vulnerable to security attacks, which may originate within or outside the network. The authors mention that IoT health devices are part of an attack taxonomy, including information, host, and network. Sun et al. [18] define the Private Device (P-device) as an entity involved in the HCPP system, such as smartphones or wearable devices. The patient uses the P-device to manage privileges on access to his health data. In Plachkinova, Andrés, and Chatterjee [20], the device must be secured, as it can leak data about the location or sensor of the patient. As the apps mentioned in their work fail to provide accurate data management information, the device can be a tool for misusing information.

The use of the data acquired from different sensors needs the implementation of several privacy and security rules. In [25] is presented a low-cost system that embeds the measurement of temperature, heart rate, respiration rate and other parameters to define the health state of the person. This system performs the networking with the healthcare professional to prevent several situations. In addition to these sensors' data, it includes the tracking of the location of the user to present several contagious. This system may be used for a preliminary diagnosis. Mobile devices are capable of acquiring different types of data in several conditions. Spain was one of the fustigated countries with this pandemic's situation, and the authors of [26] proposed the implementation of online sensing networks to provide social quarantine and reduce the contagious with the virus.

The monitoring of the COVID-19 needs the use of secured technologies, and the IEEE 802.11ah technology was used in [27] to support the prevention of the contamination with COVID-19. It can be implemented in telemonitoring technologies to provide reliable information and prevent the contact. The network should previously know which are the persons that are contaminated with the virus. The tacking of the location and movements may be performed with location, inertial, and proximity sensors that communicates the data to social networks to reduce the social contact with infected individuals. The authors of [28] studied different privacy constraints related to the real-time monitoring with the mobile devices. The monitoring with mobile devices can be considered to be a digital vaccine that help in the reducing number of contagious with massive sharing of the data.

The creation of a taxonomy was proposed by most of the related works. Alsubaei, Shiva, and Abuhussein [21] proposed a taxonomy regarding IoT layer, intruder type, compromise level, impact, attack method, CIA compromise, attack origin, attack level, and attack difficulty. As can be seen, the taxonomy embraces the security and privacy aspects of medical IoT devices. Barker et al. [10] explored three dimensions to develop a taxonomy, based on visibility, granularity, and purpose. These three dimensions focus on privacy aspects, where visibility deals with who is permitted to access the data. Granularity is focused on the characteristics of that data to direct it to the appropriate use and a dimension that deals with the data's purpose. In the work of Asaddok and Ghazali [11], the authors developed a taxonomy containing usability, security, and privacy aspects to mHealth applications. Each item of the taxonomy is derived in three or more sub-items. Silva et al. [13] developed a taxonomy or notifications on mobile devices, including communication protocols, message transmission technologies, privacy, and criteria. Plachkinova et al. [20] proposed a taxonomy for mHealth apps regarding security and privacy. The items involve app dimension, security dimension, and privacy dimension.

Table 1 presents a comparison with the related works concerning the application of the privacy aspects described above with the additional taxonomy application.

Table 1. Scope of related works

Work	Cryptography	Private Profile	Devices	Taxonomy
[12] (2007)		•		
[10] (2009)		•		•
[20] (2015)		•	•	•
[17] (2015)	•	•	•	
[18] (2015)	•	•	•	
[11] (2017)		•	•	•
[21] (2017)			•	•
[13] (2019)		•	•	•
[14] (2020)		•		•
[15] (2020)		•	•	
[16] (2020)		•		
[25] (2020)		•	•	
[26] (2020)		•	•	
[27] (2020)		•	•	
[28] (2020)		•	•	
Proposal	•	•	•	•

Even though cryptography is one of the main concerns when dealing with data privacy, descriptions of how to apply it were found explicitly only in the works of Islam et al. [17], Sun et al. [18] and Riza and Gunawan [27]. As Horst Feistel [29] said almost 50 years ago: "personal data needs protection, which can be achieved when enciphering the material". Cryptography will prevent the plaintext from being accessible to people who are not authorized to have it, whereas it is an important tool when dealing with personal data. The work of Islam et al. [17] comprises a survey of IoT in health care, including analysis regarding the security and privacy aspects. However, the authors did not expose how cryptography can be applied, instead, mentioned that some parts of the flow can be tampered by attackers to obtain the cryptographic secrets. This way, IoT systems should be designed with protections against stealing of cryptographic keys.

The work of Sun et al. [18] is focused on cryptography, as it describes a system based on this aspect. The authors designed protocols for a healthcare system in which the security aspect leverages on cryptographic tools. The HCPP allows the patient to store their medical record even on public servers, where only the patient can retrieve the information. The patient's medical record is encrypted to ensure privacy, and its content can only be retrieved by the patient and the physician when some treatment is being carried out. If by any means the patient is unable to retrieve the medical record, the system can provide the relevant information to the physician without compromising the patient's secret key. In our work, cryptography is used to prevent unauthorized access to the patient's medical records. As it can be seen in our proposed taxonomy in Figure 1, cryptography is part of the User's items, as it is a critical tool to protect the patient data. The patients' medical records should be stored and transmitted in encrypted ways, in a way that only the personnel who has the authorization and, therefore, the secret keys, can decrypt the data. Therefore, patients' medical records are encrypted and can only be accessed by the authorized staff.

In comparison to the selected works, ours stands out because it includes the indication of encryption, profile privacy, concerns on device, and the definition of the taxonomy meant to define the theme and scenario of the application more clearly. Our taxonomic definition aims to embrace the necessary aspects to be covered to enhance security measures throughout the patient's sensitive data. We developed a mobile application to validate the data flow of information from the moment patients are being admitted in the hospital until they are discharged.

The use of a mobile application that implements data privacy parameters related to the data of patients infected with COVID-19 is another contribution of this study. The data of patients may be its location, temperature, history of navigation, among others. Therefore, we consider that contagion can be identified in the first moments spent in the emergency room using basic information on the health status and the monitoring of the feverish state with the use of IoT devices. The degree of privacy applied in each user's registration process should enable identifying infected patients without the exposure of sensitive data.

To this end, we have developed a taxonomy that highlights how important it is for confidential information to be handled with care. We have included examples of privacy applications in the use of IoT devices to receive, screen, and providing patient care with a focus on the COVID-19 pandemic.

3. Taxonomy

We have developed a taxonomic definition for a better classification of the items related to the privacy parameters. A taxonomy is necessary to identify the critical aspects where security measures and policies need to be applied. Based on the goals of this paper and the comparisons made with the related works, we selected the principal parameters to manage privacy, which are divided into other levels to better embrace the desired security aspects.

As presented in works [13,30,31], a taxonomy allows the systematic organization of relevant data in the form of a hierarchy. The keywords and concepts used to define a taxonomy establish parameters throughout the information production cycle, in which distributed professionals can participate in the knowledge creation process in an organized way. This definition covers four parameters for managing privacy standards in hospital settings within the previously defined context. The selected parameters with five attributes were considered necessary for this scenario. Figure 1 shows the taxonomic definitions proposed in this paper.

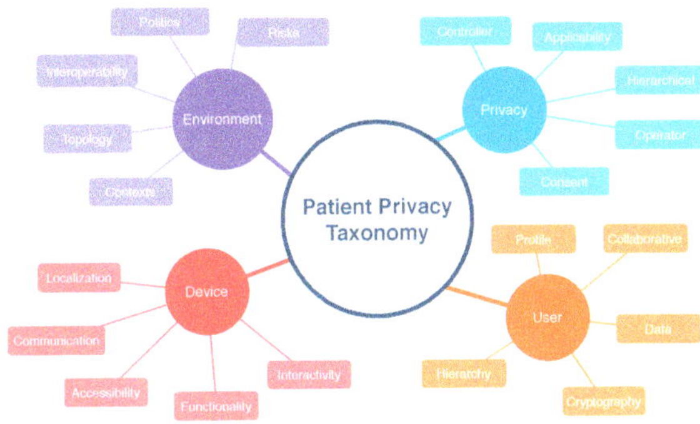

Figure 1. Proposed taxonomy.

3.1. *User Parameter*

The user parameter designates the person who provides, controls, or operates the sensitive data used in privacy handling. This parameter refers not only to the patient but also to the participants in the data's provision or control. For this parameter, we set the following attributes: profile, collaborative, hierarchy, cryptography, data. The profile attribute covers several items that will be part of the process. According to Fengou et al. [32], six entities participate in interactions taking place in the hospital environment:

- the patient himself/herself;
- the clinical network that will care for the patient, including doctors, family members, volunteers, health insurance provider, among other things;
- the hospital;
- smart home as an environment with ubiquitous equipment's capable of providing security and quality of life;
- the environment in which the patient works, the vehicle with which the patient is transferred to the clinical center.

Based on the entities listed, it can be observed that the user profile is one that must be substantiated, along with the profiles of other entities. The patient's cooperativeness in providing their registration data is fundamental for a better experience in the given setting. According to Leithardt [33], the user must provide access to their information and services, thus favoring both their expertise in using the service and the system's improvement whole. The hierarchy enables proper separation of the levels and permissions of each user type. Viswanatham and Senthilkumar [34] proposed the so-called hierarchy-based user privacy, where the information is encrypted and decrypted based on access levels and releases.

The General Data Protection Regulation (GDPR) deals with the need to protect confidential data and the inevitable risk of data theft. Encryption reinforces that all sensitive information must be covered by an acceptable security level, either at its source or at its destination. Ibraimi et al. [35] said that patient confidentiality is one of the significant obstacles in obtaining medical data, as some information is not shared for fear of it being saved in databases that do not comply with security regulations. The protection of sensitive patient information is an essential task. The Department of Health and Human Services, 2002 (HIPAA) Privacy Standard [36] deals with the security of sensitive patient information in the medical field. It is a US federal law created in 1996 to impose standards for protecting such information and preventing it from being shared without the patient's consent. Cooper et al. [37] deal with privacy and security in data mining in the medical field and cites HIPAA in information privacy matters. In 2002, they suggested that protective measures be imposed by health plans, clinical centers, and other entities involved.

3.2. Environment Parameter

The environment parameter represents the smart physical location where user data will flow between different systems and devices. For this parameter, we define the following attributes: topology, interoperability, policies, risks, hierarchy. Topology refers to the architecture of a hospital environment. Costa [38] comments that hospitals used to be built with an emphasis on the utility of the building and the technique used. The health field's processes and dynamics are often determined by how the wards, sectors, and departments that house distinct functions are arranged. In many of the methods that occur during the patient's journey through the emergency room, one or more systems are used.

Interoperability between systems is strongly present in the medical field presently. According to Lopes [39], strategies used to be designed and developed from an internal perspective of organizations, with no motivation for integration with other systems. In all the smart environments that people transit, data is shared between information systems and IoT devices. The data are a vital part of the operation of a health institution. Several policies need to be established to apply access security to these environments and define what data will be exchanged between systems and devices. According to Yildirim et al. [40], information security management is an activity that aims to implement a set of policies that help to define an acceptable level of security in these environments, minimizing the potential risks inherent in the exploitation of this information.

Risk management in hospital settings is a crucial activity for the proper functioning of the operation. According to Florence et al. [41], the risk is an estimated value that considers the probability of occurrence of damage and the severity of said damage. Therefore, procedures are meant to minimize those factors that need to be mapped, controlled, and defined. The dimension in the patient's care is large and complex. It occurs at various times and in multiple environments in the course of service, along with several interactions between the patient, other participants, and technologies. Soares et al. [9] emphasize that due to its characteristics and complexity, the hospital environment favors establishing power and asymmetrical relationships between the nursing team and patients. The asymmetry results from the patients' fragility and vulnerability in the face of health-diseases processes.

3.3. Privacy Parameter

The privacy parameter designates how each piece of information will be handled according to its characteristics. For this parameter, we define the following attributes: communication, applicability, controller, consent, operator. The transmission is linked to the type of user profile and will usually involve unsafe transmitting the information. According to Machado [42], anonymization or encryption in particular pass through the means of communication, i.e., the very existence of communication drives the need to apply security measures to data. It is a basic human right to have one's sensitive data handled with care. Thus, its applicability is significant. The General Data Protection Law (LGPD) [43], as the Brazilian Data Protection Law, aims to apply standards and laws regulating and protecting individuals' data. Without this application of standards and regulations, sensitive information could easily be used by those who should not have access to it in the first place.

A categorization determines who has the authority to decide the type of treatment that personal data will be submitted. As mentioned in the LGPD [43], the controller must obtain the consent of the individual owner or holder of the concerned data. The user may, in turn, deny or grant access to their information by a third party. The user must give their consent, a manifestation by which they agree that their information be used in a specific way for a particular purpose. As mentioned in the LGPD [43], if the controller wishes to use this data at another time, consent will be requested once more. The operator shall be responsible for carrying out the data processing determined by the controller. As mentioned in the LGPD [43], the operator is jointly and severally liable for the damages caused by data handling if the strategy does not comply with legal provisions or is not in line with the controller's instructions. The user provides their consent, and the operator is responsible for processing the information made available when for personal use or transfer to third parties.

3.4. Device

The device represents either the IoT equipment present in the smart environment that will interact with the patient's data or the wearable IoT device that will be set to monitor the patient's temperature. There may be devices that are fixed in the environment, such as surveillance cameras or devices that can be used to monitor the patient, which can be fixed or mobile. For this parameter, we define the following attributes: function, location, communication, accessibility, interactivity. The device must meet the needs of the process to which it will be directed. According to Lupiana and O'Driscolle Mtenzi [44], one of the relevant requirements for devices is their storage and processing capacity. The location attribute refers to the location where the device is installed. For Leithardt [14], the attribute that controls location must be linked to a database where all user data must be included. This database will be accessible only for updating and validating some data. The other information should be processed from the point where the user has accessed the system to provide greater security and reliability. Figure 2 shows both fixed and wearable IoT devices and how the parameters are applied. For both fixed and wearable IoT devices, all five parameters are used. The last column on the Figure 2 shows some of the possible options for each attribute.

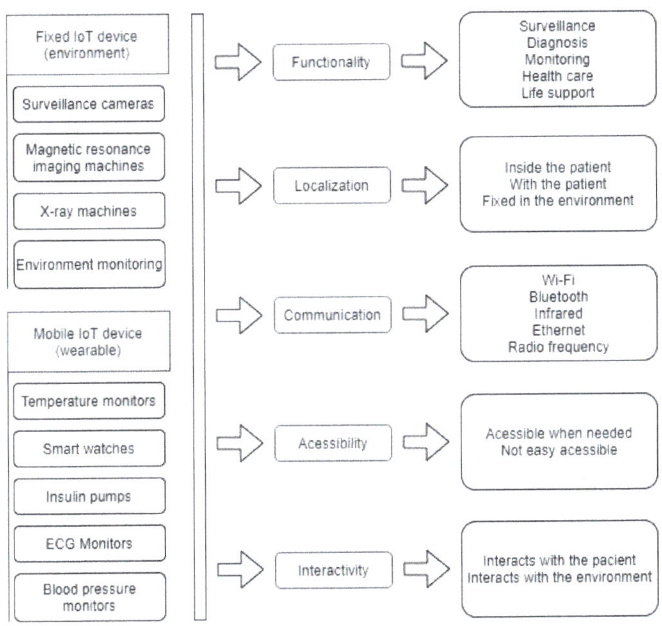

Figure 2. Devices with its parameters.

The way the device communicates with the user is addressed through the communication attribute and fits in heterogeneity, a feature that ensures information is handled evenly. According to a study presented by Pradilla, Esteve, and Palau [45], the devices are responsible for taking data acquisition through sensors, supporting data treatment with processing units, and acting in conjunction with IoT. Therefore, it is necessary to use heterogeneity in the communication protocols handled by the device and the number of services and types available. This attribute is associated with the protocols of the device, providing security in data transfer. The possibility to access the device whenever necessary is crucial, and interactivity between the device and the client must be ensured. With this in mind, we have developed a model based on the characteristics and functionalities defined in the described taxonomy.

4. Project Modeling

The model consists of use case diagrams, sequence diagrams, and context diagrams. All these notations are based on UML. The described model refers to the process from the patient's arrival at the hospital until his discharge.

4.1. Use Cases Diagrams

The first use case represents the entry of a patient into the emergency room. The patient interacts with the receptionist and performs some procedures. This use case includes some of the attributes of the proposed taxonomic definition: privacy, represented by the data which the patient grants access to and is registered in the systems; user, represented by the patient and the receptionist; environment, represented by the emergency room, shown in Figure 3.

After first care and registration, the transfer of the patient to the screening area is demonstrated in the use case pictured in Figure 4. The screening process aims to establish the urgency of the case and the risk classification. This use case includes some of the attributes present in the proposed taxonomic definition: privacy, represented by the data which the patient grants access to and is registered in the systems and the wearable IoT device; user, represented by the patient and the nurse; environment, represented by the screening room; device, represented by the wearable IoT device that will receive an identification to record the data and the classification of this patient.

And the last use case represents the patient being attended to by the doctor in the office after going through the screening process. The wearable IoT device identifies the patient so that the data is made available, and the doctor proceeds with the consultation. The doctor performs the anamnesis and records the data in the Electronic Health Record (EHR). This use case uses some of the attributes of our taxonomy as follows: privacy, represented by the data which the patient grants access to and is registered in the systems; user, represented by the patient and the doctor; environment, represented by the office; device, represented by the wearable IoT device used by the patient. This case is illustrated in Figure 5.

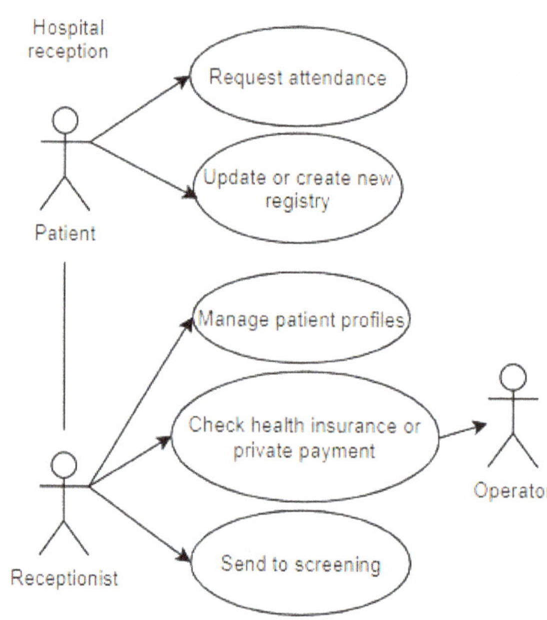

Figure 3. Reception at the Emergency Room.

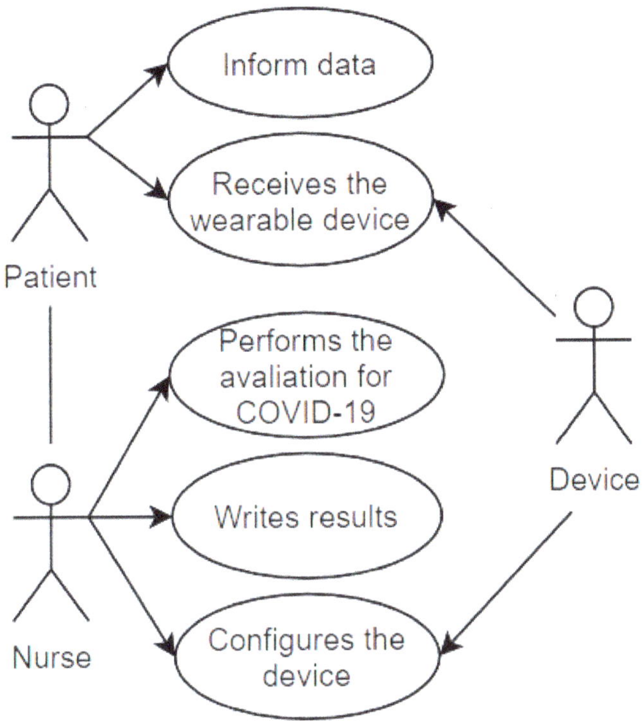

Figure 4. Screening Room.

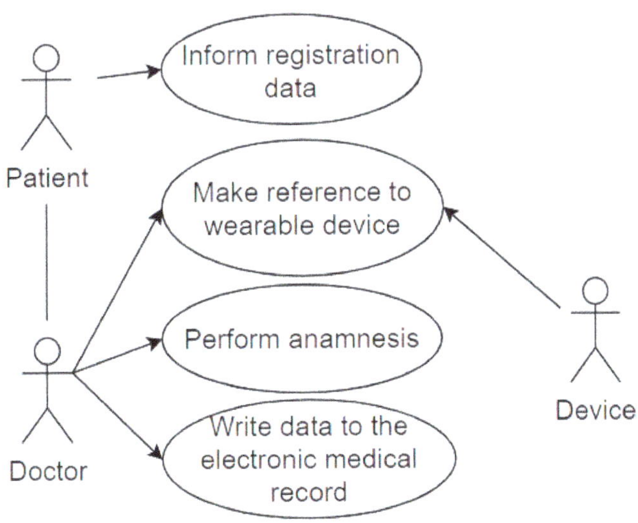

Figure 5. Reception at the Office.

Sequence diagrams of each use case were also developed. Sequence Diagram is a UML tool used to represent interactions between objects in a scenario, performed through operations or methods.

4.2. Sequence Diagrams

The sequence diagram displayed in Figure 6 represents the entry of a patient into the emergency room. It demonstrates the arrival of the patient (user) to the emergency room (environment), where they request assistance from the receptionist (user). The receptionist provides a password to the patient waiting to be called on. Upon being called on, the patient offers data for registration updates (privacy) recorded by the receptionist in the hospital system. The receptionist checks if the patient has a health plan and then records how this service's billing issue will be managed. After this procedure, the patient will be referred to as screening.

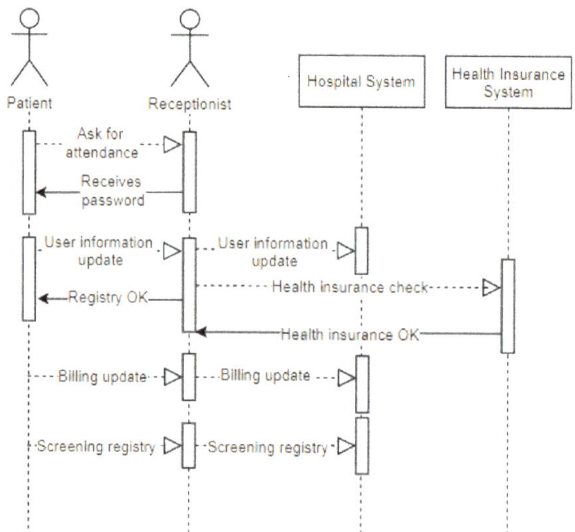

Figure 6. Sequence—Reception at the Emergency Room.

The sequence diagram displayed in Figure 7 represents the patient's entry into the screening room after completing the first stage in the emergency room. It means the arrival of the patient (user) to the screening room (environment), where they will convey their data as requested by the nurse (user). The nurse records the hospital system's data and the entry into the system that configures the wearable IoT device that will monitor the patient (device). The receptionist then hands the wearable IoT device over to the patient and starts the assessment. The patient answers the questions (privacy), and the nurse records all the information in the hospital system. All patient data is in the system, and the hospital from their wearable IoT device can track it.

The sequence diagram displayed in Figure 8 shows the patient's entry into the office after going through the screening process. It represents the arrival of the patient (user) to the office (environment), where they will convey their identification data as requested by the doctor (user). The latter records the electronic record data and refers to the patient's wearable IoT device in the hospital system. The doctor performs the anamnesis on the patient, who must answer the questions (privacy). The doctor also records this information in the patient's electronic record. The patient has already been attended to, so they are drugged and released or referred to another hospital ward based on the clinical condition's evolution.

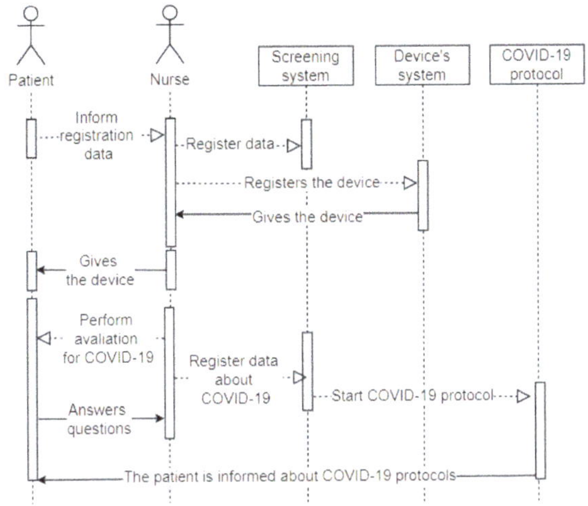

Figure 7. Sequence—Screening Room.

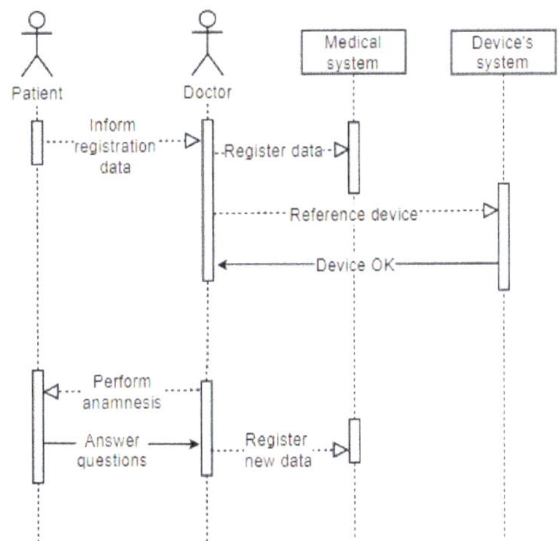

Figure 8. Sequence—Office.

4.3. Context Diagrams

The Context Diagram is a UML tool that represents the entire system as a single process. It consists of data streams that show the interfaces between the system and external entities [46]. The diagram illustrates the object of the study, the project, and its relationship to the environment. Figure 9 represents the context diagram of this project.

The patient (user) requests assistance from the receptionist (user), who will fill in the data (privacy) in the hospital system. The hospital system interacts with the operator's system that is outside the physical environment of the hospital. In the screening process, the nurse (user) conducts the questionnaire with the patient (user), entering the basic health data in the central hospital system, which interacts with the wearable IoT devices system. Finally, the doctor (user) performs the anamnesis, entering the central hospital system's consultation information. These information registration processes are focused on privacy determinations, and all processes occur in a clinical setting, explicitly represented within the context diagram.

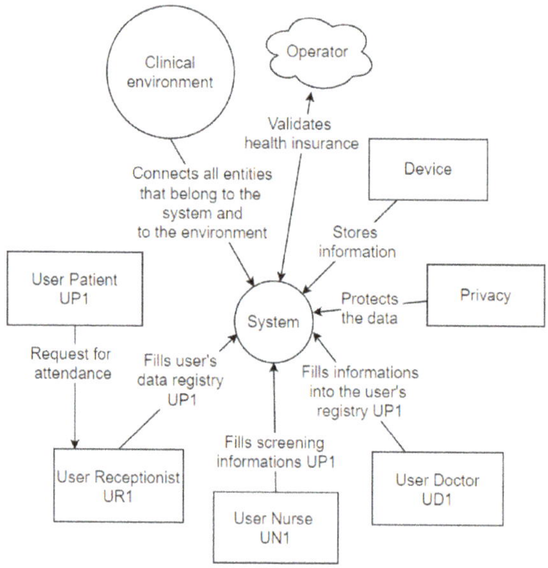

Figure 9. Context Diagram.

5. Prototype

A mobile application was developed as a prototype to illustrate the basic principles, from the admission of the patient to the emergency room and referral to the office or discharge indication. The goal of the developed mobile application is to validate some taxonomy items, for it embraces the environment (interoperability among the system and the wearable IoT device). The application was developed using NodeJS.

The application comprises an initial customer registration screen, which simulates the process of filling out the registration form upon admission to the emergency room. The prototype only contains the primary fields: name, gender, age, and address. The 'encrypt data?' checkbox has been included to select the encryption/hash algorithm. Since it is merely a prototype for demonstrating the flow of information and its security application, the hashes SHA-256 and SHA-512 were made available. In the real application, they would not serve to encrypt data because hashes are not reversible and are considered a one-way function [47]; the prototype also includes the Advanced Encryption Standard (AES) symmetric encryption algorithm. Figure 10 illustrates the first registration screen of the application with the fields mentioned above.

As shown in Figure 10, the application's flow is as follows: initially, the patient fills out a form with personal data. The data is encrypted and sent to the systems through the hospital network, as necessary. The patient, then, is sent to the screening room to answer more questions and thus help medical personnel assess their situation. He receives a wearable device to monitor his health status. All the information collected about the patient and their health status is included in their digital record. If communication with other systems is required, the information to be sent is encrypted.

The link between this device and the patient's file allows the information to be collected without a health professional's intervention. Based on the information provided by the wearable, the system makes a temperature analysis. If the patient remains in a feverish state, they are referred to the doctor's office. Since fever is one of the symptoms that prevail in detecting COVID-19, its absence can prompt a discharge. However, the lack of fever is not a guarantee that there is no infection with the virus [48], so careful monitoring is needed. In addition to the factors described, comparative tests were performed to validate the application based on the initially defined requirements in the taxonomy.

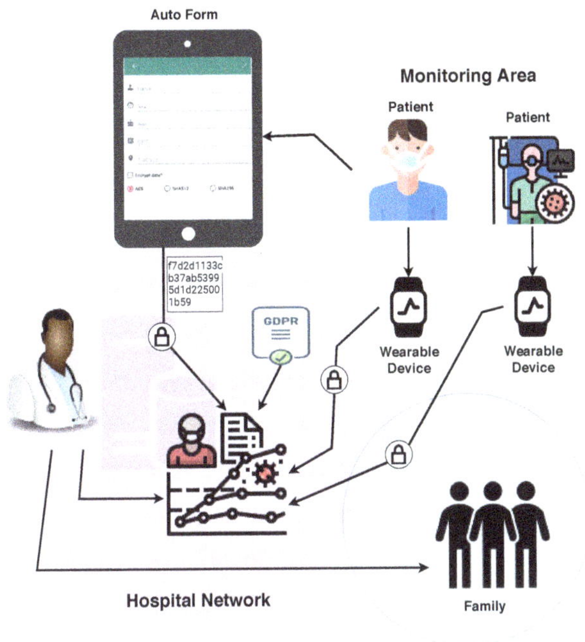

Figure 10. Application Flow.

Pseudo-Code

The algorithms applied in the development of the prototype application are described below in pseudo-code format. Pseudo-code covers the generation of a service number, temperature monitoring, and referral in case of emergency.

Algorithm 1 deals with the generation of the service number, where the patient's data will be saved in an encrypted form and forwarded to the monitoring room. In the monitoring room, the service number to be linked to the customer will be generated.

Algorithm 1: POST New medical care

```
Service Number;
Output: Attendance number
save encrypted packet data;
send to monitoring room;
```

Algorithm 2 deals with the process of monitoring temperature. The wearable IoT device collects the patient's temperature during the period defined by the medical team and sends it to the server for the monitor process. First, if it is higher than 38.5 °C, the patient is referred to the Intensive Care Unit (ICU). If it is equal to or above 37 °C for five minutes, the patient is referred to another ward for medical assistance. Finally, if it is less than 37 °C for ten minutes, the patient can be released.

Algorithm 2: Monitoring

```
The wearable IoT device send patient's' temperature for the server to monitor;
Input: Temperature
Output: Forwarding
while not forwarded do
    if temperature > 38.5 °C then
        forward to the intensive care unit;
    else if temperature >= 37 for 5 min then
        forward to the doctor;
    else if temperature < 37 for 10 min then
        forward back to home;
    else
        nothing to do;
end while
```

Algorithm 3 deals with the alert generated for the ICU in cases where the patient is classified as an emergency. If there is no emergency, the alert is generated for the doctor, informing that the patient will be referred for care.

Algorithm 3: Alert

```
Input  :ICU Data
Output:Attendance number
when patient need;
if has urgence then
    notify the intensive care unit
else
    notify doctor
end if
```

6. Tests and Results

The flow of controlled information in the application starts after the registration data has been filled in; it is also possible to apply other requirements such as encryption to the patient's data. Figure 11 illustrates the integration of basic patient information and reports that the patient was sent for temperature control. The temperature was captured and sent to the system, which will classify the feverish state, suggesting different referrals for each scenario. If the patient exhibits a feverish state and has other symptoms that may characterize COVID-19, their care must be provided in a differentiated way.

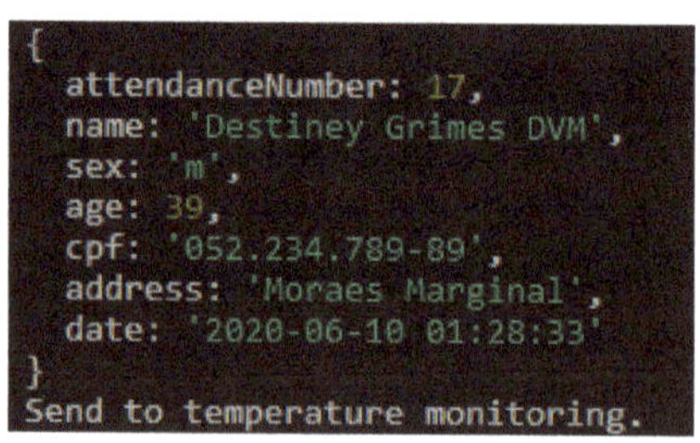

Figure 11. Saved Information.

When choosing the type of encryption in the data registration process, the data security level increases, and the information should only be made available to those who have permission. For prototype demonstration purposes, we use the AES symmetric key encryption method. The encryption application aims to secure data while transferring it to other devices. Figure 12 shows the encrypted patient registration data.


```
{
  "name":"46070d4bf934fb0d4b06d9e2c46e346944e322444900a435d7d9a95e6d7435f5",
  "cpf":"efa45c803061f485ad7a297d26253fef7764d1891e88dc8be4574d4d89f30cb2",
  "date":"2020-04-05 01:31:12",
  "address":"46070d4bf934fb0d4b06d9e2c46e346944e322444900a435d7d9a95e6d7435f5",
  "age":"c2356069e9d1e79ca924378153cfbbfb4d4416b1f99d41a2940bfdb66c5319db",
  "sex":"46070d4bf934fb0d4b06d9e2c46e346944e322444900a435d7d9a95e6d7435f5",
  "encripted":true
}
```


Figure 12. AES Encryption.

After the patient has been registered, and the information is stored safely, the data is sent to a system that continually gets updates on body temperature. With this prototype's application, we also tested the hypothesis that an IoT device can monitor the patient for changes in temperature. To test the idea, we implemented a set of random values read by the program to simulate this monitoring process. Every minute, the device will check the temperature of the patients who have entered the system and are waiting at the emergency room's reception. If their temperature can be characterized as feverish, then they are taken to the office with priority. Figure 13 describes the monitoring of a patient whose temperature remains stable, and hospital discharge is suggested.

```
Attendance number: 82. Patient Name: Estell Walter. Temperature: 36.16. Date:
Attendance number: 83. Patient Name: Estella Kutch. Temperature: 37.92. Date:
Attendance number: 84. Patient Name: Demarcus Kris. Temperature: 37.29. Date:
Attendance number: 85. Patient Name: Phoebe Buckridge. Temperature: 38.34. Dat
Attendance number: 86. Patient Name: Bernie Pollich. Temperature: 36.84. Date:
Attendance number: 87. Patient Name: Enos Yundt. Temperature: 38.04. Date: 202
Attendance number: 88. Patient Name: Lambert Yundt. Temperature: 37.49. Date:
Attendance number: 89. Patient Name: Aimee Botsford. Temperature: 37.80. Date:
Attendance number: 90. Patient Name: Brant Denesik. Temperature: 37.27. Date:
Attendance number: 91. Patient Name: Eula Beahan. Temperature: 37.49. Date: 20
Attendance number: 92. Patient Name: Ms. Bernardo Funk. Temperature: 38.49. Da
Attendance number: 93. Patient Name: Buford Hilpert. Temperature: 39.73. Date:
```

Figure 13. Patient record simulating discharge.

If the patient's state remains feverish for five minutes, a message will be sent to the doctor in charge, as shown in Figure 14. If the temperature remains stable for ten minutes, the patient will be released.

After testing and validating the application, it was possible to observe that the information flows through different devices. For the simulation environment, we experimented with only one system that communicates with a wearable device. In real applications, there could be more than one device interacting with more than one method. However, the information's fluidity would be similar: the patient's registration at the time of admission to the emergency room, the system being accessed by the screening sector to insert health status data, and the information is received from monitoring devices. At the medical consultation time, the system would receive more details regarding anamnesis, referrals for exams, or hospital discharge.

Given that the feverish state is strongly associated with a COVID-19 diagnosis, the patient should be monitored continuously and receive adequate care as long as the symptoms persist. The high contagion of the virus makes such care essential. The monitoring interval parameters, indicative of medical discharge or a possible disease carrier, are defined according to medical protocols. We emphasize that the interval and discharge suggestion present in this work are meant to simulate features.

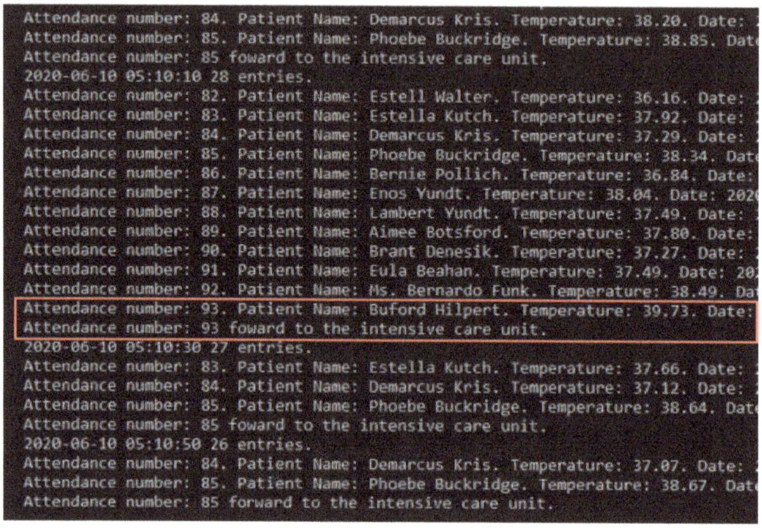

Figure 14. Patient registry simulating medical care admittance.

7. Conclusions

The COVID-19 scenario requires particular solutions for providing the emergency care process and security in the data generated in all environments. In this sense, this work proposed a taxonomy that was designed to support the development of privacy mechanisms for health environments.

The taxonomy is branched into four items containing five attributes each; all the items and their respective attributes are justifiable. For the information flow tests, we developed a prototype and application that addresses the main questions about data privacy despite being simple. The application was developed with registration data inputs and different encryption/hash to be applied according to environmental criteria. The application communicates with a wearable that monitors the patient's temperature and provides treatment in line with the patient's feverish state, guiding the referral to the doctor's office or the possibility of discharge. With the application of taxonomic definitions and the agility of medical professionals in the care of patients with suspected COVID-19, the registration data is kept confidential through encryption and privacy requirements. Temperature monitoring should be continuously done; in the case of feverish states that persist for a period defined by the entity and other symptoms suggestive of the disease, the system suggests the patient's referral without exposing personal data.

The main contribution of this research consists of the analysis of different privacy parameters with a mobile application that considers the different rules proposed in our taxonomy. There is no concrete analysis previously performed that analyzes the privacy constraints with a mobile application. Mobile technologies are commonly used by people, and it may help in the prevention of COVID-19. In addition, more search should be performed, and the taxonomy developed may be improved to be adapted with the real world.

We believe that the research we have carried out contributes to several other studies currently in progress in several countries, which propose monitoring without consent and put forward definitions of use and data privacy criteria. For future work, we are developing improvements for privacy requirements that can be adapted to different countries, thus expanding variable monitoring features to identify patients with COVID-19 and obtain new tests and results.

Author Contributions: Conceptualization, A.V.L., L.A.S., L.G.; Investigation, A.V.L., L.A.S. and V.R.Q.L. Methodology, L.A.S. and A.V.L.; Project Administration, V.R.Q.L.; Resources, I.M.P.; Supervision, V.R.Q.L. Validation, V.R.Q.L. and L.A.S.; Writing—original draft, A.V.L., R.L. and L.G. Writing—review and editing, V.R.Q.L., L.A.S., X.M., J.L.V.B. and I.M.P., Financial R.G.O. and V.R.Q.L. All authors have read and agreed to the published version of the manuscript.

Funding: The project Smart following systems, Edge Computing and IoT Consortium, CONSORCIO TC_TCUE18-20_004, CONVOCATORIA CONSORCIOTC. PLAN TCUE 2018-2020. Project managed by Fundación General de la Universidad de Salamanca and co-financed with Junta Castilla y León and FEDER funds. This work was partially supported by Fundação para a Ciência e a Tecnologia under Project UIDB/04111/2020. This work was partially funded by FCT/MEC through national funds and co-funded by FEDER–PT2020 partnership agreement under the project UIDB/50008/2020. This work was partially funded by National Funds through the FCT—Foundation for Science and Technology, I.P., within the scope of the project UIDB/00742/2020

Acknowledgments: This work was partially supported by CAPES—Financial code (001). The authors also like to acknowledge the collaboration of Computer Laboratory 7 of Instituto de Telecomunicações—IT Branch Covilhã—Portugal. Furthermore, we would like to thank the Politécnico de Viseu for their support.

Conflicts of Interest: The authors declare no conflict of interest.

Abbreviations

The following abbreviations are used in this manuscript:

AES	Advanced Encryption Standard
CIA	Confidentiality, Integrity and Availability
CPF	Cadastro de Pessoa Física
COVID-19	Coronavirus 2
COVID-19 SARS-CoV-2	Severe Acute Respiratory Syndrome Coronavirus 2
EHR	Electronic Health Record
GDPR	General Data Protection Regulation
HIPAA	Department of Health and Human Services, 2021
HCPP	Healthcare system for Patient Privacy
ICU	Intensive Care Unit
IoT	Internet of Things
LGPD	General Data Protection Law
mHealth	Mobile Health
P-device	Private Device
UML	Unified Modeling Language

References

1. Marques, G.; Saini, J.; Pires, I.M.; Miranda, N.; Pitarma, R. Internet of Things for Enhanced Living Environments, Health and Well-Being: Technologies, Architectures and Systems. In *Handbook of Wireless Sensor Networks: Issues and Challenges in Current Scenario's*; Singh, P.K., Bhargava, B.K., Paprzycki, M., Kaushal, N.C., Hong, W.C., Eds.; Springer International Publishing: Cham, Switzerland, 2020; pp. 616–631. [CrossRef]
2. Pires, I.M.; Marques, G.; Garcia, N.M.; Flórez-Revuelta, F.; Ponciano, V.; Oniani, S. A Research on the Classification and Applicability of the Mobile Health Applications. *J. Pers. Med.* **2020**, *10*, 11. [CrossRef] [PubMed]
3. Doukas, C.; Maglogiannis, I. Bringing IoT and Cloud Computing towards Pervasive Healthcare. In Proceedings of the 2012 Sixth International Conference on Innovative Mobile and Internet Services in Ubiquitous Computing, Palermo, Italy, 4–6 July 2012; pp. 922–926. [CrossRef]
4. Al-Odat, Z.A.; Srinivasan, S.K.; Al-qtiemat, E.; Dubasi, M.A.L.; Shuja, S. IoT-Based Secure Embedded Scheme for Insulin Pump Data Acquisition and Monitoring. *arXiv* **2018**, arXiv:1812.02357.
5. Farahani, B.; Firouzi, F.; Chang, V.; Badaroglu, M.; Constant, N.; Mankodiya, K. Towards fog-driven IoT eHealth: Promises and challenges of IoT in medicine and healthcare. *Future Gener. Comput. Syst.* **2017**. [CrossRef]

6. Campos, J.; Souza, V.S.A. Percepção dos Usuários do Serviço de Urgência e Emergência em relação à classificação de risco pelo protocolo de Manchester. *Rev. Unim. Cient. Montes Claros* **2014**, *16*. Available online: https://doi.org/http://www.ruc.unimontes.br/index.php/unicientifica/article/view/319/297 (accessed on 8 April 2020).
7. Rothan, H.; Siddappa, N. The epidemiology and pathogenesis of coronavirus disease (COVID-19) outbreak. *J. Autoimmun.* **2020**, *109*, 102433. [CrossRef]
8. Liang, Y.; Liang, J.; Zhou, Q.; Li, X.; Lin, F.; Deng, Z.; Zhang, B.; Li, L.; Wang, X.; Zhu, H.; et al. Prevalence and clinical features of 2019 novel coronavirus disease (COVID-19) in the Fever Clinic of a teaching hospital in Beijing: a single-center, retrospective study. *medRxiv* **2020**. [CrossRef]
9. Soares, N.V.; Dall'Agnol, C.M. Privacidade dos pacientes: uma questão para a geração do cuidado em enfermagem. *Acta Paul. Enferm.* **2011**, *24*, 683–688. [CrossRef]
10. Barker, K.; Askari, M.; Banerjee, M.; Ghazinour, K.; Mackas, B.; Majedi, M.; Pun, S.; Williams, A. A Data Privacy Taxonomy. In *British National Conference on Databases*; Springer: Berlin/Heidelberg, Germany, 2009, Volume 5588; pp. 42–54. [CrossRef]
11. Asaddok, N.; Ghazali, M. Exploring the usability, security and privacy taxonomy for mobile health applications. In Proceedings of the 2017 International Conference on Research and Innovation in Information Systems (ICRIIS), Langkawi, Malaysia, 16–17 July 2017; pp. 1–6. [CrossRef]
12. Coen-Porisini, A.; Colombo, P.; Sicari, S.; Trombetta, A. A Conceptual Model for Privacy Policies. In Proceedings of the 11th IASTED International Conference on Software Engineering and Applications (SEA '07), Cambridge, MA, USA, 19–21 November 2007; ACTA Press: Anaheim, CA, USA, 2007; pp. 570–577. [CrossRef]
13. Silva, L.A.; Leithardt, V.R.Q.; Rolim, C.O.; González, G.V.; Geyer, C.F.R.; Silva, J.S. PRISER: Managing Notification in Multiples Devices with Data Privacy Support. *Sensors* **2019**, *19*, 3098. [CrossRef]
14. Leithardt, V.; Santos, D.; Silva, L.; Viel, F.; Zeferino, C.; Silva, J. A Solution for Dynamic Management of User Profiles in IoT Environments. *IEEE Lat. Am. Trans.* **2020**, *18*, 1193–1199. [CrossRef]
15. Zwitter, A.; Gstrein, O.J. Big data, privacy and COVID-19–learning from humanitarian expertise in data protection. *J. Int. Humanit. Action* **2020**, *5*. [CrossRef]
16. Yesmin, T.; Carter, M.W. Evaluation framework for automatic privacy auditing tools for hospital data breach detections: A case study. *Int. J. Med. Inform.* **2020**, *138*. [CrossRef] [PubMed]
17. Islam, S.M.R.; Kwak, D.; Kabir, M.H.; Hossain, M.; Kwak, K. The Internet of Things for Health Care: A Comprehensive Survey. *IEEE Access* **2015**, *3*, 678–708. [CrossRef]
18. Sun, J.; Zhu, X.; Zhang, C.; Fang, Y. HCPP: Cryptography Based Secure EHR System for Patient Privacy and Emergency Healthcare. In Proceedings of the 2011 31st International Conference on Distributed Computing Systems, Minneapolis, MN, USA, 20–24 June 2011; pp. 373–382. [CrossRef]
19. Samaila, M.; Neto, M.; Fernandes, D.; Freire, M.; Inácio, P. Challenges of Securing Internet of Things Devices: A survey. *Secur. Priv.* **2018**, *1*. [CrossRef]
20. Plachkinova, M.; Andrés, S.; Chatterjee, S. A Taxonomy of mHealth Apps–Security and Privacy Concerns. In Proceedings of the 2015 48th Hawaii International Conference on System Sciences, Kauai, Hawaii, 5–8 January 2015; pp. 3187–3196. [CrossRef]
21. Alsubaei, F.; Abuhussein, A.; Shiva, S. Security and Privacy in the Internet of Medical Things: Taxonomy and Risk Assessment. In Proceedings of the 2017 IEEE 42nd Conference on Local Computer Networks Workshops (LCN Workshops), Singapore, 9–12 October 2017; pp. 112–120. [CrossRef]
22. Hankerson, D.; Menezes, A.J.; Vanstone, S. *Guide to Elliptic Curve Cryptography*; Springer Science & Business Media: Berlin, Germany, 2005; Volume 46, p. 13.
23. Yi, X.; Bertino, E.; Rao, F.Y.; Bouguettaya, A. Practical privacy-preserving user profile matching in social networks. In Proceedings of the 2016 IEEE 32nd international conference on data engineering (ICDE), Helsinki, Finland, 16–20 May 2016; pp. 373–384. [CrossRef]
24. Sivaraman, V.; Gharakheili, H.H.; Vishwanath, A.; Boreli, R.; Mehani, O. Network-level security and privacy control for smart-home IoT devices. In Proceedings of the 2015 IEEE 11th International Conference on Wireless and Mobile Computing, Networking and Communications (WiMob), Abu Dhabi, UAE, 19–21 October 2015; pp. 163–167. [CrossRef]

25. Stojanović, R.; Škraba, A.; Lutovac, B. A Headset Like Wearable Device to Track COVID-19 Symptoms. In Proceedings of the 2020 9th Mediterranean Conference on Embedded Computing (MECO), Budva, Montenegro, 8–11 June 2020; pp. 1–4.
26. Cecilia, J.M.; Cano, J.; Hernández-Orallo, E.; Calafate, C.T.; Manzoni, P. Mobile crowdsensing approaches to address the COVID-19 pandemic in Spain. *IET Smart Cities* **2020**, *2*, 58–63. [CrossRef]
27. Riza, T.A.; Gunawan, D. IEEE 802.11ah Network Challenges Supports Covid-19 Prevention Team. In Proceedings of the 2020 IEEE 10th International Conference on Electronics Information and Emergency Communication (ICEIEC), Beijing, China, 17–19 July 2020; pp. 73–76.
28. Zeinalipour-Yazti, D.; Claramunt, C. COVID-19 Mobile Contact Tracing Apps (MCTA): A Digital Vaccine or a Privacy Demolition? In Proceedings of the 2020 21st IEEE International Conference on Mobile Data Management (MDM), Versailles, France, 30 June–3 July 2020; pp. 1–4.
29. Feistel, H. Cryptography and Computer Privacy. *Sci. Am.* **1973**, *228*, 10. [CrossRef]
30. Cesconetto, J.; Augusto Silva, L.; Bortoluzzi, F.; Navarro-Cáceres, M.; Zeferino, C.A.; Leithardt, V.R.Q. PRIPRO—Privacy Profiles: User Profiling Management for Smart Environments. *Electronics* **2020**, *9*, 1519. [CrossRef]
31. Vital, L.P.; Café, L.M.A. Ontologias e taxonomias: diferenças. *Perspect. CiÊNc. Inform.* **2011**, *16*, 115–130. [CrossRef]
32. Fengou, M.; Mantas, G.; Lymperopoulos, D.; Komninos, N. Ubiquitous Health Profile Management Applying Smart Card Technology. In *International Conference on Wireless Mobile Communication and Healthcare*; Springer: Berlin/Heidelberg, Germany, 2011; Volume 83. [CrossRef]
33. Leithardt, V.; Borges, G.; Rossetto, A.; Rolim, C.; Geyer, C.; Correia, L.; Nunes, D.; Sá Silva, J. A Privacy Taxonomy for the Management of Ubiquitous Environments. *J. Commun. Comput.* **2013**, *10*, 1529–1553. [CrossRef]
34. Senthilkumar, S.; Viswanatham, V.M. HB-PPAC: Hierarchy-based privacy preserving access control technique in public cloud. *Int. J. High Perform. Comput. Netw.* **2017**, *10*, 13. [CrossRef]
35. Ibraimi, L.; Asim, M.; Petković, M. Secure Management of Personal Health Records by Applying Attribute-Based Encryption. In Proceedings of the 6th International Workshop on Wearable, Micro, and Nano Technologies for Personalized Health, Oslo, Norway, 24–26 June 2009; pp. 71–74. [CrossRef]
36. Centers for Medicare & Medicaid Services. The Health Insurance Portability and Accountability Act of 1996 (HIPAA). 1996. Available online: http://www.cms.hhs.gov/hipaa/ (accessed on 25 August 2020).
37. Cooper, T.; Collman, J. Managing Information Security and Privacy in Healthcare Data Mining. In *Medical Informatics: Knowledge Management and Data Mining in Biomedicine*; Chen, H., Fuller, S.S., Friedman, C., Hersh, W., Eds.; Springer: Boston, MA, USA, 2005; pp. 95–137. [CrossRef]
38. Costa, R.G.R. Apontamentos para a arquitetura hospitalar no Brasil: Entre o tradicional e o moderno. *Hist. Cienc. Saude-Manguinhos* **2011**, *18*, 53–66. [CrossRef]
39. Lopes, S. Data Privacy in Interoperability Contexts–The Area of Health. Ph.D. Thesis, Universidade de Évora, Évora, Portugal, 2016.
40. Yeniman Yildirim, E.; Akalp, G.; Aytac, S.; Bayram, N. Factors Influencing Information Security Management in Small- and Medium-Sized Enterprises: A Case Study from Turkey. *Int. J. Inf. Manag.* **2011**, *31*, 360–365. [CrossRef]
41. Florence, G.; Calil, S.J. Uma nova perspectiva no controle dos riscos da utilização de tecnologia médico-hospitalar. *MultiCiência* **2005**, *5*, 1–14.
42. Machado, D.; Doneda, D. Proteção de Dados Pessoais e Criptografia: Tecnologias Criptográficas Entre Anonimização e Pseudonimização de Dados. *Rev. Trib.* **2019**, *998*, 99–125.
43. Brasil. Lei n. 13.079, de 14 de Agosto de 2018. Lei Geral de Proteção de Dados Pessoais (LGPD). 2018. Available online: http://www.planalto.gov.br/ccivil_03/_ato2015-2018/2018/lei/L13709.htm (accessed on 30 Agust 2020).
44. Lupiana, D.; O'Driscoll, C.; Mtenzi, F. Taxonomy for ubiquitous computing environments. In Proceedings of the 2009 First International Conference on Networked Digital Technologies, Ostrava, Czech Republic, 28–31 July 2009; pp. 469–475. [CrossRef]
45. Pradilla, J.; Esteve, M.; Palau, C. SOSFul: Sensor Observation Service (SOS) for Internet of Things (IoT). *IEEE Lat. Am. Trans.* **2018**, *16*, 1276–1283. [CrossRef]
46. Fowler, M. *UML Essencial: Um Breve Guia Para Linguagem Padrao*; Bookman: Orange, CA, USA, 2005.

47. Menezes, A.J.; Vanstone, S.A.; Oorschot, P.C.V. *Handbook of Applied Cryptography*, 1st ed.; CRC Press, Inc.: Boca Raton, FL, USA, 1996.
48. Singhal, T. A Review of Coronavirus Disease-2019 (COVID-19). *Indian J. Pediatr.* **2020**, *87*, 281–286. [CrossRef] [PubMed]

sensors

MDPI

Article

An Experimental Study on the Validity and Reliability of a Smartphone Application to Acquire Temporal Variables during the Single Sit-to-Stand Test with Older Adults

Diogo Luís Marques [1], Henrique Pereira Neiva [1,2], Ivan Miguel Pires [3,4,5], Eftim Zdravevski [6], Martin Mihajlov [7], Nuno M. Garcia [3], Juan Diego Ruiz-Cárdenas [8], Daniel Almeida Marinho [1,2] and Mário Cardoso Marques [1,2,*]

1 Department of Sport Sciences, University of Beira Interior, 6201-001 Covilhã, Portugal; diogo.marques@ubi.pt (D.L.M.); hpn@ubi.pt (H.P.N.); dmarinho@ubi.pt (D.A.M.)
2 Research Center in Sports Sciences, Health Sciences and Human Development, CIDESD, 6201-001 Covilhã, Portugal
3 Instituto de Telecomunicações, Universidade da Beira Interior, 6200-001 Covilhã, Portugal; impires@it.ubi.pt (I.M.P.); ngarcia@di.ubi.pt (N.M.G.)
4 Computer Science Department, Polytechnic Institute of Viseu, 3504-510 Viseu, Portugal
5 Health Sciences Research Unit: Nursing, School of Health, Polytechnic Institute of Viseu, 3504-510 Viseu, Portugal
6 Faculty of Computer Science and Engineering, University Ss Cyril and Methodius, 1000 Skopje, North Macedonia; eftim.zdravevski@finki.ukim.mk
7 Laboratory for Open Systems and Networks, Jozef Stefan Institute, 1000 Ljubljana, Slovenia; martin@e5.ijs.si
8 Physiotherapy Department, Faculty of Health Sciences, Catholic University of Murcia, 30107 Murcia, Spain; jdruiz@ucam.edu
* Correspondence: mmarques@ubi.pt

Citation: Marques, D.L.; Neiva, H.P.; Pires, I.M.; Zdravevski, E.; Mihajlov, M.; Garcia, N.M.; Ruiz-Cárdenas, J.D.; Marinho, D.A.; Marques, M.C. An Experimental Study on the Validity and Reliability of a Smartphone Application to Acquire Temporal Variables during the Single Sit-to-Stand Test with Older Adults. *Sensors* **2021**, *21*, 2050. https://doi.org/10.3390/s21062050

Academic Editors: Rebeca P. Díaz Redondo and Jesús Ureña

Received: 22 January 2021
Accepted: 11 March 2021
Published: 15 March 2021

Abstract: Smartphone sensors have often been proposed as pervasive measurement systems to assess mobility in older adults due to their ease of use and low-cost. This study analyzes a smartphone-based application's validity and reliability to quantify temporal variables during the single sit-to-stand test with institutionalized older adults. Forty older adults (20 women and 20 men; 78.9 ± 8.6 years) volunteered to participate in this study. All participants performed the single sit-to-stand test. Each sit-to-stand repetition was performed after an acoustic signal was emitted by the smartphone app. All data were acquired simultaneously with a smartphone and a digital video camera. The measured temporal variables were stand-up time and total time. The relative reliability and systematic bias inter-device were assessed using the intraclass correlation coefficient (ICC) and Bland-Altman plots. In contrast, absolute reliability was assessed using the standard error of measurement and coefficient of variation (CV). Inter-device concurrent validity was assessed through correlation analysis. The absolute percent error (APE) and the accuracy were also calculated. The results showed excellent reliability (ICC = 0.92–0.97; CV = 1.85–3.03) and very strong relationships inter-devices for the stand-up time ($r = 0.94$) and the total time ($r = 0.98$). The APE was lower than 6%, and the accuracy was higher than 94%. Based on our data, the findings suggest that the smartphone application is valid and reliable to collect the stand-up time and total time during the single sit-to-stand test with older adults.

Keywords: mobile application; accelerometer sensor; stand-up time; total time; aging

1. Introduction

As the population's age increases in industrialized countries [1], older adults' care is critical for their well-being. Consequently, the evaluation and quantification of daily activities are essential for determining health status changes and, subsequently, detecting early signs of loss of autonomy [2]. Standing up from a sitting position and its counterpart transition, sitting down from a standing position, are the two most common daily motor activities that could be critical indicators for older adults' functional autonomy [3,4]. The sit-to-stand movement is one of the most challenging activities in terms of mechanics [5]. It requires the optimization of several kinematic tasks, including coordination, balance, mobility, muscular strength, and power output [6]. Within the population of older adults, increased sit-to-stand time is associated with a high risk of fall occurrence [7–9], decreased leg muscle power and strength [10,11], slow walking speed [12–14], and mobility disability [15]. Usually, the sit-to-stand test's time combined with the subject's age and previous medical history (e.g., recovering from an injury or surgery) helps identify the fall risk, assessing functional lower extremity strength, transitional movements, and balance.

Recently, approaches for quantifying mobility emerged that rely on inexpensive sensor technologies [16]. Remarkably, smartphone use has been suggested as a useful tool to objectively monitor and improve patients' health and fitness [17], which has been verified in research [18] and clinical practice [19]. Smartphone sensor technology has become sufficiently reliable and accurate to substitute specific biomechanics lab equipment and portable devices used in functional mobility research. Some authors showed that a smartphone with a motion sensor could be used as a low-cost integration device to evaluate the patient's balance and mobility [20]. Other authors demonstrated that the smartphones' accelerometer could measure kinematic tremor frequencies equivalent to electromyography's tremor frequency [21]. Wile et al. [22] utilized a smartwatch to differentiate the symptoms in patients with Parkinson's disease and related tremor diseases by calculating the first four harmonics' signal power.

Several clinical tests, such as the timed-up and go test and sit-to-stand test, have been developed to evaluate physical performance and mobility-related to every-day tasks [8,23]. The sit-to-stand test is a widely adopted clinical test used to evaluate older adults' functionality [24]. Initially designed to measure the lower extremities' functional capacity [25], it has been applied and investigated in different populations to assess the rehabilitation process and functional performance in older people with varied medical conditions [26]. During the sit-to-stand test, researchers and clinicians commonly measure the time spent to perform a fixed number of repetitions or the number of repetitions performed during a specific time [4,14]. Commonly, the evaluators use a chronometer to measure the time during the sit-to-stand test [27,28]. Despite its low cost and ease of use, chronometers present some limitations, mainly associated with human error (e.g., reaction time delay and position judgment) [29–31]. Therefore, to overcome the limitations mentioned above, clinicians and researchers may opt for alternative and reliable technologies to measure biomechanical parameters during the sit-to-stand test with aged populations [32,33]. Several authors have analyzed smartphone applications to quantify kinematic variables during the sit-to-stand test using high-speed video recordings [34,35]. Other studies utilized the triaxial accelerometer sensor embedded in the mobile smartphone [30,31,36]. For example, González-Rojas et al. [37] characterized the time measurement of sit-to-stand transitions by transforming the relative acceleration signal, recorded by a triaxial accelerometer. Cerrito et al. [30] validated a smartphone-based app using the accelerometer sensor to quantify the sit-to-stand test movement in older adults. These authors captured vertical ground reaction forces and vertical acceleration simultaneously using two force plates (reference standard) and a mobile smartphone. The total movement duration, peak force, rate of force development, and peak power were measured. Chan et al. [31] also developed a mobile application to calculate the time during the five-repetition sit-to-stand and timed-up and go tests in older women. The mobile application also includes a beep sound to cue the participants to initiate the test, which aims to eliminate potential human errors when using a chronometer, including the reaction time delay.

As frailty is viewed as a transitional state from robustness to functional decline, identifying a pre-frailty state can alleviate or postpone the consequences of this syndrome [38]. Within this context, temporal variables have been used as predictors of frailty in several older adults' studies. For example, Hausdorff et al. [39] measured the stride time, swing time, stance time, and percentage stance time. Fallers compared with non-fallers revealed higher standard deviations and coefficients of variation across all variables. In the study by Zhou et al. [40], gait is deconstructed into clinically observable spatial-temporal variables to establish a quantitative model to classify fallers and non-fallers.

Considering that the sit-to-stand test is strongly recommended in clinical and research settings to assess functional independence, detect frailty, and sarcopenia in aged populations [15,41], the practicability of using smartphones for assessing the sit-to-stand in older adults can be a logical advancement of an inherent concept. Therefore, as stated above, measuring reliably and accurately temporal variables during the sit-to-stand test with older adults, including the stand-up time and the total time, is determinant to identify those who present functional impairments and designing individual clinical interventions to improve functionality [42,43]. Due to the difficulties of performing the different measurements with older adults, developing an easy-use solution is essential for taking different preventive actions. Therefore, this study aimed to analyze a smartphone application's validity and reliability to acquire temporal descriptors during the single sit-to-stand test with institutionalized older adults, including the stand-up time and total time. We hypothesized that the smartphone application would be valid and reliable for measuring the temporal variables during the single sit-to-stand test with older adults. This study's novelty consists of creating a scientifically valid and reliable mobile application for the independent and automatic measurement of temporal variables during the single sit-to-stand test with older adults. To our best knowledge, only one study using a mobile app to quantify the sit-to-stand test with older adults exhibited the data in real-time through graphics [30]. However, analyzing the data through graphics might not be a practical approach in clinical contexts due to its complexity and time-consuming. An essential factor to bear in mind is that clinicians and researchers want to immediately access the results when the test ends to provide feedback in real-time for the participants. The automatic method will also ensure that the data were properly collected. Therefore, with our mobile app, the possibility of automatically recording, processing, and presenting the results in the smartphone screen when the test ends entails a new clinical approach to assess the single sit-to-stand performance with older adults.

2. Methods

As part of the research on developing solutions for Ambient Assisted Living (AAL) [44–46], the scope of this study consists of using technological equipment that embeds inertial sensors that acquire different data types to measure and identify human movements [47–50].

2.1. Study Design

This study was a cross-sectional design aiming to analyze a smartphone application's validity and reliability to capture temporal descriptors during the single sit-to-stand test with institutionalized older adults. A digital video camera was used further to validate the correct execution of the sit-to-stand movement, and the results presented in this study only considered the valid ones. The sit-to-stand test is quick and easy to administer and presents practical utility in clinical and research settings to evaluate functional independence in older adults [7,8]. The experimental procedures were carried out over ten weeks. Each session was performed between 10:00 and 11:00 a.m. in the same location (room temperature 22–24 °C). In the first week, we familiarized the participants with the testing procedures. We also measured the body mass (TANITA BC-601, Tokyo, Japan) and height (Portable Stadiometer SECA, Hamburg, Germany). Then, from the second to the tenth week, the participants performed one testing session per week. In each session, we assessed a group of four to five participants in the sit-to-stand test.

2.2. Participants

Forty institutionalized older adults (20 men and 20 women) volunteered to participate in this study. Inclusion criteria were age $\geq$ 65 years old, men and women, capable of stand-up from a chair independently with the arms crossed over the chest, and willingness to participate in the experimental procedures. Exclusion criteria were severe physical and cognitive impairment (i.e., Barthel index score < 60 and mini-mental state examination score < 20), deafness, musculoskeletal injuries in the previous three months, and terminal illness (life expectancy < 6 months). Table 1 presents the characteristics of the participants. All participants gave their informed consent for inclusion before they participated in the study. The study was conducted following the Declaration of Helsinki. The Ethics Committee approved the protocol of the University of Beira Interior (code: CE-UBI-Pj-2019-019).

Table 1. Participants' characteristics.

	n	Age (Years)	Body Mass (kg)	Height (m)	BMI (kg/m^2)
Women	20	81.9 ± 8.1	65.4 ± 11.6	1.49 ± 0.1	29.4 ± 5.4
Men	20	76.0 ± 8.2	78.0 ± 15.5	1.66 ± 0.1	28.4 ± 4.7
Total	40	78.9 ± 8.6	71.7 ± 15.0	1.57 ± 0.1	28.9 ± 5.0

Data are mean ± standard deviation.

2.3. *Sit-to-Stand Test*

For this study, we used the single sit-to-stand test, starting with a 10 min general warm-up consisting of light walking and mobility exercises, as described in Marques et al. [51]. The participants were equipped with a smartphone placed inside a waistband (Sports Waistband Universal Phone Holder), which was in turn attached to the waist. We placed the mobile phone on the waist because the center of gravity is located around the abdomen [52]. The waistband was tightened to avoid slight movements of the mobile phone that would adversely affect data capture. The participants sat on an armless chair (height = 0.49 cm) with the back straight and the arms crossed over the chest. We did not allow the participants to lean back on the chair neither to assume a perched position. All participants were instructed to maintain a 90° hip and knee flexion, which the operator closely monitored during the test. After 10 s of the smartphone application's activation, an acoustic signal cued the participants to stand up and sit down on the chair while maintaining the arms crossed over the chest, thus performing a single sit-to-stand movement. When they finished the movement, the participant rested on the chair with the arms crossed over the chest for 15 s. The subsequent single sit-to-stand movement was performed after hearing another acoustic signal. After six single sit-to-stand repetitions, the participants had a 3-min rest before repeating the test four more times. This procedure was necessary to ensure that enough and correct data (i.e., having enough rest and without previous body movement before the beep) was collected for post-analysis. Before each sit-to-stand transition, we instructed the participants to perform the repetitions as fast as possible immediately after hearing the acoustic signal. In all trials, a researcher was standing next to the participants to ensure safety during the movement transitions. Figure 1 illustrates the sit-to-stand testing procedure.

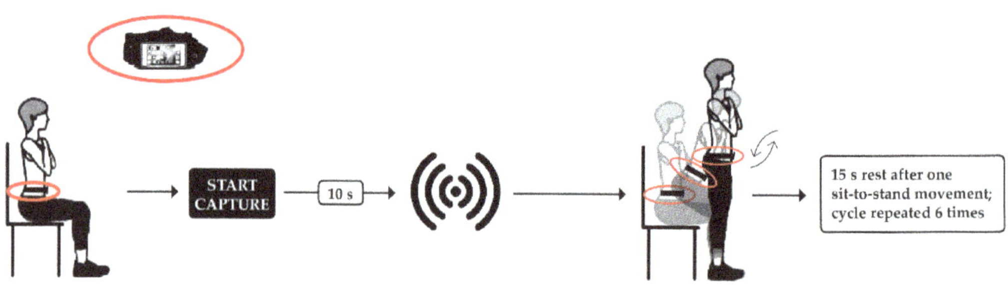

Figure 1. Illustration of the single sit-to-stand test.

2.4. Data Acquisition

A smartphone application and a digital video camera acquired the data simultaneously. The latter device was considered the reference criterion [53–55]. The smartphone model was the Xiaomi Mi A1. This device embeds a triaxial accelerometer (model Bosch BMI120), which acquires the data at a sampling frequency of 200 Hz. As described before, we placed the smartphone inside a waistband, which was then attached to the participants' waist. The digital video camera (Canon LEGRIA HF R46, Tokyo, Japan) was positioned perpendicular to the field of view (distance = 3 m) and attached to a stationary tripod (height = 1.2 m). We recorded the participants from the sagittal plane at a sampling frequency of 25 frames per second. Although the sampling frequency between devices was different (200 vs. 25 Hz), this was not considered a significant limitation [53]. In fact, previous studies with older adults comparing the accelerometer data vs. video camera data during postural transitions (e.g., sit-to-stand or gait analysis) used different sampling frequencies between devices [54–56]. Therefore, according to the scientific literature, when comparing handheld devices (e.g., mobile phones) vs. machines (e.g., video cameras), it is impossible to achieve synchronization or sampling frequency equality [53].

Regarding using a 25 Hz video camera for analysis, it is essential to note that one frame's error corresponds to an error of 0.04 s when using this sampling frequency. Table 2 shows that the stand-up time and total time values are around 1.62 and 2.75 s, respectively. These results indicate that the video camera's error is around 2.5% in the stand-up time (i.e., 0.04/1.68) and 1.5% in the total time (i.e., 0.04/2.75), in the worst-case scenario which means a minor measurement error. Therefore, as stated by Winter [57], except for high-speed running and athletic movements, slower movement analyses (e.g., walking) can be reliably done with minor errors using a 25 Hz video camera. Previous studies with older adults used a sampling frequency of 25 Hz to analyze kinematic data during movement transitions, such as sit-to-stand, stand-to-sit, or sit-to-walk [54–56,58,59], which reinforces the validity and reliability of using this frequency for movement analysis.

Table 2. Relative reliability and relationship inter-devices.

Variable	App (s) (Mean ± SD)	Video (s) (Mean ± SD)	Cronbach's α	ICC (95% CI)	Correlation (ϱ) (95% CI)
Stand-up Time (n = 842)	1.68 ± 0.29	1.62 ± 0.30	0.97	0.92 (0.82–0.96)	0.94 (0.94–0.95) ***
Total Time (n = 892)	2.81 ± 0.50	2.75 ± 0.50	0.99	0.97 (0.93–0.98)	0.98 (0.97–0.98) ***

CI: confidence interval; ICC: intra-class correlation coefficient; ϱ: Spearman's rank correlation coefficient; *** p-value < 0.001.

2.5. Data Analysis

2.5.1. Mobile Application

The accelerometer data were acquired with a mobile application, which automatically pre-processes the raw data and measures the stand-up time and total time (Figure 2). These measures are related to the occurrence of events, such as when standing up starts or ends. The mobile application was developed with Android Studio 4.1., with Java SE 12. It was used for the automatic detection of different events during the sit-to-stand test. It was developed and adjusted, considering the previous study [60], by this study's research team. The application will be available in the market after validation. Contrary to other studies, and to avoid the mobile device's incorrect positioning on the waist, we used the Euclidean norm of the accelerometer's outputs. After, the data was filtered, and the different calculations were applied. The different measurements can be performed locally without an Internet connection. The results are presented as soon as the test ends. The stand-up time starts from the acoustic signal until the minimum (negative) acceleration value is reached before the maximum (positive) acceleration value. The time frame between the acoustic signal and the maximum (positive) acceleration value is defined as the total time (Figure 2).

For the measurement of the stand-up time and total time, we researched in the literature how these variables are automatically detected considering the accelerometer data [41,42].

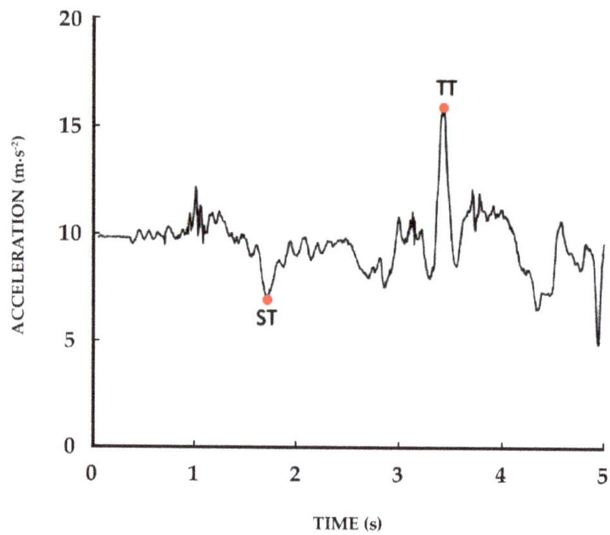

Figure 2. Signal plot of the acceleration during one sit-to-stand repetition (i.e., the complete cycle of stand-up and sit down on the chair); ST: stand-up time; TT: total time.

2.5.2. Video-Camera Recordings

The video recording files were transferred to a personal laptop and then analyzed using Adobe Premiere Pro (version 14.4.0, Adobe Systems, San Jose, CA, USA). We analyzed all video files frame by frame. The first frame was considered the start of the acoustic signal. After that, we calculated the stand-up time and total time. The stand-up time was defined as the moment from the acoustic signal until the participant was stand-up with the legs fully extended and an upright torso. The total time was measured from the acoustic signal until the participant returned to the seated position, the moment of contact with the chair, with vertical velocity decreased to zero. We identified that the person was entirely sat by monitoring subsequent frames and ensuring that the last frame corresponded to the moment they were fully seated. We converted the data to seconds by dividing the frame number by 25 frames per second. The repetitions were invalid if participants moved any segment of the body the instant before the acoustic signal or did not complete the sit-to-stand cycle. Therefore, we only selected valid repetitions for further analysis.

2.6. Statistical Analysis

The calculation of the interquartile range of the mean difference between devices in each temporal variable enabled the detection of outliers. If data were higher than 1.5 or lower than −1.5 times the Inter-quartile Range (IQR), it was removed [53]. The intra-class correlation coefficient (ICC with 95% confidence intervals [CI]) analyzed the level of agreement or relative reliability inter-device [61]. The ICC model was the two-way random-effects, absolute agreement, single rater/measurement [$ICC_{(2,1)}$] [61]. Cronbach's alpha analyzed internal consistency. ICC values were interpreted as: <0.50, poor; 0.50–0.75, moderate; 0.75–0.90, good; >0.90, excellent [61]. Bland-Altman plots with 95% limits of agreement (LOA) (mean difference $\pm$ 1.96 $\times$ standard deviation [SD] of the differences) analyzed the systematic bias/differences between devices [62]. The Kendall Rank Correlation Coefficient (τ) between the absolute differences and the mean of both devices analyzed the degree of heteroscedasticity. If $\tau > 0.1$, the data were considered heteroscedastic and transformed by logarithms to the base 10 (log10) [63]. Linear regressions and Spearman's Rank Correlation Coefficients (ρ) analyzed the concurrent validity inter-device. The magnitude of correlation was interpreted as: 0.00–0.10, negligible; 0.10–0.39, weak; 0.40–0.69, moderate; 0.70–0.89, strong; 0.90–1.00, very strong [64]. The assumption of homoscedasticity was analyzed by inspecting the standardized residuals' scatter plots against the standardized predicted values. The absolute reliability was analyzed by estimating the standard error of measurement (SEM = SD of the difference between the smartphone application and video camera scores divided by the $\sqrt{2}$), the coefficient of variation (CV = (SEM/mean of both devices) × 100), and minimal detectable change (MDC = $\sqrt{2}$ × SEM × 1.96) [65]. CV values < 5% were considered acceptable [66]. The absolute percent error of the measurements (APE = ((|smartphone application − video camera)/video camera|) × 100) [67], and the accuracy ((video camera − (|video camera − smartphone application|)/video camera) × 100) were also calculated. An APE < 10% was considered acceptable [67]. We conducted a sample size calculation based on an expected reliability level of 0.90 and a minimum acceptable reliability level of 0.80. With an alpha value of 0.05 and 6 repetitions per participant, a minimum sample size of 32 was required to obtain a power of 80% [68]. The significance level was set at $p < 0.05$. All data were analyzed using Microsoft Office Excel 2016 and SPSS version 27 (SPSS Inc., Chicago, IL, USA). Figures were designed using the GraphPad Prism version 7.0 (GraphPad Software Inc., San Diego, CA, USA).

3. Results

Table 2 presents the relative reliability and relationship between devices. The results obtained through manual based on the video camera recordings correspond to traditional methods when an operator observes the patient performing the test and times his/her movements with a chronometer. It is expected that the careful analysis frame-by-frame to measure the time of the test is more accurate than an operator measuring the time with a chronometer. Therefore, the results provided in Tables 2 and 3 and Figures 3 and 4 comparing the video camera results and the smartphone-based approach correspond to a comparison between a traditional and a smartphone-based approach as well. The stand-up time and total time showed excellent relative reliability and very strong significant relationships ($p < 0.001$) inter-devices.

Table 3. Absolute reliability and accuracy inter-devices.

	Stand-Up Time (n = 842)	**Total Time (n = 892)**
SEM (s)	0.05	0.05
CV (%)	3.03	1.85
MDC (s)	0.14	0.14
APE (%)	5.79	3.90
Accuracy (%)	94.21	96.10

APE: absolute percent error calculated as: ((|smartphone application − video camera)/video camera|) × 100); Accuracy calculated as: ((video camera − (|video camera − smartphone application|)/video camera) × 100); CV: coefficient of variation; MDC: minimal detectable change; SEM: standard error of measurement.

Figure 3 shows the Bland-Altman plots of agreement between the mobile application and video-camera for the stand-up time (A) and total time (B).

Figure 4 shows the linear regression between the mobile application and the video-camera for all variables. The line of 45° indicates the amount of difference inter-devices in the measurement of the variables. Both the stand-up time and total time fall nearby the line of 45°. The resulting linear regression equation is provided for both variables.

Table 3 presents the absolute reliability and accuracy between devices. The stand-up time and total time showed CV values lower than 4%, revealing excellent absolute reliability. In both variables, the APE values were lower than 6%, and the accuracy was higher than 94%, thus revealing a high accuracy level.

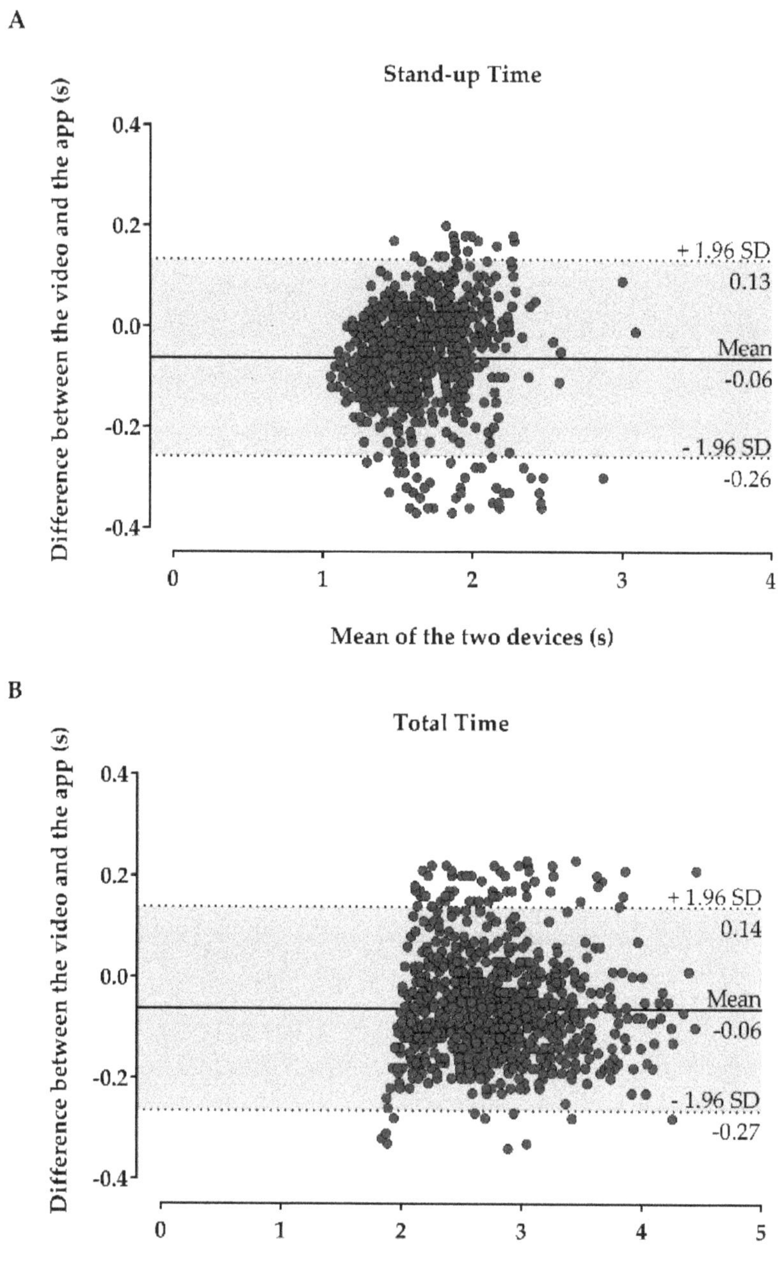

Figure 3. Bland-Altman plots with 95% limits of agreement (mean difference $\pm$ 1.96 $\times$ standard deviation [SD] of the differences) between the mobile application and video-camera for the stand-up time (**A**) and total time (**B**); the solid lines in the middle of the plots represent the mean difference/bias, while the upper and lower dotted lines represent the upper and lower LOA.

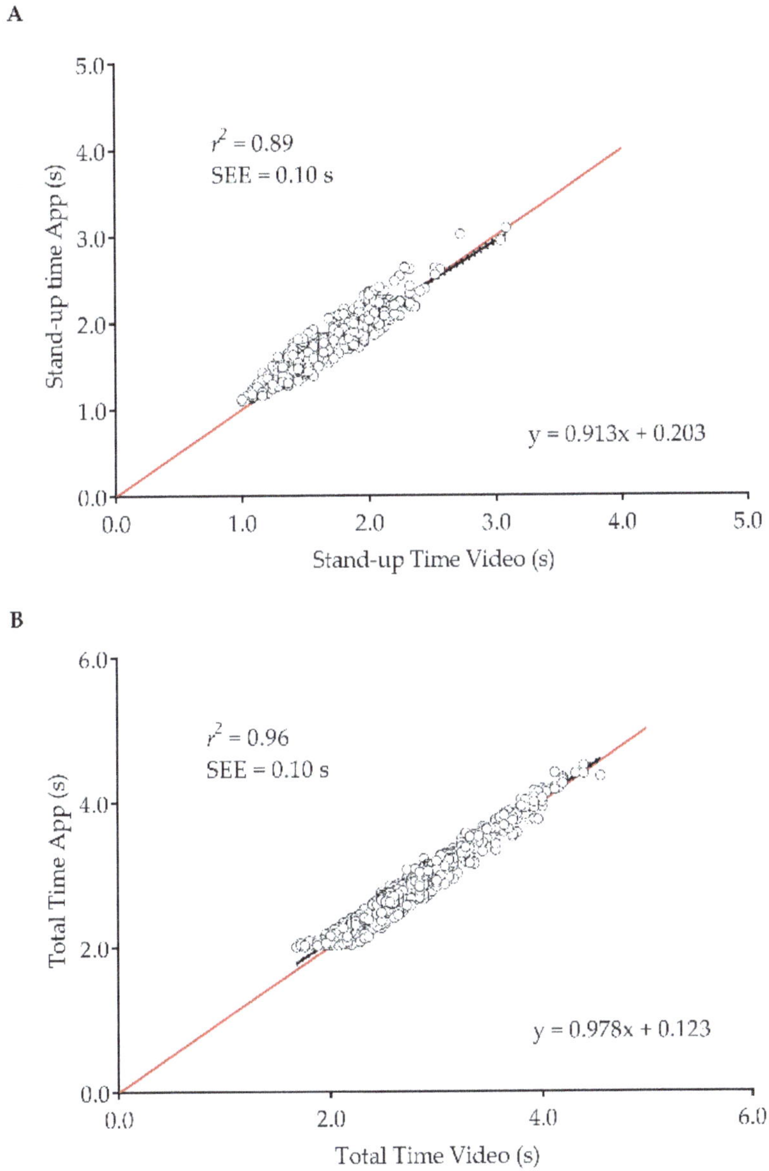

Figure 4. Linear regression between the mobile application and video-camera for the stand-up time (**A**) and total time (**B**); r^2: coefficient of determination; the black lines indicate the regression line, while the red lines indicate the line of 45°; dotted lines indicate 95% confidence intervals.

4. Discussion

In this study, we analyzed the validity of a mobile smartphone application to quantify the stand-up time and total time during the single sit-to-stand test with institutionalized older adults. The results revealed excellent reliability, high accuracy, and very strong relationships between devices in both temporal variables. These results agree with our central hypothesis, meaning that the mobile application is valid and reliable for measuring temporal variables during the single sit-to-stand test with institutionalized older adults.

Regarding stand-up time, only two studies reported acquiring this temporal variable with a mobile application during the sit-to-stand test with older adults. In a study with stroke patients [36] of both sexes (67.50 ± 13.18 years), the authors could observe a mean stand-up time of 1.95 s (SD = 0.08), which is 16% longer than the value observed in our study (1.68 ± 0.29 s). With our study focusing on older adults without any medical conditions that would affect mobility during the sit-to-stand test, these differences are expected. In the second study [30], which included community-dwelling older adults (73.5 ± 10.4 years), the reported mean stand-up time of 1.66 s (SD = 0.42) is close to our results. Possible reasons for these similarities might be associated with using a triaxial accelerometer embedded in the smartphone, a similar system to develop the mobile application, the participants' age, and the maximal intended speed during the sit-to-stand transfer. Other studies that captured the stand-up time through a mobile application [35] reported significantly lower times (0.47 ± 0.09 s). However, the results are incomparable as the participant sample pool included adults of a significantly wider age range (21–87 years), and the stand-up time in this study was defined as the rising phase of the sit-to-stand movement without taking into account the preparatory phase, i.e., when the trunk is shifted forward prior to seat-off. Additionally, the mobile application developed in that study quantified the sit-to-stand test based on high-speed video recordings and not through the triaxial accelerometer incorporated in the smartphone. Having only three studies related to the sit-to-stand test and its analysis using mobile devices suggests that this is an active area of research. Furthermore, the inconsistent results reported by different studies due to the reasons mentioned above justify the work performed in our research, which is in a more controlled and homogeneous age group of older adults, lacking in other studies.

Studies that used body-fixed sensors instead of a mobile device to record the stand-up time with older adults observed that the time ranged between 1.81 to 2.17 s [69–72]. Furthermore, when the participants were instructed to perform the movement as fast as possible, the time decreased to 1.74 s (SD = 0.33) [69] and 1.7 s (SD = 0.80) [42]. These results are relevantly like the stand-up time values presented in our study, mainly when the sit-to-stand transfer is performed at the maximal intended velocity. Considering the findings above, the accelerometer data acquired with a hybrid sensor or a mobile application seems to present similar results in the sit-to-stand test with older adults.

Regarding the total time (i.e., the complete measurement of the stand-up/sit-down cycle), to our best knowledge, only one study measured this variable with a mobile application [36]. Merchán-Baeza et al. [36], in a study with stroke patients, reported a total time of 4.09 ± 0.07 s. This time is 45% higher than the total time presented in our study (2.81 ± 0.50 s). However, this result is expected as none of our participants had suffered from a stroke. Hence, the observation of a faster time for sit-to-stand transitions in older adults without mobility disability when compared to stroke patients can be anticipated.

Our mobile application demonstrated high accuracy levels and minor errors to capture temporal variables during the single sit-to-stand test with older adults, reinforcing its validity and reliability. Although several related studies mentioned accurate mobile apps to measure temporal variables during the sit-to-stand test with older adults [31,35], none reported the accuracy values like ours, which does not allow valid comparisons with our results.

We want to note that the paper discusses a mobile app designed primarily for use in clinical settings or in controlled settings when the patient has help from caregivers, medical personnel, or family members to set up the application and place the mobile device properly. If used in an uncontrolled environment, then the test results can be invalid. However, even in such limiting circumstances, not having to visit a medical center to perform the test is quite valuable for older adults with mobility and dexterity problems.

Future studies should consider the following limitations and perform new analyses in the sit-to-stand test with aged populations to strengthen this field's knowledge. Firstly, the sit-to-stand test analysis can be strengthened by capturing other biomechanical variables. For example, developing an algorithm to calculate the velocity, force, and power generated during the sit-to-stand transitions will provide insightful information for researchers and clinicians. Secondly, determining the mobile application's intra-device reliability by repeating the experiment over different trials will help analyze the results' consistency. Thirdly, analyzing the mobile app's validity and reliability considering smartphones with different sampling frequencies will help understand if its use can be generalized among several devices. Finally, applying other field-based tests such as upper and lower body strength tests, the timed-up and go, or walking speed tests will enable researchers to analyze their relationship with the temporal variables collected during the sit-to-stand test.

Another promising avenue for research is to also include capabilities in the mobile application to identify whether the test was performed correctly, such as due to incorrect body positioning or performing incomplete movements. For the current research, we assume that the participants were trained to perform the test when asked to install the mobile application. Additionally, the mobile app will also include informative videos and tutorials about the test's procedures.

5. Conclusions

For older adults, sit-to-stand tasks are an essential facet of independence and well-being. Therefore, improved quantification of the sit-to-stand test is warranted. It can provide important information that can help improve the quality of life of older adults. The smartphone application presented in this study is suitable for valid and reliable measurements of temporal variables during the single sit-to-stand test with institutionalized older adults, specifically, stand-up time and total time. Researchers and clinicians commonly use these variables for different purposes, such as identifying frailty and analyzing the effects of different training interventions on these variables (i.e., do they improve the time to stand-up from the chair after a training program?). Therefore, having a valid and reliable instrument like our mobile application to measure these variables is clinically essential for capturing accurate data. The smartphone application can also be used in contexts where budget, space, time, and equipment are limited. It is also essential to note that, as the test ends, the results are presented on the smartphone screen in real-time, meaning that the evaluators can immediately access the data and provide reliable feedback regarding the test's performance. As a result, there is no need to use other materials more sensitive to human error like chronometers to capture the data. Finally, the data is also stored in the mobile phone and cloud, enabling follow-up analysis.

In the future, these tests should be evaluated by multidisciplinary teams comprised of coaches, physiotherapists, physicians, nurses, and technicians to identify potential issues that might have been neglected during this study. A pilot test in a broader population performed for a prolonged period should be conducted to evaluate the long-term effects of exercise or rehabilitation on older adults' sit-to-stand performance. As a result, it would help determine whether the current practice should be modified or updated and under which conditions.

Author Contributions: Conceptualization, methodology, software, validation, formal analysis, investigation, writing—original draft preparation, writing—review and editing; D.L.M., H.P.N., I.M.P., E.Z., M.M., N.M.G., J.D.R.-C., D.A.M. and M.C.M. All authors have read and agreed to the published version of the manuscript.

Funding: This research was funded by Portuguese Foundation for Science and Technology, I.P., grant number **SFRH/BD/147608/2019** and project number **UIDB/04045/2020**. This work is also funded by FCT/MEC through national funds and co-funded by FEDER—PT2020 partnership agreement under the project **UIDB/50008/2020** (*Este trabalho é financiado pela FCT/MEC através de fundos nacionais e cofinanciado pelo FEDER, no âmbito do Acordo de Parceria PT2020 no âmbito do projeto UIDB/50008/2020*), as well as by National Funds through the FCT—Foundation for Science and Technology, I.P., within the scope of the project **UIDB/00742/2020**.

Institutional Review Board Statement: The study was conducted according to the guidelines of the Declaration of Helsinki and approved by the Ethics Committee of University of Beira Interior (protocol code CE-UBI-Pj-2019-019; 27/05/2019).

Informed Consent Statement: Informed consent was obtained from all subjects involved in the study.

Data Availability Statement: Data is available in Mendeley data at http://dx.doi.org/10.17632/335rmgrfw2.4 io (accessed on 29 September 2020).

Acknowledgments: We wish to express our sincere gratitude to the care staff of the Santa Casa da Misericórdia do Fundão and thank all the participants involved in this study. We also would like to thank João Leal, full professor in the University of Agder in Norway, for his contribution to programming the mobile application. This research was funded by Portuguese Foundation for Science and Technology, I.P., grant number **SFRH/BD/147608/2019** and project number **UIDB/04045/2020.** This work is funded by FCT/MEC through national funds and co-funded by FEDER—PT2020 partnership agreement under the project **UIDB/50008/2020** (*Este trabalho é financiado pela FCT/MEC através de fundos nacionais e cofinanciado pelo FEDER, no âmbito do Acordo de Parceria PT2020 no âmbito do projeto UIDB/50008/2020*). This work is also funded by National Funds through the FCT—Foundation for Science and Technology, I.P., within the scope of the project **UIDB/00742/2020.** This article is based upon work from COST Action IC1303–AAPELE–Architectures, Algorithms and Protocols for Enhanced Living Environments and COST Action CA16226–SHELD-ON–Indoor living space improvement: Smart Habitat for the Elderly, supported by COST (European Cooperation in Science and Technology). More information in www.cost.eu. Furthermore, we would like to thank the Politécnico de Viseu for their support.

Conflicts of Interest: The authors declare no conflict of interest.

References

1. Bloom, D.E.; Canning, D.; Lubet, A. Global Population Aging: Facts, Challenges, Solutions & Perspectives. *Chall. Solut. Perspect. Daedalus* **2015**, *144*, 80–92.
2. Noury, N.; Quach, K.-A.; Berenguer, M.; Bouzi, M.-J.; Teyssier, H. A feasibility study of using a smartphone to monitor mobility in elderly. In Proceedings of the 2012 IEEE 14th International Conference on e-Health Networking, Applications and Services (Healthcom), Beijing, China, 11–13 October 2012.
3. Dall, P.M.; Kerr, A. Frequency of the sit to stand task: An observational study of free-living adults. *Appl. Ergon.* **2010**, *41*, 58–61. [CrossRef]
4. Frykberg, G.E.; Häger, C.K. Movement analysis of sit-to-stand—research informing clinical practice. *Phys. Ther. Rev.* **2015**, *20*, 156–167. [CrossRef]
5. Yoshioka, S.; Nagano, A.; Hay, D.C.; Fukashiro, S. The minimum required muscle force for a sit-to-stand task. *J. Biomech.* **2012**, *45*, 699–705. [CrossRef] [PubMed]
6. Cadore, E.L.; Izquierdo, M. New Strategies for the Concurrent Strength-, Power-, and Endurance-Training Prescription in Elderly Individuals. *J. Am. Med. Dir. Assoc.* **2013**, *14*, 623–624. [CrossRef] [PubMed]
7. Tiedemann, A.; Shimada, H.; Sherrington, C.; Murray, S.; Lord, S. The comparative ability of eight functional mobility tests for predicting falls in community-dwelling older people. *Age Ageing* **2008**, *37*, 430–435. [CrossRef]
8. Buatois, S.; Miljkovic, D.; Manckoundia, P.; Gueguen, R.; Miget, P.; Vanã§on, G.; Perrin, P.; Benetos, A. FIVE TIMES SIT TO STAND TEST IS A PREDICTOR OF RECURRENT FALLS IN HEALTHY COMMUNITY-LIVING SUBJECTS AGED 65 AND OLDER. *J. Am. Geriatr. Soc.* **2008**, *56*, 1575–1577. [CrossRef] [PubMed]
9. Lusardi, M.M.; Fritz, S.; Middleton, A.; Allison, L.; Wingood, M.; Phillips, E.; Criss, M.; Verma, S.; Osborne, J.; Chui, K.K. Determining Risk of Falls in Community Dwelling Older Adults: A Systematic Review and Meta-analysis Using Posttest Probability. *J. Geriatr. Phys. Ther.* **2017**, *40*, 1–36. [CrossRef]
10. Minematsu, A.; Hazaki, K.; Harano, A.; Okamoto, N. Association between muscle strength and physical performance in Japanese elderly: The Fujiwara-kyo Study. *JCGG* **2018**, *9*, 44–51. [CrossRef]
11. Glenn, J.M.; Gray, M.; Binns, A. Relationship of Sit-to-Stand Lower-Body Power With Functional Fitness Measures Among Older Adults With and Without Sarcopenia. *J. Geriatr. Phys. Ther.* **2017**, *40*, 42–50. [CrossRef] [PubMed]
12. Schaubert, K.L.; Bohannon, R.W. Reliability and Validity of Three Strength Measures Obtained From Community-Dwelling Elderly Persons. *J. Strength Cond. Res.* **2005**, *19*, 717–720. [CrossRef] [PubMed]
13. Suwannarat, P.; Kaewsanmung, S.; Thaweewannakij, T.; Amatachaya, S. The use of functional performance tests by primary health-care providers to determine walking ability with and without awalking device in community-dwelling elderly. *Physiother. Theory Pr.* **2021**, *37*, 64–72. [CrossRef]
14. Yanagawa, N.; Shimomitsu, T.; Kawanishi, M.; Fukunaga, T.; Kanehisa, H. Relationship between performances of 10-time-repeated sit-to-stand and maximal walking tests in non-disabled older women. *J. Physiol. Anthr.* **2016**, *36*, 2. [CrossRef] [PubMed]

15. Wang, C.-Y.; Yeh, C.-J.; Hu, M.-H. Mobility-related performance tests to predict mobility disability at 2-year follow-up in community-dwelling older adults. *Arch. Gerontol. Geriatr.* **2011**, *52*, 1–4. [CrossRef]
16. Pinho, A.S.; Salazar, A.P.; Hennig, E.M.; Spessato, B.C.; Domingo, A.; Pagnussat, A.S. Can We Rely on Mobile Devices and Other Gadgets to Assess the Postural Balance of Healthy Individuals? A Systematic Review. *Sensors* **2019**, *19*, 2972. [CrossRef] [PubMed]
17. Higgins, J.P. Smartphone Applications for Patients' Health and Fitness. *Am. J. Med.* **2016**, *129*, 11–19. [CrossRef] [PubMed]
18. Dorsey, E.R.; Chan, Y.-F.Y.; McConnell, M.V.; Shaw, S.Y.; Trister, A.D.; Friend, S.H. The Use of Smartphones for Health Research. *Acad. Med.* **2017**, *92*, 157–160. [CrossRef]
19. Moore, S.; Jayewardene, D. The use of smartphones in clinical practice. *Nurs. Manag. (Springhouse)* **2014**, *21*, 18–22. [CrossRef]
20. Horak, F.; King, L.; Mancini, M. Role of Body-Worn Movement Monitor Technology for Balance and Gait Rehabilitation. *Phys. Ther.* **2015**, *95*, 461–470. [CrossRef]
21. Joundi, R.A.; Brittain, J.-S.; Jenkinson, N.; Green, A.L.; Aziz, T. Rapid tremor frequency assessment with the iPhone accelerometer. *Park. Relat. Disord.* **2011**, *17*, 288–290. [CrossRef] [PubMed]
22. Wile, D.J.; Ranawaya, R.; Kiss, Z.H. Smart watch accelerometry for analysis and diagnosis of tremor. *J. Neurosci. Methods* **2014**, *230*, 1–4. [CrossRef]
23. Weiss, A.; Herman, T.; Plotnik, M.; Brozgol, M.; Giladi, N.; Hausdorff, J.M. An instrumented timed up and go: The added value of an accelerometer for identifying fall risk in idiopathic fallers. *Physiol. Meas.* **2011**, *32*, 2003–2018. [CrossRef] [PubMed]
24. Zhang, F.; Ferrucci, L.; Culham, E.; Metter, E.J.; Guralnik, J.; Deshpande, N. Performance on Five Times Sit-to-Stand Task as a Predictor of Subsequent Falls and Disability in Older Persons. *J. Aging Health* **2012**, *25*, 478–492. [CrossRef] [PubMed]
25. Csuka, M.; Mccarty, D.J. Simple method for measurement of lower extremity muscle strength. *Am. J. Med.* **1985**, *78*, 77–81. [CrossRef]
26. Morita, A.; Bisca, G.W.; Machado, F.V.C.; A Hernandes, N.; Pitta, F.; Probst, V.S. Best Protocol for the Sit-to-Stand Test in Subjects With COPD. *Respir. Care* **2018**, *63*, 1040–1049. [CrossRef] [PubMed]
27. Marques, D.L.; Neiva, H.P.; Marinho, D.A.; Marques, M.C. Novel Resistance Training Approach to Monitoring the Volume in Older Adults: The Role of Movement Velocity. *Int. J. Environ. Res. Public Health* **2020**, *17*, 7557. [CrossRef] [PubMed]
28. Alcazar, J.; Losa-Reyna, J.; Rodriguez-Lopez, C.; Alfaro-Acha, A.; Rodriguez-Mañas, L.; Ara, I.; García-García, F.J.; Alegre, L.M. The sit-to-stand muscle power test: An easy, inexpensive and portable procedure to assess muscle power in older people. *Exp. Gerontol.* **2018**, *112*, 38–43. [CrossRef] [PubMed]
29. Janssen, W.G.M.; Bussmann, J.B.J.; Horemans, H.L.D.; Stam, H.J. Validity of accelerometry in assessing the duration of the sit-to-stand movement. *Med. Biol. Eng. Comput.* **2008**, *46*, 879–887. [CrossRef] [PubMed]
30. Cerrito, A.; Bichsel, L.; Radlinger, L.; Schmid, S. Reliability and validity of a smartphone-based application for the quantification of the sit-to-stand movement in healthy seniors. *Gait Posture* **2015**, *41*, 409–413. [CrossRef]
31. Chan, M.H.; Keung, D.T.; Lui, S.Y.; Cheung, R.T. A validation study of a smartphone application for functional mobility assessment of the elderly. *Hong Kong Physiother. J.* **2016**, *35*, 1–4. [CrossRef] [PubMed]
32. Pires, I.; Felizardo, V.; Pombo, N.; Garcia, N.M. Limitations of Energy Expenditure Calculation Based on a Mobile Phone Accelerometer. In Proceedings of the 2017 International Conference on High Performance Computing & Simulation (HPCS), Genoa, Italy, 19–21 July 2017.
33. Pires, I.M.; Garcia, N.M.; Pombo, N.; Florez-Revuelta, F. Limitations of the Use of Mobile Devices and Smart Environments for the Monitoring of Ageing People. In Proceedings of the 4th International on Information and Communication Technologies for Ageing Well and e-Health, Lisbon, Portugal, 27 March 2018. [CrossRef]
34. Orange, S.T.; Metcalfe, J.W.; Liefeith, A.; Jordan, A.R. Validity of various portable devices to measure sit-to-stand velocity and power in older adults. *Gait Posture* **2020**, *76*, 409–414. [CrossRef] [PubMed]
35. Ruiz-Cárdenas, J.D.; Rodríguez-Juan, J.J.; Smart, R.R.; Jakobi, J.M.; Jones, G.R. Validity and reliability of an iPhone App to assess time, velocity and leg power during a sit-to-stand functional performance test. *Gait Posture* **2018**, *59*, 261–266. [CrossRef]
36. Merchán-Baeza, J.A.; González-Sánchez, M.; Cuesta-Vargas, A.I. Using Smartphones to Collect Quantitative Data on Lower Limb Functionality in People Who Have Suffered a Stroke. *J. Stroke Cerebrovasc. Dis.* **2018**, *27*, 3555–3562. [CrossRef] [PubMed]
37. Rojas, H.A.G.; Cuevas, P.C.; Figueras, E.E.Z.; Foix, S.C.; Egea, A.J.S. Time measurement characterization of stand-to-sit and sit-to-stand transitions by using a smartphone. *Med. Biol. Eng. Comput.* **2017**, *56*, 879–888. [CrossRef] [PubMed]
38. Dent, E.; Morley, J.E.; Cruz-Jentoft, A.J.; Woodhouse, L.; Rodríguez-Mañas, L.; Fried, L.P.; Woo, J.; Aprahamian, I.; Sanford, A.; Lundy, J.; et al. Physical Frailty: ICFSR International Clinical Practice Guidelines for Identification and Management. *J. Nutr. Health Aging* **2019**, *23*, 771–787. [CrossRef] [PubMed]
39. Hausdorff, J.M.; Edelberg, H.K.; Mitchell, S.L.; Goldberger, A.L.; Wei, J.Y. Increased gait unsteadiness in community-dwelling elderly fallers. *Arch. Phys. Med. Rehabil.* **1997**, *78*, 278–283. [CrossRef]
40. Zhou, Y.; Rehman, R.Z.U.; Hansen, C.; Maetzler, W.; Del Din, S.; Rochester, L.; Hortobágyi, T.; Lamoth, C.J.C. Classification of Neurological Patients to Identify Fallers Based on Spatial-Temporal Gait Characteristics Measured by a Wearable Device. *Sensors* **2020**, *20*, 4098. [CrossRef] [PubMed]
41. Millor, N.; Lecumberri, P.; Gómez, M.; Martínez-Ramírez, A.; Izquierdo, M. An evaluation of the 30-s chair stand test in older adults: Frailty detection based on kinematic parameters from a single inertial unit. *J. Neuroeng. Rehabil.* **2013**, *10*, 86. [CrossRef]

42. Van Lummel, R.C.; Walgaard, S.; Maier, A.B.; Ainsworth, E.; Beek, P.J.; van Dieën, J.H. The Instrumented Sit-to-Stand Test (iSTS) Has Greater Clinical Relevance than the Manually Recorded Sit-to-Stand Test in Older Adults. *PLoS ONE* **2016**, *11*, e0157968. [CrossRef]
43. Boukadida, A.; Piotte, F.; Dehail, P.; Nadeau, S. Determinants of sit-to-stand tasks in individuals with hemiparesis post stroke: A review. *Ann. Phys. Rehabil. Med.* **2015**, *58*, 167–172. [CrossRef]
44. Marques, G.; Saini, J.; Pires, I.M.; Miranda, N.; Pitarma, R. Internet of Things for Enhanced Living Environments, Health and Well-Being: Technologies, Architectures and Systems. *Adv. Intell. Syst. Comput.* **2020**, *1132*, 616–631. [CrossRef]
45. Goleva, R.I.; Garcia, N.M.; Mavromoustakis, C.X.; Dobre, C.; Mastorakis, G.; Stainov, R.; Chorbev, I.; Trajkovik, V. AAL and ELE Platform Architecture. *Ambient Assist. Living Enhanc. Living Environ.* **2017**, 171–209. [CrossRef]
46. Garcia, N.M.; Rodrigues, J.J.P.C. *Ambient Assisted Living*; CRC Press: Boca Raton, FL, USA, 2015; ISBN 9781439869857.
47. Pires, I.M.; Teixeira, M.C.; Pombo, N.; Garcia, N.M.; Flórez-Revuelta, F.; Spinsante, S.; Goleva, R.; Zdravevski, E. Android Library for Recognition of Activities of Daily Living: Implementation Considerations, Challenges, and Solutions. *Open Bioinform. J.* **2018**, *11*, 61–88. [CrossRef]
48. Pires, I.M.; Garcia, N.M.; Teixeira, M.C.C. Calculation of Jump Flight Time using a Mobile Device. In Proceedings of the 8th International Conference on Health Informatics, Lisabon, Portugal, 12–15 January 2015.
49. Sousa, P.S.; Sabugueiro, D.; Felizardo, V.; Couto, R.; Pires, I.; Garcia, N.M. mHealth Sensors and Applications for Personal Aid. In *Mobile Health*; Springer International Publishing: Berlin/Heidelberg, Germany, 2015; Volume 5, pp. 265–281.
50. Pires, I.M.; Hussain, F.; Garcia, N.M.; Zdravevski, E. Improving Human Activity Monitoring by Imputation of Missing Sensory Data: Experimental Study. *Futur. Internet* **2020**, *12*, 155. [CrossRef]
51. Marques, D.L.; Neiva, H.P.; Pires, I.M.; Marinho, D.A.; Marques, M.C. Accelerometer data from the performance of sit-to-stand test by elderly people. *Data Brief.* **2020**, *33*, 106328. [CrossRef] [PubMed]
52. Yamada, T.; Demura, S.; Takahashi, K. Center of gravity transfer velocity during sit-to-stand is closely related to physical functions regarding fall experience of the elderly living in community dwelling. *Health* **2013**, *5*, 2097–2103. [CrossRef]
53. Viecelli, C.; Graf, D.; Aguayo, D.; Hafen, E.; Füchslin, R.M. Using smartphone accelerometer data to obtain scientific mechanical-biological descriptors of resistance exercise training. *PLoS ONE* **2020**, *15*, e0235156. [CrossRef] [PubMed]
54. Atrsaei, A.; Dadashi, F.; Hansen, C.; Warmerdam, E.; Mariani, B.; Maetzler, W.; Aminian, K. Postural transitions detection and characterization in healthy and patient populations using a single waist sensor. *J. Neuroeng. Rehabil.* **2020**, *17*, 70. [CrossRef] [PubMed]
55. Stack, E.; Agarwal, V.; King, R.; Burnett, M.; Tahavori, F.; Janko, B.; Harwin, W.; Ashburn, A.; Kunkel, D. Identifying balance impairments in people with Parkinson's disease using video and wearable sensors. *Gait Posture* **2018**, *62*, 321–326. [CrossRef]
56. Bourke, A.K.; Ihlen, E.A.F.; Bergquist, R.; Wik, P.B.; Vereijken, B.; Helbostad, J.L. A Physical Activity Reference Data-Set Recorded from Older Adults Using Body-Worn Inertial Sensors and Video Technology—The ADAPT Study Data-Set. *Sensors* **2017**, *17*, 559. [CrossRef]
57. Winter, D.A. Signal Processing. In *Biomechanics and Motor Control of Human Movement*; John Wiley & Sons: Hoboken, NJ, USA, 2009; pp. 14–44. ISBN 978-0-470-39818-0.
58. Gandomkar, Z.; Bahrami, F. Method to classify elderly subjects as fallers and non-fallers based on gait energy image. *Heal. Technol. Lett.* **2014**, *1*, 110–114. [CrossRef]
59. Nan, Y.; Lovell, N.H.; Redmond, S.J.; Wang, K.; Delbaere, K.; Van Schooten, K.S. Deep Learning for Activity Recognition in Older People Using a Pocket-Worn Smartphone. *Sensors* **2020**, *20*, 7195. [CrossRef] [PubMed]
60. Pires, I.M.; Marques, D.; Pombo, N.; Garcia, N.M.; Marques, M.C.; Florez-Revuelta, F. Measurement of the Reaction Time in the 30-S Chair Stand Test using the Accelerometer Sensor Available in off-the-Shelf Mobile Devices. In Proceedings of the 4th International on Information and Communication Technologies for Ageing Well and e-Health, Lisbon, Portugal, 27 March 2018. [CrossRef]
61. Koo, T.K.; Li, M.Y. A Guideline of Selecting and Reporting Intraclass Correlation Coefficients for Reliability Research. *J. Chiropr. Med.* **2016**, *15*, 155–163. [CrossRef] [PubMed]
62. Bland, J.M.; Altman, D.G. Measuring agreement in method comparison studies. *Stat. Methods Med. Res.* **1999**, *8*, 135–160. [CrossRef] [PubMed]
63. Bland, J.M.; Altman, D.G. Statistics Notes: Measurement error proportional to the mean. *BMJ* **1996**, *313*, 106. [CrossRef] [PubMed]
64. Schober, P.; Boer, C.; Schwarte, L.A. Correlation Coefficients: Appropriate use and interpretation. *Anesth. Analg.* **2018**, *126*, 1763–1768. [CrossRef]
65. De Vet, H.C.W.; Terwee, C.B.; Mokkink, L.B.; Knol, D.L. *Measurement in Medicine*; Cambridge University Press: Cambridge, UK, 2011; ISBN 9780521118200.
66. Courel-Ibáñez, J.; Martínez-Cava, A.; Morán-Navarro, R.; Escribano-Peñas, P.; Chavarren-Cabrero, J.; González-Badillo, J.J.; Pallarés, J.G. Reproducibility and Repeatability of Five Different Technologies for Bar Velocity Measurement in Resistance Training. *Ann. Biomed. Eng.* **2019**, *47*, 1523–1538. [CrossRef] [PubMed]
67. Montoye, A.H.K.; Mitrzyk, J.R.; Molesky, M.J. Comparative Accuracy of a Wrist-Worn Activity Tracker and a Smart Shirt for Physical Activity Assessment. *Meas. Phys. Educ. Exerc. Sci.* **2017**, *21*, 201–211. [CrossRef]
68. Arifin, W.N. Sample Size Calculator (Version 2.0) [Spreadsheet File]. Available online: http://wnarifin.github.io (accessed on 29 September 2020).

69. Regterschot, G.R.H.; Folkersma, M.; Zhang, W.; Baldus, H.; Stevens, M.; Zijlstra, W. Sensitivity of sensor-based sit-to-stand peak power to the effects of training leg strength, leg power and balance in older adults. *Gait Posture* **2014**, *39*, 303–307. [CrossRef]
70. Regterschot, G.R.H.; Zhang, W.; Baldus, H.; Stevens, M.; Zijlstra, W. Accuracy and concurrent validity of a sensor-based analysis of sit-to-stand movements in older adults. *Gait Posture* **2016**, *45*, 198–203. [CrossRef] [PubMed]
71. Zijlstra, A.; Mancini, M.; Lindemann, U.; Chiari, L.; Zijlstra, W. Sit-stand and stand-sit transitions in older adults and patients with Parkinson's disease: Event detection based on motion sensors versus force plates. *J. Neuroeng. Rehabil.* **2012**, *9*, 75. [CrossRef] [PubMed]
72. Van Lummel, R.C.; Ainsworth, E.; Lindemann, U.; Zijlstra, W.; Chiari, L.; Van Campen, P.; Hausdorff, J.M. Automated approach for quantifying the repeated sit-to-stand using one body fixed sensor in young and older adults. *Gait Posture* **2013**, *38*, 153–156. [CrossRef] [PubMed]

sensors

MDPI

Article

Using Direct Acyclic Graphs to Enhance Skeleton-Based Action Recognition with a Linear-Map Convolution Neural Network

Tan-Hsu Tan [1], Jin-Hao Hus [1], Shing-Hong Liu [2,*], Yung-Fa Huang [3] and Munkhjargal Gochoo [4]

1 Department of Electrical Engineering, National Taipei University of Technology, Taipei 10608, Taiwan; thtan@ntut.edu.tw (T.-H.T.); t103310022@ntut.edu.tw (J.-H.H.)
2 Department of Computer Science and Information Engineering, Chaoyang University of Technology, Taichung 413310, Taiwan
3 Department of Information and Communication Engineering, Chaoyang University of Technology, Taichung 413310, Taiwan; yfahuang@cyut.edu.tw
4 Department of Computer Science & Software Engineering, College of Information Technology, United Arab Emirates University, Al Ain P.O. Box 15551, Abu Dhabi, United Arab Emirates; mgochoo@uaeu.ac.ae
* Correspondence: shliu@cyut.edu.tw; Tel.: +886-4-233-230-000-7811

Abstract: Research on the human activity recognition could be utilized for the monitoring of elderly people living alone to reduce the cost of home care. Video sensors can be easily deployed in the different zones of houses to achieve monitoring. The goal of this study is to employ a linear-map convolutional neural network (CNN) to perform action recognition with RGB videos. To reduce the amount of the training data, the posture information is represented by skeleton data extracted from the 300 frames of one film. The two-stream method was applied to increase the accuracy of recognition by using the spatial and motion features of skeleton sequences. The relations of adjacent skeletal joints were employed to build the direct acyclic graph (DAG) matrices, source matrix, and target matrix. Two features were transferred by DAG matrices and expanded as color texture images. The linear-map CNN had a two-dimensional linear map at the beginning of each layer to adjust the number of channels. A two-dimensional CNN was used to recognize the actions. We applied the RGB videos from the action recognition datasets of the NTU RGB+D database, which was established by the Rapid-Rich Object Search Lab, to execute model training and performance evaluation. The experimental results show that the obtained precision, recall, specificity, F1-score, and accuracy were 86.9%, 86.1%, 99.9%, 86.3%, and 99.5%, respectively, in the cross-subject source, and 94.8%, 94.7%, 99.9%, 94.7%, and 99.9%, respectively, in the cross-view source. An important contribution of this work is that by using the skeleton sequences to produce the spatial and motion features and the DAG matrix to enhance the relation of adjacent skeletal joints, the computation speed was faster than the traditional schemes that utilize single frame image convolution. Therefore, this work exhibits the practical potential of real-life action recognition.

Keywords: linear-map convolutional neural network; direct acyclic graph; action recognition; spatial feature; temporal feature

Citation: Tan, T.-H.; Hus, J.-H.; Liu, S.-H.; Huang, Y.-F.; Gochoo, M. Using Direct Acyclic Graphs to Enhance Skeleton-Based Action Recognition with a Linear-Map Convolution Neural Network. *Sensors* **2021**, *21*, 3112. https://doi.org/10.3390/s21093112

Academic Editor: Ivan Miguel Serrano Pires

Received: 30 March 2021
Accepted: 27 April 2021
Published: 29 April 2021

1. Introduction

Recently, the lifespans of the world's population are increasing, and society is gradually aging. According to the report of the United Nations [1], the number of elderly people (over 65) in the world in 2019 was 703 million, and this is estimated to double to 1.5 billion by 2050. From 1990 to 2019, the proportion of the global population over 65 years old increased from 6% to 9%, and the proportion of the elderly population is expected to further increase to 16% by 2050. In Taiwan, the report of the National Development Council indicated that the elderly population with age over 65 will exceed 20% of the national population at 2026 [2].

Taiwan will enter a super-aged society in 2026. This means that the labor manpower will gradually decrease in the future. Thus, the cost of home care for elders will significantly increase. In homecare, the monitoring of elderly people living alone is a major issue. The behaviors of their activities have a high relation with their physical and mental health [3,4]. Therefore, how to use artificial intelligence (AI) techniques to reduce the cost of home care is an important challenge.

The recognition of body activities has two major techniques. One is physical sensors, like accelerometers [5,6], gyroscopes [7], and strain gauges [8], which have the advantage that they can be worn on the body to monitor dangerous activities throughout the day and the disadvantage that few activities can be recognized. Therefore, the physical sensors are typically not used to identify the daily activities. Another technique is the charge-coupled device (CCD) camera [9,10], which has the advantage of being able to recognize many daily activities and the disadvantage that it can only be used to monitor the activities of people in a local area. Thus, it is suitable to be used in a home environment.

Many previous studies have used deep learning techniques to recognize daily activities, including two-stream convolutional neural networks (CNNs), long short-term memory networks (LSTMNs), and three-dimensional CNNs (3D CNNs). For the two-stream CNN, Karpathy et al. used context stream and fovea streams to train a CNN [11]. The two streams proposed by Simonyan et al. were spatial and temporal streams, which represent the static and dynamic frames of each action's film [12]; however, the spending time for each action was different.

Thus, Wang et al. proposed a time segment network to normalize the spatial and temporal streams [13]. Jiang et al. used two streams as the input combined with CNN and LSTMN [14]. Ji et al. proposed 3D CNN to obtain the features of the spatial and temporal streams [15]. The two-stream methods using image and optical flow to represent the spatial and temporal streams had a better performance for recognizing activities compared with the one-stream methods. However, the weakness is that its doubled data amount requires more time to train the model.

Studies have used skeletal data as the common input feature for human action recognition [16–18], where the 3D skeletal data were typically obtained by use of the depth camera. In these studies, the number of recognized actions was less than 20 [17], and the skeletal data had to be processed to extract the features. Machine learning methods were used in these studies. The spatiotemporal information of skeleton sequences was exploited using recurrent neural networks (RNNs) [19,20].

Both the amount of data and recognized actions were less than in the video datasets [16–20]. An RNN tends to overemphasize the temporal information and ignore the spatial information, leading to low accuracy. However, the advantage of methods employing skeletal data is that these requires less training data and training time compared to those using image data. Hou et al. used a CNN to recognize actions with skeletal features [21]. Therefore, an effective method to encode the spatiotemporal information of a skeleton sequence into color texture images that could be recognized by a CNN is a relevant issue.

A directed acyclic graph (DAG) consists of a combination of nodes and edges. Each node points to another node by an edge. These directions do not become a circle graph that will end at the extremities. DAGs are usually used to represent causal relations amongst variables, and they are also used extensively to determine which variables need to be controlled for confounding in order to estimate causal effects [22]. The physical posture of people can be described by the positions of the skeletal joints. The adjacent joints have a causal relation when the body is moving. Thus, we can define the DAG of skeletal joints to explain the relations of physical skeletons.

This study aims to recognize the daily activities with films recorded by CCD cameras. To reduce the large amount of data for model training, we transferred body images to physical postures with an open system, AlphaPose [23]. The posture information is the skeleton sequences captured from the films of actions to build the spatial and motion features. These features all include both the spatial and temporal characteristics of actions. The relations of adjacent joints were used to build direct acyclic graph (DAG) matrices, the source matrix, and the target matrix.

These features are expanded by the DAG matrices as color texture images. The linear-map CNN has a two-dimensional linear map at the beginning of each layer to adjust the number of channels. Then, a two-dimensional CNN is used to recognize the actions. A structure with two streams was used to increase the accuracy of the action recognition. The datasets (NTU RGB+D) used in this study is an open source supported by Rapid-Rich Object Search Lab, National Technological University, Singapore [18].

In our work, a total of 49 actions, including daily actions, medical conditions, and mutual actions, were considered for action recognition. The total number of films was 46,452 for the cross-subject and cross-view sources. Of the cross-subject sources, 32,928 films were used for training, and 13,524 films were used for testing. Of the cross-view sources, 30,968 films were used for training, and 15,484 were used for testing. The experimental results show that the performance of our method was better than those in the previous studies.

2. Methods

Figure 1 shows the flowchart of action recognition in this study, which has three phases. In the feature phase, RGB images were processed by AlphaPose [23] to obtain the coordinate values of the skeletal joints of the subject in an image as a vector. A film contained 300 images that were used to build a posture matrix as the feature. We defined the spatial features and motion features by the coordinate values of the skeletal joints for each film. Spatial features are the position information of skeletons and joints, and motion features are the optical-flow information of skeletons and joints.

Each feature was expanded into two features of source and target by DAGs. In the recognition phase, a 10-layer linear-map CNN was used to recognize the activities. The cross-subject and cross-view evaluations were used to test the performance of this linear-map CNN. In the output phase, the results of the spatial and motion features were fused to show the recognized actions.

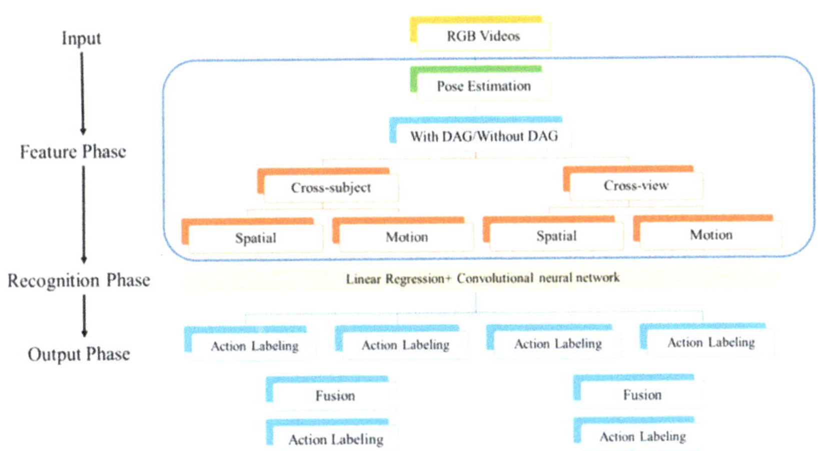

Figure 1. The flowchart of the action recognition in this study.

2.1. NTU RGB+D Dataset

The datasets of action recognition supported by the Rapid-Rich Object Search Lab, National Technological University, Singapore [24] were used in the study. There were 56,880 files, including 60 action classes. Each file consists of RGB, depth, and skeleton data of human actions. All actions were recorded by three Kinect V2 cameras. The size of the RGB images was 1920 × 1080. There were 40 classes for daily actions, 9 classes for medical conditions, and 11 classes for mutual actions. Forty distinct subjects were invited for this data collection.

The physical activities of only a single person were recognized, and the sample size of 46,452, including the 49 physical activities, was used in this study. To ensure standard evaluations for all the reported results on the benchmark, two types of action classification evaluation (cross-subject evaluation, and cross-view evaluation) were used [24]. In the cross-subject evaluation, the sample sizes for training and testing sets were 40,320 and 16,560, respectively. In the cross-view evaluation, the sample sizes for training and testing sets were 37,920 and 18,960, respectively.

2.2. Spatial and Motion Features

The RGB images were processed by the AlphaPose [23] to obtain the coordinate values of skeleton joints of people in the image, and the format is shown in Equation (1).

$$\text{pose} = \{(x_0, y_0, c_0, \cdots, x_M, y_M, c_M) \parallel M = 17\} \quad (1)$$

where x_i and y_i are the coordinate values of ith joint, c_i is the confidence score, and M is the index of the joints. According to the coordinate values of the joints, the spatial and motion variables were defined, as shown in Table 1. n is the index of the frames. The spatial variables are the joint data (v_i), and skeleton data ($s_{i,j}$). The motion variables are the motion data of the joints and skeleton (m_{vi} and m_{si}). Thus, there are four features in this study, F_v, F_s, F_{mv}, and F_{ms}. Table 2 shows the indexes and definition of 18 joints and 17 edges in the body. The 17 edges, e_i, are defined as the relations between two adjacent joints, i-1th and ith.

Table 1. The indexes of the joints and relations between every two adjacent joints at the 17 edges.

Spatial Variables	
Joint Data	**Skeleton Data**
$v_i(n) = (x_i(n), y_i(n))$	$s_{i,j}(n) = \left(x_i(n) - x_j(n), y_i(n) - y_j(n)\right)$
Motion Variables	
Joint-Motion Data	**Skeleton-Motion Data**
$mv_i(n) = v_i(n+1) - v_i(n)$	$ms_i = s_{i,i+1}(n+1) - s_{i,i+1}(n)$

2.3. Directed Acyclic Graph

DAG was used to describe the relations of 18 joints. The nodes of DAG represent the joints, and the flows represent the edges. Each edge has a source joint and a target joint. Thus, two DAG matrices, the source matrix and target matrix, can be defined. If the ith joint is the source point of the jth edge, the element (j, i) of the source matrix is set as 1. Otherwise, the element (j, i) is set as 0.

The target matrix is set as the source matrix. Then, each row of the source and target matrices is normalized to avoid overvalues. To match the size of feature, e(0, 0) = 1 is defined as a virtual edge. The sizes of the source and target matrices are 18 × 18. Figure 2a is the source matrix, S, color-none is 0, color-black is 1, and color-original is 0.25. Figure 2b is the target matrix, T.

Table 2. The indexes and positions of the joints and edges for the skeletal data.

Index	Joint Position	Index	Joint Position	Indexe of Joints and Edges
0	Nose	9	Right knee	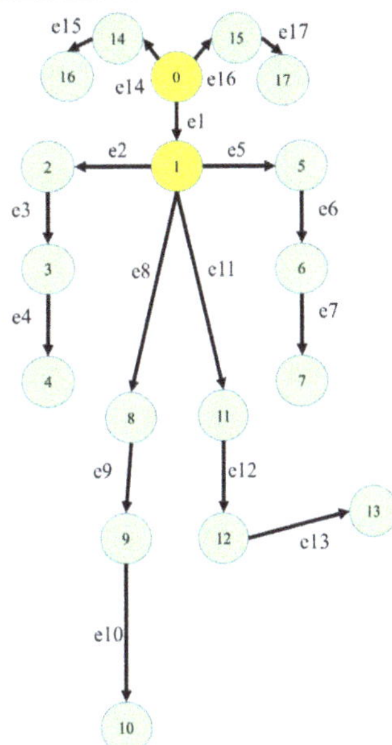
1	Neck	10	Right ankle	
2	Right shoulder	11	Left femur	
3	Right elbow	12	Left knee	
4	Right wrist	13	Left ankle	
5	Left shoulder	14	Right eye	
6	Left elbow	15	Left eye	
7	Left wrist	16	Right ear	
8	Right femur	17	Left ear	

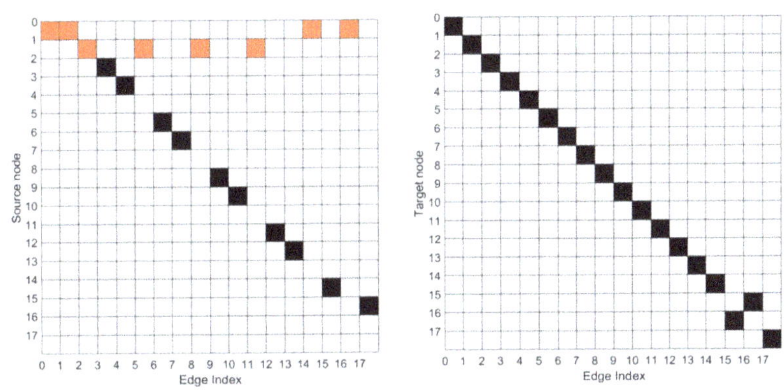

(**a**) (**b**)

Figure 2. (**a**) Source matrix, (**b**) Target matrix. Color-none is 0, color-black is 1, and color-orange is 0.25.

2.4. Input Features

We used 300 frames for every film. The joint data built the joint feature (F_v), the skeleton data built the skeleton feature (F_s), the joint-motion data built the joint-motion feature (F_{mv}), and the skeleton-motion data built the skeleton-motion feature (F_{ms}). Figure 3 shows the contents of a feature with x and y values of a data. Thus, the information of the film was reduced to four 600 × 18 matrices for four features, F_v, F_s, F_{mv}, and F_{ms}. The F_v was expanded by the DAG matrix into two features, F_{vin} and F_{vout}.

$$F_{vin} = F_v \times S^T \quad (2)$$

$$F_{vout} = F_v \times T^T \quad (3)$$

F_s was expanded by the DAG matrix to F_{sin} and F_{sout}; F_{mv} was expanded to F_{mvin} and F_{mvout}; and F_{ms} was expanded to F_{msin} and F_{msout}. Table 3 shows the contents of the spatial feature ($F_{spatial}$) and motion feature (F_{motion}). $F_{saptial}$ is the combination of $F_{spatial\text{-}joint}$ and $F_{spatial\_skeleton}$. F_{motion} is the combination of $F_{motion\text{-}joint}$ and $F_{motion\_skeleton}$. We used the two features ($F_{spatial}$ and F_{motion}) to evaluate the performance of the linear-map CNN.

Table 3. The channel contents of the input features.

$F_{spatial}$ (1200,18,3)	$F_{spatial\_joint}$ (600,18,3) $F_{spatial\_skeleton}$ (600,18,3)	(F_v, F_{sin}, F_{sout}) (F_s, F_{vin}, F_{vout})
F_{motion} (1200,18,3)	F_{motion_joint} (600,18,3) $F_{motion\_skeleton}$ (600,18,3)	(F_{mv}, F_{msin}, F_{msout}) (F_{ms}, F_{mvin}, F_{mvout})

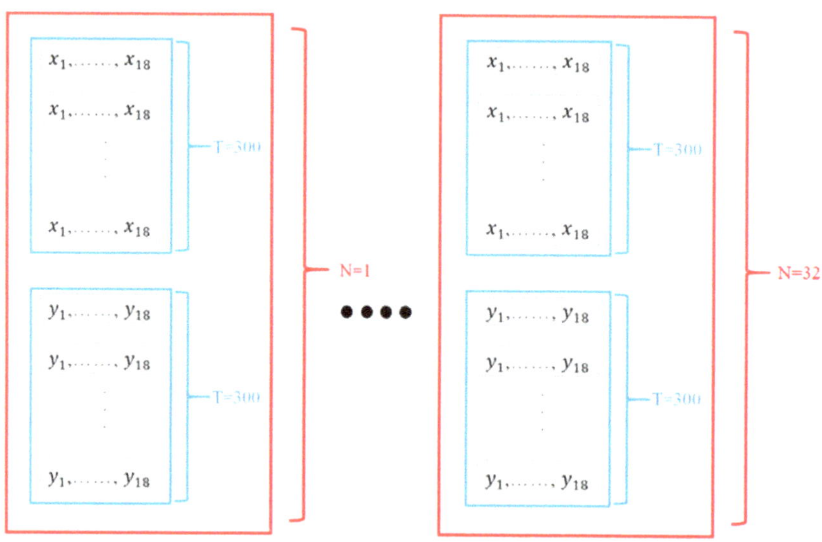

Figure 3. One feature built by the joint data, skeleton data, joint-motion data, or skeleton-motion data, with the size of 600 × 18.

2.5. Linear-Map CNN

Figure 4 shows the structure of a 10-layer linear-map CNN. The linear-map was used to adjust the number of channels at the beginning of each layer. Batch normalization (BN) can overcome the disappearance of the learning gradient and, thus, use a larger learning rate. In the CNN, the kernel size is a 9 × 1 matrix, the stride is (1,1), and the padding is (4,0). In the input feature, columns represent the different joints, and rows represent the time sequence of the actions. The relation of the adjacent joints was enhanced by the DAG matrix. Thus, the kernel of convolution is a 9 × 1 matrix. Table 4 shows the detailed information of the linear-map CNN. The output layer has 49 nodes representing the 49 action classes. The optimal method was momentum. The batch number is 32.

Table 4. The parameters of linear-map CNN in each layer.

	Linear	BN	AF	Conv2d	BN	AF
L1	In feature (600,18,3), Out features (600,18,64)	64	ReLu	(600,18,64), kernel = (9,1), stride = (1,1), padding = (4,0)	64	ReLu
L2	In feature (600,18,192), Out features (600,18,64)	64	ReLu	(600,18,64), kernel = (9,1), stride = (1,1), padding = (4,0)	64	ReLu
L3	In feature (600,18,192), Out features (600,18,64)	64	ReLu	(600,18,64), kernel = (9,1), stride = (1,1), padding = (4,0)	64	ReLu
L4	In feature (600,18,192), Out features (600,18,64)	64	ReLu	(600,18,64), kernel = (9,1), stride = (1,1), padding = (4,0)	64	ReLu
L5	In feature (600,18,192), Out features (600,18,128)	128	ReLu	(600,18,128), kernel = (9,1), stride = (2,1), padding = (4,0)	128	ReLu
L6	In feature (300,18,384), Out features (300,18,128)	128	ReLu	(300,18,128), kernel = (9,1), stride = (1,1), padding = (4,0)	128	ReLu
L7	In feature (300,18,384), Out features (300,18,128)	128	ReLu	(300,18,128), kernel = (9,1), stride = (1,1), padding = (4,0)	128	ReLu
L8	In feature (300,18,384), Out features (300,18,256)	256	ReLu	(300,18,256), kernel = (9,1), stride = (2,1), padding = (4,0)	256	ReLu
L9	In feature (150,18,768), Out features (150,18,256)	256	ReLu	(150,18,256), kernel = (9,1), stride = (1,1), padding = (4,0)	256	ReLu
L10	In feature (150,18,768), Out features (150,18,256)	256	ReLu	(150,18,256), kernel = (9,1), stride = (1,1), padding = (4,0)	256	ReLu
Flatten	Input 512 Output 49		Softmax			

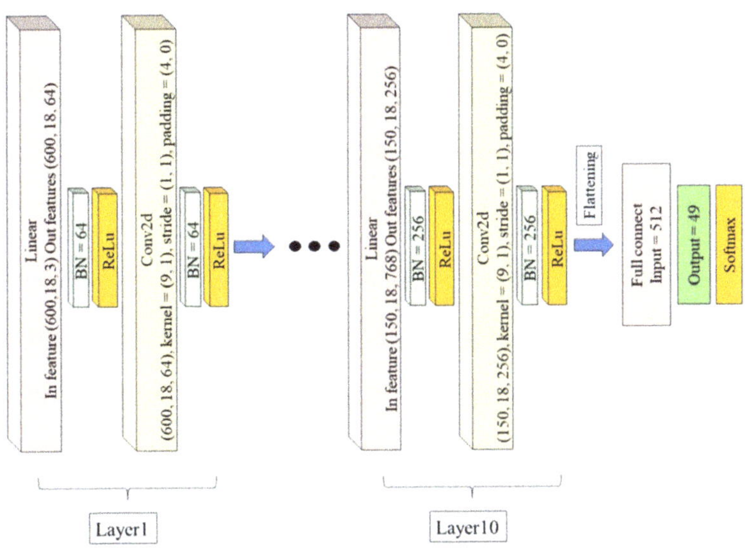

Figure 4. The structure of the 10-layer linear-map CNN.

2.6. Statistical Analysis

According to our proposed method, a film is considered as true positive (TP) when the classification action is correctly identified; false positive (FP) when the classification action is incorrectly identified; true negative (TN) when the action classification is correctly rejected, and false-negative (FN) when the action classification is incorrectly rejected. Here, the performance of the proposed method was evaluated using these parameters,

$$Precision(\%) = \frac{TP}{TP + FP} \times 100\%, \quad (4)$$

$$Recall(\%) = \frac{TP}{TP + FN} \times 100\%, \tag{5}$$

$$Specificity(\%) = \frac{TN}{TN + FP} \times 100\%, \tag{6}$$

$$F_1score = \frac{2 \times Precision \times Recall}{Precision + Recall} \times 100\%, \tag{7}$$

$$Accuracy(\%) = \frac{TP + TN}{TP + TN + FP + FN} \times 100\%. \tag{8}$$

3. Results

In this study, the hardware employed was CPU Intel Core i7-8700 and GPU GeForce GTX1080. The operating system was Ubuntu 16.04LTS software, the development system was Anaconda 3 at python 3.7 version, the tool of deep learning was Pytorch 1.10, and the compiler was Jupyter Notebook. We evaluated the performance of DAG with the cross-subject and cross-view sources, and four features ($F_{spatial\_joint}$, $F_{spatial\text{-}skeleton}$, F_{motion_joint}, and $F_{motion\_skeleton}$). At last, we used the two-stream concept, class score fusion for $F_{spatial}$ and F_{motion}, to evaluate the performance of the proposed method with cross-subject and cross-view sources.

Table 5 shows the results without the DAG transfer. The best feature is $F_{spatial}$ under the cross-subject source, which resulted in an accuracy and F1-score of 99.3% and 82.8%, respectively. There were 10 actions with recall rates below 70%: 0, 4, 9, 10, 11, 16, 28, 29, 43, and 48. The worst feature is F_{motion} under the cross-subject source, its accuracy and F1-score are 99.2% and 79.6%, respectively. There are 11 actions with recall rates below 70%: 3, 10, 11, 16, 24, 28, 29, 31, 36, 43, and 45. Table 6 shows the results with the DAG transfer. The best feature was $F_{spatial}$ under the cross-view source; its accuracy and F1-score were 99.9% and 96.2%, respectively.

Only four actions, 10, 11, 28, and 29, had recall rates below 70%. The worst feature was F_{motion} under the cross-subject source, which obtained an accuracy and F1-score of 99.1% and 79.1%, respectively. There were 10 actions with recall rates below 70%: 2, 10, 11, 16, 28, 29, 31, 43, 44, and 45. We found that the DAG transfer could significantly improve the recognition rate of different actions, not only for spatial features but also for motion features. Table 7 shows the results of class score fusion with and without DAG transfer. We found that the performance of DAG transfer used in the cross-view source was better than used in the cross-subject source, with an accuracy of 99.9% vs. 99.5% and F1-score of 94.7% vs. 86.3%. The recall rates for all 49 actions were not below 70%.

We used the two-dimensional joint and skeleton features to perform the training and testing of the linear-map CNN, which could reduce the running time more than those using two- or three-dimensional joint and skeleton images. Table 8 shows the training and testing time with and without DAG transfer. We found that the GPU could process about 30 frames/second (fps) in the training phase, and process about 125 fps in the testing phase. Although the DAG transfer required time to process, the delay time was about 30 min in the training phase. The maximum testing time was 141 s.

Table 5. The results without the DAG transfer.

	Feature (Epochs)	Precision (100%)	Recall (100%)	Specificity (100%)	F1-Score (100%)	Accuracy (100%)	Actions (Recall < 70%)
Cross-subject	$F_{spatial}$ (70)	84.2%	82.8%	99.7%	82.8%	99.3%	0, 4, 9, 10, 11, 16, 28, 29, 43, 48
	F_{motion} (65)	80.8%	79.7%	99.6%	79.6%	99.2%	3, 10, 11, 16, 24, 28, 29, 31, 36, 43, 45
Cross-view	$F_{spatial}$ (90)	86.4%	82.0%	99.7%	81.6%	99.3%	4, 9, 11, 16, 18, 28, 43, 44
	F_{motion} (20)	82.9%	79.7%	99.7%	79.7%	99.2%	4, 9, 10, 11, 28, 29, 31, 32

Table 6. The results with the DAG transfer.

	Feature (Epochs)	Precision (100%)	Recall (100%)	Specificity (100%)	F1-Score (100%)	Accuracy (100%)	Actions (Recall < 70%)
Cross-subject	$F_{spatial}$ (65)	86.9%	86.0%	99.9%	86.2%	99.4%	10, 11, 28
	F_{motion} (35)	80.6%	78.7%	99.7%	79.1%	99.1%	2, 10, 11, 16, 28, 29, 31, 43, 44, 45
Cross-view	$F_{spatial}$ (65)	94.3%	94.3%	99.9%	94.2%	99.9%	10, 11, 28, 29
	F_{motion} (65)	90.9%	90.2%	99.9%	90.4%	99.7%	10, 11, 28, 29

Table 7. The results of class score fusion with and without DAG transfer.

		Precision (100%)	Recall (100%)	Specificity (100%)	F1-Score (100%)	Accuracy (100%)	Actions (Recall < 70%)
Cross-subject	Non GAD	85.7%	85.0%	99.8%	85.0%	99.4%	10, 11, 16, 28, 29, 31
	GAD	86.9%	86.1%	99.9%	86.3%	99.5%	10, 11, 28, 29
Cross-view	Non GAD	89.8%	87.3%	99.8%	87.3%	99.6%	5, 10, 12, 29
	GAD	94.8%	94.7%	99.9%	94.7%	99.9%	

Table 8. The training and testing time with and without DAG transfer.

Without DAG					
	Date Source	Feature	Epoch	Time: Day: Hr.: Min.:Sec.	Fps
Train	Cross-subject	Spatial	120	1:12:41:00	30
		Motion	92	1:04::04:00	30
Train	Cross-view	Spatial	120	1:11:39:00	30
		Motion	92	1:02:55:00	30
Test	Cross-subject	Spatial		0:00:02:01	125
		Motion		0:00:01:57	125
Test	Cross-view	Spatial		0:00:02:18	125
		Motion		0:00:02:19	125

Table 8. *Cont.*

With DAG					
Train	Cross-subject	Spatial	120	1:13:27:00	30
		Motion	92	1:04:29:00	30
	Cross-view	Spatial	120	1:11:05:00	30
		Motion	92	1:03:30:00	30
Test	Cross-subject	Spatial		0:00:02:03	125
		Motion		0:00:02:03	125
	Cross-view	Spatial		0:00:02:21	125
		Motion		0:00:02:19	125

4. Discussion

In this study, we used DAG transfer and the two-stream method to improve the accuracy of action recognition. When the input features were transferred with the DAG matrices, the precision, recall, specificity, F1-score, and accuracy were improved by 1.2%, 1.1%, 0.1%, 1.3%, and 0.1%, respectively, for the cross-subject source, and were improved by 9.1%, 7.4%, 0.1%, 7.4%, and 0.5%, respectively, for the cross-view source. In the two-stream method, previous studies have typically used the spatial and temporal, or optical flow features to perform the active score fusion [11–14]. They also proved that the performance of two streams was better than one stream.

We used joint and skeleton sequences as the spatial motion features that had temporal characteristics. We utilized 300 frames to describe an action. Thus, the spatial feature of one action included the spatial and temporal characters. However, the motion relations of the joints and skeletons were different from one action to another. Therefore, we defined the motion variables, mv_i and ms_i, as shown in Table 1, to establish the motion features.

Our results also show that the precision, recall, specificity, F1-score, and accuracy of the two streams were improved by 0% vs. 6.3%, 0.1% vs. 7.4%, 0% vs. 0.2%, 0.1% vs. 7.1%, and 0.1% vs. 0.4%, over spatial and motion streams with the DAG transfer in the cross-subject source, and improved by 0.5% vs. 3.9%, 0.4% vs. 4.5%, 0% vs. 0%, 0.5% vs. 4.3%, and 0% vs. 0.2% in the cross-view source.

The comparison of our results with the previous studies under the recall rate is shown in Table 9. These studies all used cross-subject and cross-view sources from the NTU RGB+D database to recognize actions and also used three-dimensional characteristics of each posture as the input features [19,25–31]. Our method had the best recall rates of 86.1% and 94.7% in the cross-subject and cross-view sources.

We analyzed the actions with lower recall rates in the cross-subject and cross-view sources in Tables 5–7. The four actions that often had lower recall rates were A10 (reading), A11 (writing), A28 (phone call), and A29 (playing with laptop). Figure 5a is the posture of reading, and Figure 5b is the posture of writing. The subject is standing up, looking down, and holding something. The difference between the two images is only in the gestures of two hands. However, according to the description of the body posture in Table 2, only the right and left wrist joints are marked, which cannot show the gestures of two hands.

The postures of the subject making a phone call (in Figure 5c) and using the laptop (in Figure 5d)) had the same problem. The difference between the two images was also in the gestures of the two hands. These actions were difficult to recognize using spatial features, such as the movement trajectories of the arms, elbows, and wrists. Thus, the results of our method with the DAG transfer and two-stream method in Table 7 show that no action had a lower recall rate in the cross-view source.

Table 9. These studies all used cross-subject and cross-view sources in the NTU RGB+D database to recognize the actions, which also used the three-dimensional characteristics of each posture as the input features [19,25–31]. Our method had the best recall rates in the cross-subject and cross-view sources at 86.1% and 94.7%.

Methods	Recall Rate (%)	
	Cross-Subject	Cross-View
Lie Group [25]	50.1	52.8
HBRNN [19]	59.1	64.0
Deep RNN [26]	59.29	64.09
Deep LSTM [26]	60.7	67.3
Part-aware LSTM [26]	62.9	70.3
ST-LSTM + Trust Gate [27]	69.2	77.7
Two-stream RNN [28]	71.3	79.5
Clips + CNN + MTLN [29]	79.6	84.8
ST-GCN [30]	81.5	88.3
SR-TSL [31]	84.8	92.4
Proposed DAG + linear-map CNN	86.1	94.7

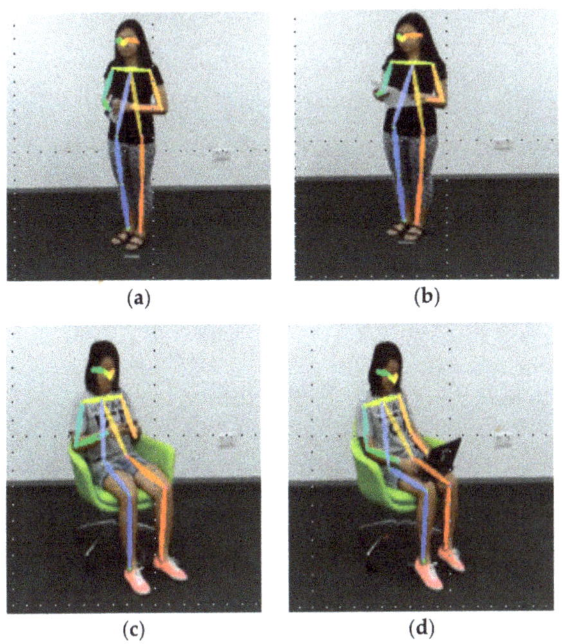

Figure 5. (**a**) the posture of reading, (**b**) the posture of writing, (**c**) the posture of making a phone call, (**d**) the posture of using the laptop.

5. Conclusions

The large scale of the collected data in the NTU RGB+D database enabled us to apply the posture-driven learning method for action recognition. The posture information represented by the skeleton data was obtained from the 300 frames of the film. The joint and skeleton sequences were used to build spatial and motion features that included the spatial and temporal characteristics of the actions. The relations of the adjacent skeletal joints were used to build the DAG matrices.

The spatial or motion features were expanded by DAG matrices as color texture images. The expanded features all indicated that the relations between adjacent joints were enhanced. Our method effectively reduced the amount of data for training the linear-map CNN, and its performance was superior to the previous schemes using deep learning methods. Notably, since the computation speed can reach around 125 fps in the testing phase with GPU, our scheme could be used to monitor the daily activities of elders in real time in home care applications.

Author Contributions: Conceptualization, T.-H.T. and S.-H.L.; Data curation, J.-H.H.; Investigation, T.-H.T.; Methodology, J.-H.H.; Project administration, T.-H.T.; Software, J.-H.H.; Supervision, Y.-F.H. and M.G.; Validation, Y.-F.H. and M.G.; Writing original draft, S.-H.L.; Writing review and editing, S.-H.L., Y.-F.H., and M.G. All authors have read and agreed to the published version of the manuscript.

Funding: This research was funded by the Ministry of Science and Technology, Taiwan, under grants MOST 109-2221-E-324-002-MY2 and MOST 109-2221-E-027-97.

Institutional Review Board Statement: Not applicable.

Informed Consent Statement: Not applicable.

Conflicts of Interest: The authors declare no conflict of interest.

References

1. United Nations, Department of Economic Social Affairs. *World Population Ageing 2019*; United Nations: New York, NY, USA, 2019.
2. Available online: https://www.ndc.gov.tw/Content_List.aspx?n=695E69E28C6AC7F3 (accessed on 28 April 2021).
3. Nyboe, L.; Lund, H. Low levels of physical activity in patients with severe mental illness. *Nord. J. Psychiatry* **2013**, *67*, 43–46. [CrossRef]
4. Oliveira, J.; Ribeiro, F.; Gomes, H. Effects of a home-based cardiac rehabilitation program on the physical activity levels of patients with coronary artery disease. *J. Cardiopulm. Rehabil. Prev.* **2008**, *28*, 392–396. [CrossRef]
5. Liu, S.-H.; Cheng, W.-C. Fall detection with the support vector machine during scripted and continuous unscripted activities. *Sensors* **2012**, *12*, 12301–12316. [CrossRef] [PubMed]
6. Liu, S.-H.; Chang, Y.-J. Using accelerometers for physical actions recognition by a neural fuzzy network. *Telemed. e-Health* **2009**, *15*, 867–876. [CrossRef]
7. Tufek, N.; Yalcin, M.; Altintas, M.; Kalaoglu, F.; Li, Y.; Bahadir, S.K. Human action recognition using deep learning methods on limited sensory data. *IEEE Sens. J.* **2020**, *20*, 3101–3112. [CrossRef]
8. Wong, W.Y.; Wong, M.S.; Lo, K.H. Clinical applications of sensors for human posture and movement analysis: A review. *Prosthet. Orthot. Int.* **2007**, *31*, 62–75. [CrossRef]
9. Feichtenhofer, C.; Pinz, A.; Zisserman, A. Convolutional two-stream network fusion for video action recognition. In Proceedings of the IEEE Conference Computer Vision and Pattern Recognition, Las Vegas, NV, USA, 27–30 June 2016; pp. 1933–1941.
10. Fernando, B.; Gavves, E.; Oramas, M.J.; Ghodrati, A.; Tuytelaars, T. Modeling video evolution for action recognition. In Proceedings of the IEEE Conference Computer Vision and Pattern Recognition, Boston, MA, USA, 7–12 June 2015; pp. 5378–5387.
11. Karpathy, A.; Toderici, G.; Shetty, S.; Leung, T.; Sukthankar, R.; Fei, L. Large-scale video classification with convolutional neural networks. In Proceedings of the IEEE Conference Computer Vision and Pattern Recognition, Columbus, OH, USA, 23–28 June 2014; pp. 1725–1732.
12. Simonyan, K.; Zisserman, A. Two-stream convolutional networks for action recognition in videos. *arXiv Prepr.* **2014**, arXiv:1406.2199.
13. Wang, L.; Xiong, Y.; Wang, Z.; Qiao, Y.; Lin, D.; Tang, X.; Gool, L.V. Temporal segment networks: Towards good practices for deep action recognition. In Proceedings of the European Conference on Computer Vision, Amsterdam, The Netherlands, 11–14 October 2016; pp. 20–36.
14. Jiang, Y.; Wu, Z.; Tang, J.; Li, Z.; Xue, X.; Chang, S. Modeling multimodal clues in a hybrid deep learning framework for video classification. *IEEE Trans. Multimed.* **2018**, *20*, 3137–3147. [CrossRef]
15. Ji, S.; Xu, W.; Yang, M.; Yu, K. 3D convolutional neural networks for human action recognition. *IEEE Trans. Pattern Anal. Mach. Intell.* **2013**, *35*, 221–231. [CrossRef] [PubMed]
16. Chaudhry, R.; Ofli, F.; Kurillo, G.; Bajcsy, R.; Vidal, R. Bioinspired dynamic 3D discriminative skeletal features for human action recognition. In Proceedings of the IEEE Conference on Computer Vision and Pattern Recognition, Portland, OR, USA, 23–28 June 2013; pp. 471–478.
17. Vemulapalli, R.; Arrate, F.; Chellappa, R. R3DG features: Relative 3D geometry-based skeletal representations for human action recognition. *Comput. Vis. Image Understand.* **2016**, *152*, 155–166. [CrossRef]
18. Wang, P.; Li, W.; Gao, Z.; Zhang, J.; Tang, C.; Ogunbona, P.O. Action recognition from depth maps using deep convolutional neural networks. *IEEE Trans. Hum.–Mach. Syst.* **2016**, *46*, 498–509. [CrossRef]

19. Du, Y.; Wang, W.; Wang, L. Hierarchical recurrent neural network for skeleton based action recognition. In Proceedings of the IEEE Conference on Computer Vision and Pattern Recognition, Boston, MA, USA, 7–12 June 2015; pp. 1110–1118.
20. Veeriah, V.; Zhuang, N.; Qi, G.-J. Differential recurrent neural networks for action recognition. In Proceedings of the IEEE Conference on Computer Vision and Pattern Recognition, Boston, MA, USA, 7–12 June 2015; pp. 4041–4049.
21. Hou, Y.; Li, Z.; Wang, P.; Li, W. Skeleton optical spectra-based action recognition using convolutional neural networks. *IEEE Trans. Circ. Syst. Video Technol.* **2018**, *28*, 807–811. [CrossRef]
22. VanderWeele, T.J.; Robins, J.M. Four types of effect modification: A classification based on directed acyclic graphs. *Epidemiology* **2007**, *18*, 561–568. [CrossRef] [PubMed]
23. Fang, H.; Xie, S.; Tai, Y.; Lu, C. RMPE: Regional multi-person pose estimation. In Proceedings of the IEEE International Conference on Computer Vision, Venice, Italy, 22–29 October 2017; pp. 2353–2362.
24. Rapid-Rich Object Search (ROSE) Lab. 2019. Available online: http://rose1.ntu.edu.sg/datasets/actionrecognition.asp (accessed on 28 April 2021).
25. Vemulapalli, R.; Arrate, F.; Chellappa, R. Human action recognition by representing 3d skeletons as points in a lie group. In Proceedings of the IEEE Conference on Computer Vision and Pattern Recognition, Columbus, OH, USA, 23–28 June 2014; pp. 588–595.
26. Shahroudy, A.; Liu, J.; Ng, T.; Wang, G. NTU RGB+D: A large scale dataset for 3d human activity analysis. In Proceedings of the IEEE Conference on Computer Vision and Pattern Recognition, Las Vegas, NV, USA, 27–30 June 2016; pp. 1010–1019.
27. Liu, J.; Shahroudy, A.; Xu, D.; Wang, G. Spatial-temporal LSTM with trust gates for 3D human action recognition. In Proceedings of the European Conference on Computer Vision, Amsterdam, The Netherlands, 11–14 October 2016; pp. 816–833.
28. Wang, H.; Wang, L. Modeling temporal dynamics and spatial configurations of actions using two-stream recurrent neural networks. In Proceedings of the IEEE Conference on Computer Vision and Pattern Recognition, Honolulu, HI, USA, 21–26 July 2017; pp. 499–508.
29. Ke, Q.; Bennamoun, M.; An, S.; Sohel, F.; Boussaid, F. A new representation of skeleton sequences for 3D action recognition. In Proceedings of the IEEE Conference on Computer Vision and Pattern Recognition, Honolulu, HI, USA, 21–26 July 2017.
30. Yan, S.; Xiong, Y.; Lin, D. Spatial temporal graph convolutional networks for skeleton-based action recognition. In Proceedings of the AAAI Conference on Artificial Intelligence, New Orleans, LA, USA, 2–7 February 2018; pp. 7444–7452.
31. Si, C.; Jing, Y.; Wang, W.; Wang, L.; Tan, T. Skeleton-based action recognition with spatial reasoning and temporal stack learning. In Proceedings of the European Conference on Computer Vision, Munich, Germany, 8–14 September 2018; pp. 106–121.

sensors

MDPI

Article

Usability, User Experience, and Acceptance Evaluation of CAPACITY: A Technological Ecosystem for Remote Follow-Up of Frailty

Rodrigo Pérez-Rodríguez [1,2,3,*], Elena Villalba-Mora [2,4], Myriam Valdés-Aragonés [2,5], Xavier Ferre [2], Cristian Moral [2], Marta Mas-Romero [6], Pedro Abizanda-Soler [3,6,7] and Leocadio Rodríguez-Mañas [2,3,5]

1 Biomedical Research Foundation, Getafe University Hospital, 28905 Getafe, Spain
2 Centre for Biomedical Technology (CTB), Universidad Politécnica de Madrid (UPM), Pozuelo de Alarcón, 28223 Madrid, Spain; elena.villalba@ctb.upm.es (E.V.-M.); myriam.valdes@salud.madrid.org (M.V.-A.); xavier.ferre@ctb.upm.es (X.F.); cristian.moral@ctb.upm.es (C.M.); leocadio.rodriguez@salud.madrid.org (L.R.-M.)
3 CIBER of Frailty and Healthy Aging (CIBERFES), 28001 Madrid, Spain; pabizanda@sescam.jccm.es
4 Centro de Investigación Biomédica en Red en Bioingenieria, Biomateriales y Nanomedicina (CIBER-BBN), 28029 Madrid, Spain
5 Geriatrics Service, Getafe University Hospital, 28095 Getafe, Spain
6 Geriatrics Service, Albacete University Hospital, 02006 Albacete, Spain; mmasr@sescam.jccm.es
7 Faculty of Medicine, University of Castilla-La Mancha, 02008 Albacete, Spain
* Correspondence: rprodrigo@salud.madrid.org

Citation: Pérez-Rodríguez, R.; Villalba-Mora, E.; Valdés-Aragonés, M.; Ferre, X.; Moral, C.; Mas-Romero, M.; Abizanda-Soler, P.; Rodríguez-Mañas, L. Usability, User Experience, and Acceptance Evaluation of CAPACITY: A Technological Ecosystem for Remote Follow-Up of Frailty. *Sensors* **2021**, *21*, 6458. https://doi.org/10.3390/s21196458

Academic Editor: Ivan Miguel Serrano Pires

Received: 7 September 2021
Accepted: 26 September 2021
Published: 27 September 2021

Abstract: Frailty predisposes older persons to adverse events, and information and communication technologies can play a crucial role to prevent them. CAPACITY provides a means to remotely monitor variables with high predictive power for adverse events, enabling preventative personalized early interventions. This study aims at evaluating the usability, user experience, and acceptance of a novel mobile system to prevent disability. Usability was assessed using the system usability scale (SUS); user experience using the user experience questionnaire (UEQ); and acceptance with the technology acceptance model (TAM) and a customized quantitative questionnaire. Data were collected at baseline (recruitment), and after three and six months of use. Forty-six participants used CAPACITY for six months; nine dropped out, leaving a final sample of 37 subjects. SUS reached a maximum averaged value of 83.68 after six months of use; no statistically significant values have been found to demonstrate that usability improves with use, probably because of a ceiling effect. UEQ, obtained averages scores higher or very close to 2 in all categories. TAM reached a maximum of 51.54 points, showing an improvement trend. Results indicate the success of the participatory methodology, and support user centered design as a key methodology to design technologies for frail older persons. Involving potential end users and giving them voice during the design stage maximizes usability and acceptance.

Keywords: frailty; home monitoring; user-centered design; usability; user experience; acceptance

1. Introduction

1.1. Research Context

Intrinsic capacity, according to the World Health Organization (WHO), is the combination of the physical and mental (including psychological) capacities of the individuals. Intrinsic capacity is thus part of functional ability together with the environment and the interactions with it. The concept of frailty is closely related and complementary to intrinsic capacity. Frailty can be defined as a stage of age-related decline reducing intrinsic capacity and functional reserve of older persons, thus predisposing them to adverse events (mortality and disability, among others). These days, there is a pressing need to develop comprehensive community-based approaches and to introduce interventions to prevent functional decline [1].

The risk of developing chronic conditions, including disability and dependency, increases with age [2,3], and is changing the classical approach to manage functionally declining older persons. Considering that functional decline is accompanied by a loss in functional reserve, it is very unlikely that disability is reversed. In this way, healthcare systems need to move towards person-centered approaches that anticipate the earliest stages of functional decline (i.e., frailty) to prevent disability, since becoming frail can be delayed, slowed, or even reversed.

Estimated prevalence of frailty is 18% (95% CI: 15–21%), and it seems to be correlated with age, gender (female), and socio-economic factors such as lower education and wealth [4]. Good news is that frailty is reversible, but to achieve it, it is of paramount importance to fight inactivity and sedentariness [5]. Scientific literature supports activity-centered interventions to delay and even reverse frailty and disability [6–11]. Furthermore, interventions on nutrition, such as modifying habits, increasing protein and micronutrient intake, are also recommended [12,13], as well as interventions on inadequate drug prescriptions [14–16]. And finally, it is also important that the physiological and social aspects are not left apart [11].

A frail older person usually shows decreased neurological and muscle function [17], normally accompanied by an accelerated involuntary weight loss and a decline in the skeletal muscle [18]. Moreover, considering the results published in a relatively recent systematic review, in the 30.6% of the studies that were analyzed, associations between gait speed, disability, frailty, sedentary lifestyle, falls, muscular weakness, diseases, body fat, cognitive impairment, mortality, stress, lower life satisfaction, lower quality of life, and poor performance in quantitative parameters of gait were found [19].

Ageing in Place purses that older persons continue living at their homes as they age [20], which brings along important economic benefits given the reduction of the institutionalized care [21]; information and communication technologies (ICTs) can play a crucial role to promote it [22]. For instance, having fresh and periodic information on variables associated to poor health outcomes (e.g., gait speed, muscle power, and involuntary weight loss) can be a great asset to trigger early interventions to prevent disability and dependency. Smart home technologies [23–27], wearable sensors [28,29] or mHealth technology [30] may enable continuous, ubiquitous and transparent monitoring of the independent older adult, supporting the traditional geriatric approach to identify older people at risk of disability. Notwithstanding, more effort is still needed to assess not only how reliable and valid the ICT-based approached to measure frailty are, but also deeply study the associated ethical, technical, and economic issues [31].

Nevertheless, the lack of consensus in terms of technology acceptance by older persons must be considered. Several authors have reached the conclusion that older persons are not interested in innovative technologies [32,33], while others state that older people have already accepted new technologies, mainly because they have been proven useful in meeting their information needs, especially in health [34]. Yet it seems that the use of ICTs by the older population is strongly linked to physical limitations (e.g., abilities, chronic illnesses, etc.), mental limitations (e.g., fear of damaging the technology, electric shocks, making mistakes, etc.), educational limitations (e.g., low levels of literacy, limited electronic literacy, learning barriers, etc.), structural limitations (e.g., design of the appliance), instructional limitations (e.g., instructions on how to use a technology are hard to follow) and to a limited access to the technology (e.g., financial costs) [35,36].

The design of those technologies to be used by older persons must be done according to their characteristics. Methodologies such as user-centered design (UCD) [37] and participatory design (PD) are a good alternative to develop right solutions for a specific audience [33,38], since they help designers better understand the environment of use. Older people are usually excluded from product design activities since they are stigmatized as people reluctant to engage with technology, and this is probably one of the primary causes that prevent older people becoming loyal users of technological solutions [38,39].

The current demographic challenge is forcing researchers to focus on discovering feasible alternative ways of providing healthcare to the older population who are at an increased risk of suffering adverse events [40]. And, as it has been mentioned earlier, ICTs may help identifying early risk indicators for adverse events, providing a means for self-managing them. To this extent, its use in the context of frailty still at the very beginning [41].

1.2. Objective

The main objective of this work is to evaluate the usability, user experience (UX), and acceptance of older persons' interaction layer of CAPACITY, a frailty home-monitoring system aimed at the prevention of disability.

The manuscript is structured as follows. First, the CAPACITY ecosystem is presented as a modular infrastructure to monitor frailty and prevent disability. Second, the specific methodology followed in this work is described to later present and discuss the obtained results. Finally, conclusions are extracted and future work proposed.

2. Materials and Methods

2.1. Overview of the CAPACITY Technological Ecosystem

CAPACITY is a technological ecosystem aimed at preventing disability among the older population by detecting and intervening regarding frailty; it also provides a substrate to connect all relevant people in the care process (see Figure 1). Using CAPACITY, the older population can be remotely supervised by community care professionals. So, in case worrying declines are detected, specialists (i.e., geriatricians) can be included in the loop. Intervention provided to older persons is grounded on three main pillars: the VIVIFRAIL physical activity program (declared as a success story by the European Commission) [42], personalized nutritional recommendations, and a program to detect risk of polypharmacy.

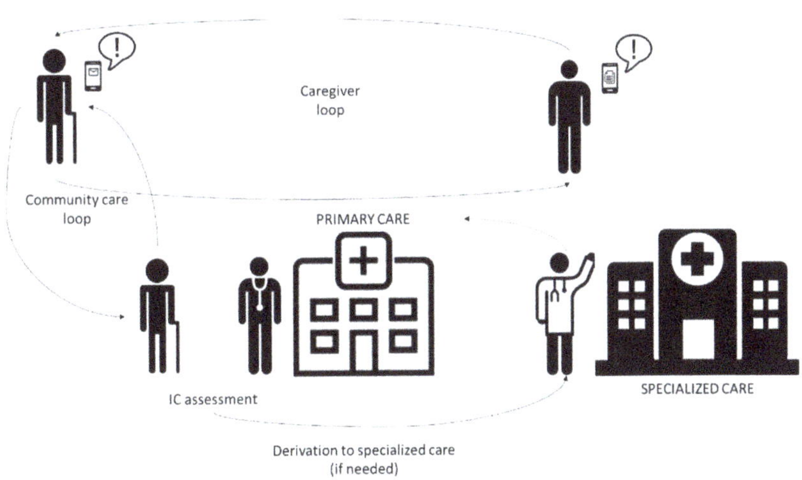

Figure 1. CAPACITY providers and interactions.

A Randomized Clinical Trial (RCT; ClinicalTrials.gov Identifier: NCT03707145) has allowed demonstrating that CAPACITY is an effective tool for a fast improvement of the frailty status as well as to reduce the use of healthcare resources [41].

2.2. CAPACITY Interaction System

CAPACITY services are offered through a set of user-adapted mobile applications. The functionalities offered to older persons are:

- Unceasing intrinsic capacity follow-up that enables triggering potential deterioration alarms;
- Access to a customized therapeutic plan (intervention), given the peculiarities and needs of the older person;
- Retrieving their own evolution;
- Communication with formal carers via asynchronous channels; and
- Notifications on pertinent alarms related to health.

Apart from helping older persons, CAPACITY also offers different functionalities to other relevant stakeholders, namely primary and specialized care professionals, and informal caregivers, as shown in Figure 2. Work published [43] contains a wider description of all functionalities and services offered by the CAPACITY technological ecosystem to all involved people. In any case, this work solely focuses on the older persons as they are the center of the care.

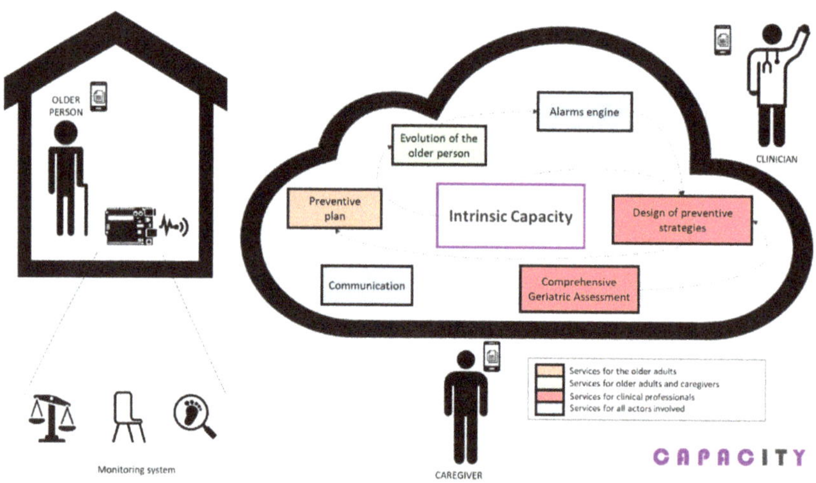

Figure 2. Conceptual architecture of CAPACITY.

Older persons being followed by CAPACITY need to use a home monitoring kit aimed at measuring variables with high predictive value for adverse events. This monitoring system consists in a gait-speed sensor [44], a sensor to indirectly (through the chair stand test) measure power in the lower limbs [45], and a wireless commercial weight scale to measure involuntary weight loss. Figure 3 illustrates the prototypes of the sensors originally designed for CAPACITY.

The interaction with the home monitoring kit is handled by a mobile application that acts as a guiding element to the older person, as a data concentrator (bluetooth connection with the monitoring kit; see Figure 3 and Supplementary Material for details), and as data input point, not only enabling the older adult using the sensors but also completing a set of questionnaires to enrich the information handled by the clinical professionals. These questionnaires are adapted versions of the Frailty Phenotype criteria [46], Mini Nutritional Assessment (MNA) [47], Barthel Index [48], FRAIL Scale [49], and the Functional Activities Questionnaire (FAQ) [50].

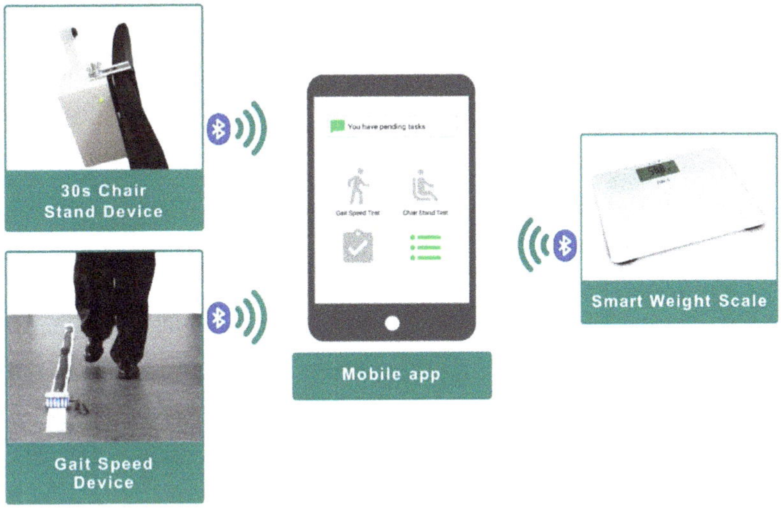

Figure 3. Home-monitoring kit.

This interaction system was iteratively designed under a user-centered approach. Different prototypes were created and tested, first in a laboratory environment, later in a clinical environment, and, finally, at the final users' dwellings. In the two last cases, the system was evaluated by older users. The outcomes of each iteration allowed designers improve and adapt the interaction system to the needs, preferences, and context of use of the older adults.

Figure 4 shows how the interaction system evolved during the process. An in-depth description of this iterative process and the resulting interaction system can be found in [51].

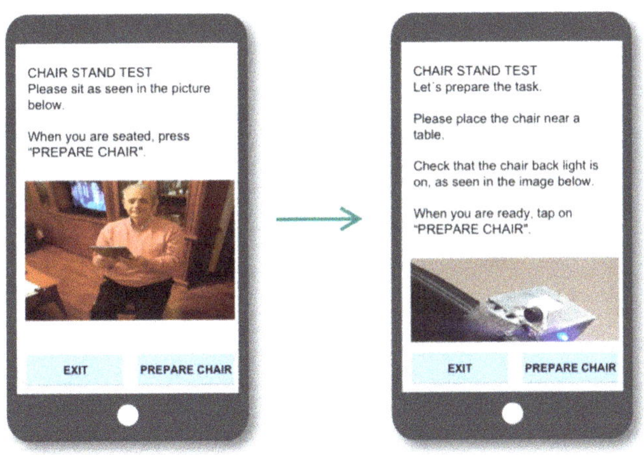

(a)

Figure 4. *Cont.*

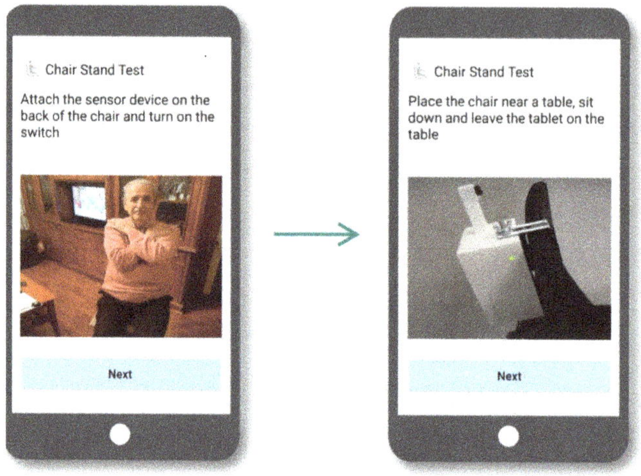

(b)

Figure 4. (a) First and (b) second—final—prototypes.

Figure 5 shows the workflow that needs to be followed to interact with any of the components of the monitoring kit. The process starts with the app notifying a pending measurement (prescribed by the physician as part of a personalized follow-up) and the user pushing the corresponding button to start it. Then, the older person is provided with a short explanatory video showing how the measurement will be performed. Once the user is ready, that is explicated by pressing a specific button, the actual measurement takes place; this part is fully guided by voice commands and accompanying pictures (e.g., 'please switch the sensor on', 'please sit on a chair', 'the process will start after the countdown', etc.). Transparently to the user, the app and the sensor establish a bluetooth communication channel used to register the datum. Once the process is over, the older person receives a confirmation with some feedback related to the measurement.

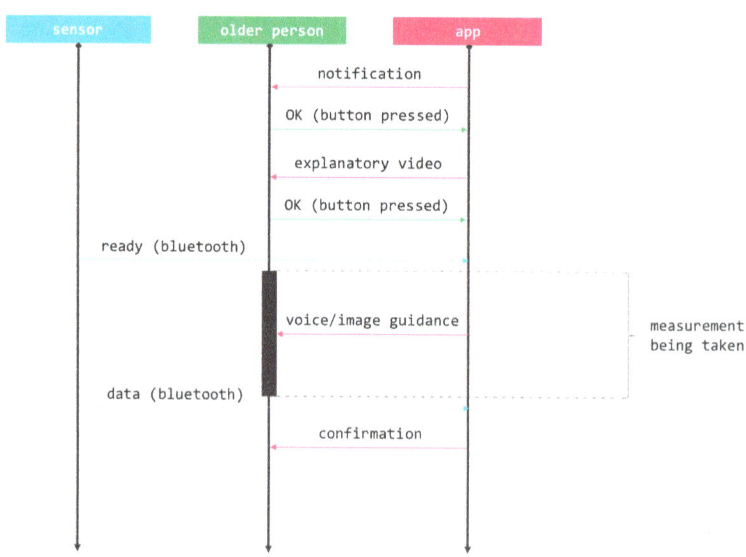

Figure 5. CAPACITY's workflow to collect data from the home monitoring kit.

2.3. Assessment Tools

The usability, UX, and user acceptance related to the CAPACITY technological ecosystem has been assessed.

Usability, defined as the extent to which a product can be used by specified users to achieve specified goals with effectiveness, efficiency, and satisfaction in a specified context of use [52], has been assessed using the system usability scale [53,54]. SUS is a short 10-item Likert questionnaire that provides a measure of people's subjective perceptions of the usability of a system; concretely, SUS focuses on learnability and usability, which are indeed correlated [55]. These 10 items can be evaluated from '1—fully disagree' to '5—fully-agree'. The total score ranges from 0 to 100. Although SUS is a simple tool, a study carried out by Tullis [56], who compared the effectiveness and accuracy of five questionnaires for assessing usability across different sample sizes, reached the conclusion that it is a reliable scale, especially when the sample is over 12 users.

UX has been assessed with the user experience questionnaire [57]. UEQ does not provide an overall score, but a score related to six different categories: attractiveness, perspicuity, efficiency, dependability, stimulation, and novelty. The score for each category is calculated by averaging the different items within it; each item's value ranges from −3 to 3, where extreme values represent two opposite concepts (e.g., attractive vs. unattractive). UEQ was included as an evaluation tool to complement the domains' SUS addresses.

Acceptance has been evaluated using the technology acceptance model [58] (adapted to the use case. See Supplementary Material for details) and a customized short quantitative questionnaire. TAM evaluates, throughout 12 items (answers range from '1—fully disagree' to '5—fully-agree'), two different categories: perceived usefulness and perceived ease of use. The maximum score is 60 (30 points in each category); final scores are calculated by averaging. To further investigate acceptance, a customized acceptance interview consisting in a Likert-type scale (from 1 to 5, the same as SUS) assessing all three components of the home monitoring kit was used. This interview had the following structure:

- Q1: The information the device provides motivates me to have a healthier lifestyle;
- Q2: The device makes me feel cared for;
- Q3: Using the device is a burden for me;
- Q4: The device enables me to control my own health; and
- Q5: I would use it.

For all the scales and questionnaires, Spanish versions were used (local language).

2.4. Recruitment and Data Collection

The impact of administering a multicomponent intervention partially supported by the CAPACITY ecosystem was assessed by conducting a pilot, prospective, randomized, and blind study. The pilot study lasted 12 months: 6 months were dedicated to recruitment, and 6 months to intervention. Within this wider experiment, where the primary endpoint was to investigate whether the proposed technology helped preventing or reversing frailty, usability, UX, and user acceptance were evaluated as secondary endpoints (along with others). The pilot study was carried out simultaneously in two institutions: Getafe University Hospital and Albacete University Hospital.

Participation criteria were:

- Inclusion criteria:
    - +70 years old;
    - Living at home;
    - Barthel index [48] ≥90; and
    - Being pre-frail or frail.
- Exclusion criteria:
    - Inadequate home infrastructure impeding the installation of the technology;
    - Inability to understand how to use CAPACITY;
    - Medical condition incompatible with the VIVIFRAIL physical activity program;

- History of drug/alcohol abuse;
- Psychiatric disorders;
- Living with a participant; and
- Parti9cipating in another interventional clinical study.

Pre-frail participants were those meeting two Frailty Phenotype criteria [46] and suffering from at least four comorbidities, since they are the ones with the highest risk for developing frailty. Frail individuals were those meeting at least three Frailty Phenotype criteria and having at least four comorbidities.

Two research groups (arms) were defined. A control group receiving usual geriatric care and an intervention group who received the same multicomponent intervention but partially supported by the CAPACITY system. Stratified randomization by age (70–85, +85), sex (male, female), diagnosis (pre-frail, frail) and educational level (non-formal education, higher education, others) was applied to ensure groups were balanced.

Sample size could not be empirically calculated due to a lack of similar studies aiming to the same primary endpoint, so it was established to 90. Reasons behind this decision were:

- There were two different groups of interest (i.e., pre-frail and frail older persons);
- Given the usual standards, a recruitment objective of 20 subjects per interest group and research arm was set, for a total of 80; and
- Researchers assumed that a potential dropout rate of 10–15% over the previous calculation, so the target sample size was increased to 90.

Data reported in this manuscript are restricted to those participants who were randomly allocated into the intervention group (n = 46), since they were the only ones that used the CAPACITY system during the six months of intervention. The modules that supported the intervention were: (1) monitoring system, (2) evolution of the older person (e.g., access to follow-up information collected by the home-monitoring kit), and (3) basic asynchronous communication. All technological components were preconfigured prior to the delivery to the participating older persons (i.e., a tablet was delivered with the app already installed and configured to receive data from the home monitoring kit), so they only had to follow notifications and instructions. Besides, older participants received an initial training during the installation of the technology in their homes. This face-to-face training was delivered once and lasted approximately one hour. During the session, a user manual was provided that was used as a reference to show all functionalities to the older person, who had to repeat what was learnt (e.g., how to measure gait speed or complete a questionnaire). After this session, a telephone line remained open during the weekdays at working hours to attend any consultation or issue coming from the older participants.

Data related to usability and acceptance were collected at baseline and after three and six months of intervention. SUS and TAM were registered in all three sampling points while UEQ and the ad hoc acceptance questionnaires were only administered in the last data collection point to enrich the collected data with UX information and prospective acceptance.

3. Results

A total of 46 older persons used the CAPACITY technological solution to undergo an intervention aimed at preventing/reversing frailty; 14 were male (30.43%) and 32 female; mean age was 82.11 (SD = 5.42) years old. Regarding educational level, 20 of participants using technology did not have formal education (43.48%), 20 had primary studies (43.48%), 5 received secondary education (10.87%), and 1 received higher education (2.17%). Finally, most of the participants (30 persons -65.22%-) did not have any previous experience with technology (i.e., smart phones and internet) while 9 of them (19.56%) used it in a daily basis; the remaining part made an occasional use of the technology (3 subjects -6.52%-) or had used it once or twice before this study (4 subjects -8.70%-). Table 1 shows the description of the population that participated in the study.

Table 1. Description of the older population that participated in the study.

		N	Proportion (%)
Sex	Female	32	69.57
	Male	14	30.43
Educational level	No formal education	20	43.48
	Secondary education	20	43.48
	Primary education	5	10.87
	Higher education	1	2.17
Experience with technology	No experience	30	65.22
	Once or twice	4	8.70
	Occasional use	3	6.52
	Daily experience	9	19.56

Nine participants dropped out from the study during the follow-up period, raising a final sample for analysis of 37 subjects. All subjects completed the questionnaires about usability, UX, and acceptance evaluation at the second visit (three months) and at the end of the follow-up; 25 participants completed it at baseline.

Table 2 depicts the adherence to the monitoring plan calculated as the average commitment to the measurements that the users needed to perform as part of their treatment. Full adherence (100%) means that all participants performed all prescribed measurements. Table 2 also shows the default periodicity for the different measurements, but it must be taken into consideration that additional measurements could be requested. For all measurements a push notification was sent to the user through the app.

Table 2. Average adherence to monitoring plan.

	Chair Stand	Weight Scale	Gait Speed	Frailty Phenotype Criteria	Barthel Index	FAQ	FRAIL Scale	MNA
Default measurement periodicity (weeks)	2	2	2	4	4	4	4	8
Average adherence (%)	96.38	97.44	95.70	100	99.54	95.95	94.44	94.91
SD	0.12	0.07	0.08	0.00	0.03	0.08	0.23	0.19

Table 3 shows the usability results. Usability obtained averaged SUS values of 80.11/100 (SD = 13.66) at baseline, 83.31/100 (SD = 15.07) at month 3 and 83.68/100 (SD = 1.62) at the end of the study.

Table 4 depicts the results regarding UX, that was only assessed at the end of the intervention. Averaged values of 2.20/3 (SD = 0.64) in terms of attractiveness, 2.30/3 (SD = 0.73) in perspicuity, 1.99/3 (SD = 0.75) in efficiency, 2.16/3 (SD = 0.66) in dependability, 2.05/3 (SD = 0.72) in stimulation, and, finally, 2.09/3 (SD = 0.98) in novelty were obtained.

Table 5 contains those results corresponding to assessing the acceptance of the CAPACITY solution in terms of TAM, that show an improving trend ($p = 0.15$) starting with a value of 49.00/60 (SD = 8.24) at baseline, that gets to 50.68/60 (SD = 6.68) at month 3 and reaches 51.54 (SD = 6.97) at month 6. On the other hand, Table 6 presents the results related to the administration of the ad-hoc quantitative questionnaires, that evaluate individually each component of the home-monitoring kit.

Table 3. SUS results.

SUS Item	Baseline	Month 3	Month 6
Q1	4.32	4.43	4.32
Q2	2.00	1.65	1.57
Q3	4.48	4.46	4.49
Q4	2.88	2.43	1.92
Q5	4.76	4.38	4.24
Q6	1.48	1.65	1.65
Q7	4.04	3.95	4.00
Q8	1.16	1.19	1.16
Q9	4.84	4.84	4.76
Q10	2.40	2.11	2.35
Mean	80.11	83.31	83.68
SD	13.66	15.07	11.62
$P_{\text{baseline-m3}}$	0.57		
$P_{\text{baseline-m6}}$	0.49		
$P_{\text{m3-m6}}$	0.88		

Table 4. Categorized UEQ results.

UEQ Category	Mean	SD	Confidence	Conf. Interval ($p = 0.05$)	
Attractiveness	2.20	0.64	0.21	1.99	2.40
Perspicuity	2.30	0.73	0.23	2.06	2.53
Efficiency	1.99	0.75	0.24	1.75	2.23
Dependability	2.16	0.66	0.21	1.95	2.38
Stimulation	2.05	0.72	0.23	1.82	2.28
Novelty	2.09	0.98	0.32	1.78	2.41

Table 5. TAM results.

	TAM item	Baseline	Month 3	Month 6
Perceived usefulness	Q1	4.16	4.35	4.16
	Q2	3.88	3.89	4.03
	Q3	3.84	4.16	4.24
	Q4	4.08	4.22	4.11
	Q5	4.12	4.27	4.30
	Q6	4.56	4.51	4.24
	Mean	24.64	25.41	25.08
	SD	4.79	3.68	4.48
	$P_{\text{baseline-m3}}$	0.64		
	$P_{\text{baseline-m6}}$	0.84		
	$P_{\text{m3-m6}}$	0.40		
Perceived ease-of-use	Q1	3.72	4.30	4.24
	Q2	3.88	4.35	4.51
	Q3	4.36	4.30	4.65
	Q4	4.52	4.11	4.46
	Q5	3.72	4.00	4.16
	Q6	4.16	4.22	4.43
	Mean	24.36	25.27	26.46
	SD	4.86	5.16	3.77
	$P_{\text{baseline-m3}}$	0.25		
	$P_{\text{baseline-m6}}$	0.02		
	$P_{\text{m3-m6}}$	0.36		

Table 5. *Cont.*

TAM item	Baseline	Month 3	Month 6
Mean	49.00	50.68	51.54
SD	8.24	6.68	6.97
$p_{baseline\text{-}m3}$	0.26		
$p_{baseline\text{-}m6}$	0.15		
$p_{m3\text{-}m6}$	0.83		

Table 6. Acceptance results (ad-hoc questionnaires).

Sensor		Q1	Q2	Q3	Q4	Q5
Gait speed	Mean	4.19	4.53	1.44	4.00	4.44
	Std. dev	0.75	0.61	0.84	0.99	0.88
Chair stand	Mean	4.61	4.64	1.08	4.53	4.69
	Std. dev	0.64	0.68	0.37	0.70	0.79
Weight	Mean	4.69	4.61	1.19	4.61	4.92
	Std. dev	0.58	0.90	0.82	0.69	0.28

SUS and TAM data were analyzed according to the educational level (i.e., non-formal education, primary education, secondary education, or higher education), living conditions (i.e., alone, with younger relatives, or with other older person), daily help received (i.e., from nobody, from a younger relative, from other older person, or from social services), previous experience with technology (i.e., no experience, used it once or twice before, occasionally used, or daily use), and frailty diagnosis (i.e., pre-frail, or frail). Table 7 shows the evolution of the reported SUS and TAM according to the category labels.

Table 7. SUS and TAM evolution per category label.

	SUS at Baseline	SUS at M3	SUS at M6	TAM at Baseline	TAM at M3	TAM at M6
Non-formal education	75.75	80.16	82.33	46.55	48.56	50.40
Primary education	83.25	84.12	82.94	48.33	51.53	51.12
Secondary education	92.50	85.63	84.50	56.67	55.50	56.40
Higher education	80.00	-	-	53.00	-	-
p	0.31	0.69	0.94	0.24	0.14	0.24
Living alone	80.21	80.33	82.14	49.08	49.47	49.07
Living with younger relatives	90.00	88.75	90.63	50.50	54.50	54.75
Living with other older person	80.91	83.06	81.84	48.64	50.83	52.68
p	0.65	0.60	0.37	0.96	0.41	0.21
Help from nobody	84.04	87.61	84.89	48.77	52.05	51.50
Help from a younger relative	78.21	75.31	80.71	48.29	47.13	50.43
Help from other older person	78.13	77.50	79.64	50.75	50.50	53.29
Help from social services	80.00	60.00	77.50	50.00	50.00	48.00
p	0.79	0.052	0.66	0.97	0.37	0.84

Table 7. *Cont.*

	SUS at Baseline	SUS at M3	SUS at M6	TAM at Baseline	TAM at M3	TAM at M6
No previous experience with technology	79.17	82.60	80.30	49.94	50.50	50.88
Used technology once or twice	92.50	85.83	94.17	52.00	52.67	56.33
Occasional use of technology	85.63	84.38	90.71	46.75	51.13	52.14
Daily use of technology	80.00	70.00	71.25	35.00	48.00	50.50
p	0.54	0.66	0.017	0.31	0.96	0.65
Pre-frail	80.00	84.50	83.89	49.69	51.40	53.00
Frail	82.71	82.27	82.59	48.25	50.56	51.07
p	0.62	0.66	0.57	0.67	0.86	0.87

Statistically significance related to the evolution= of the reported SUS and TAM within the categories described above has been analyzed. Only those older persons living with a younger relative showed a marginal but significant improvement in the reported SUS between baseline and month 3 ($p = 0.049$).

4. Discussion

This research study shows that the CAPACITY technological ecosystem has a very high-performance in term of usability, UX, and acceptance. Results have been obtained in a real-world scenario, where pre-frail and frail older persons used CAPACITY as the main vehicle to avoid transiting to disability.

Usage information demonstrates a high adoption rate, with an average adherence to the monitoring plan very close to or matching 100% for all components of the monitoring plan (i.e., use of sensors and completion of questionnaires). This endorses the validity of data collected in terms of usability, UX, and acceptance, since these are based on an intensive use of the system under assessment. However, the high usage of the system is not fully correlated with the expected use; for instance, during the experimentation the physicians detected that some users were not complying with the temporality of the monitoring plan (i.e., some measurements were missing, and they had to reach out to the older person to remind him or her). This has a twofold interpretation: on the one hand, sometimes notifications are neglected by the users, implying that new strategies should be found to promote prompt responses; but, on the other hand, the information that is constantly being provided to the clinical team allows a closer follow-up of the older persons, enabling early actuation of potentially worrying situations.

Based on the data collected by Sauro [59], the mean SUS score across a large number of usability studies is 68. If that value is used as a reference, the mean SUS obtained in all sampling points is highly above average. Furthermore, according to the Sauro–Lewis SUS grading curve [60], obtained score would be qualified as an A, with the last measurement really close of reaching A+, set at 84.1. So we can state that the evaluated user interaction is perceived as very good, almost excellent [61]. However, although usability seems to improve with use, obtained paired data Student's test does not demonstrate that this improvement is statistically significant; a plausible explanation for this non-significant result could be linked to a ceiling effect, probably associated to an insufficient sample size. A further analysis by category showed that those older users living with a younger person marginally but significantly improved reported SUS between baseline and month 3 ($p = 0.049$), but this isolated result does not entitle to draw any solid conclusion since no other significant differences were observed.

Although usability is very high, which implies the UCD process was highly successful, it is not the highest possible, so there is still room for improvement. Most of the averaged items scored very close to the edges of the scale, which is good for the evaluation of the system, but some others deviate a bit from the expected value and are those which are susceptible to be improved. SUS items Q4, Q7, and Q10 are the ones lowering the overall score (without significant changes along the follow-up, except for Q4, that seems to show an improvement trend). The description of these items is:

- Q4: I think that I would need the support of a technical person to be able to use this system;
- Q7: I would imagine that most people would learn to use this system very quickly; and
- Q10: I needed to learn a lot of things before I could get going with this system.

These relatively low evaluations in these items could be linked to the unfamiliarity or insecurity of the older adults who used the technology during the intervention (65.22% of the sample did not have any experience with technology). Actually, obtained results indicate that, after six months of use, there are statistically significant differences in the reported SUS depending on the previous experience with technology ($p = 0.017$), which implies that its relationship with reported SUS needs to be further investigated.

UEQ does not provide an overall score for the UX but an individual score for each category. Scores between -0.8 and 0.8 usually represent a neutral evaluation, while values over 0.8 represent a positive evaluation; values below -0.8 represent a negative evaluation [62]. Obtained results are exceptional since all categories received averaged scores higher to or really close to 2, and given the fact that extreme UEQ values are very uncommon [63]). Furthermore, lower bounds of all confidence intervals per category are significantly above the minimum threshold established to be considered as positive evaluations ($p = 0.05$).

UX results in terms of UEQ have been benchmarked using a dataset with data from 9905 responses corresponding to 246 studies [64]; however, and given the fact that product categories have been not considered, this benchmarking can only be used as a first indicator to assess the UX of the system under study. Figure 6 represents the result of the benchmarking; in all six categories CAPACITY ranks over average (i.e., top 10%).

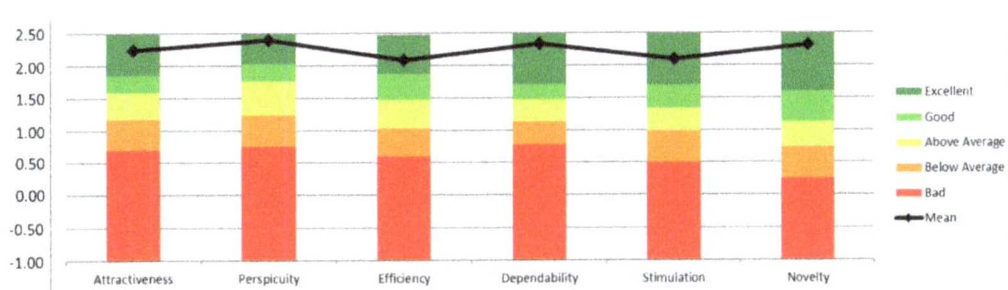

Figure 6. CAPACITY benchmarking according to UEQ.

It is important to analyze whether the UEQ respondents have provided random answers to endorse validity of the obtained results. In this case, given the specific characteristics of the population who participated in the study (i.e., older persons with poor education background and digital literacy), some inconsistencies (i.e., all items in a scale should measure a similar UX quality aspect; if there is a big difference—>3—this is an indicator for a problematic data pattern) in the provided answers have been found for several respondents. These inconsistent answers can be due to misunderstanding of one or several items. One respondent was inconsistent in three categories, six in two categories, and eight in one category. According to Schrepp [62], answers to UEQ with two or more inconsistencies should be considered suspicious. No significant changes are observed when the doubtful information is removed from the analysis: all six categories stay with averaged values above two; in the same way, all categories remain qualified as excellent in the benchmarking.

Acceptance in terms of TAM reached a maximum score of 51.54 in the last sampling point, also showing an increasing trend amongst data collection points (p = 0.15 from baseline to month 6); given the small sample size of this research work, and the fact that statistical significance is a function of both the sample size and the magnitude of the estimated effect, p-values lower than 0.2 could be considered statistically significant [65,66]. Furthermore, Student's t-test was significant (p = 0.02) for the positive evolution of the perceived ease of use between baseline and month 6. All individual items obtained averaged values above 4 in the last sampling point. The item that took the longest in reaching a value of 4 was Q2, under the category of 'perceived usefulness'; this item relates to whether users perceived that CAPACITY contributed to his or her daily life independency. A possible explanation could be related to the fact that results associated to a physical intervention are not perceived immediately. In any case, this seems to be more related to the clinical aspects of the project rather than to the technological ones. On the other hand, the evaluation of each individual device (i.e., each component of the home monitoring kit) indicates very high acceptance: all questions pursuing a value of 5 got an average value over 4, while those aiming at 1 got values below 1.5. Acceptance results not only suggest that the older population would accept using and having CAPACITY devices at home as a way of being constantly monitored in terms of function, but also that all components of the monitoring kit are perceived as empowerment tools to motivate having a healthier lifestyle and to control his or her own health.

Not many RCTs exist in the field of ICTs applied to frailty management, which prevents the availability of a wider number of works focused on assessing the usability of technologies for treating frail population [41]. Works analyzing usability-related aspects of technology during real interventions also report satisfactory results [67,68], however, since the design procedure followed is scarcely described, the sample size is significantly smaller than the one presented in this paper, and data related to adherence are not optimal (i.e., far from 100% adherence as reported in this paper), those results should be interpreted with caution. On the other hand, the majority of the research addressing how older persons interact with technology is done in controlled environments and under the supervision of domain experts [69–73]. Most of the published related research use standardized tools such as those used in this work, which is aligned with the methods followed in the current approach; moreover, despite the heterogeneous approaches in terms of the target application, ranging from rehabilitation [67–69] to exergames [70], monitoring cognitive impairment [71], fall risk [72], or evaluating available health apps [73], the used interaction instruments, including mobile devices [67,69,72,73], personal computers [68,71], or custom prototypes [70], and the diverse characteristics of the target population, that in some cases have a previous experience with the technology to be used [68] and in some others cannot use it without help [71], the UCD approach is a commonality backing almost all approaches from a methodological perspective.

5. Conclusions

The objective of this research work was to investigate the usability, UX, and acceptance of CAPACITY, a technological ecosystem to prevent disability. This objective has been achieved, obtaining very satisfactory results in all domains under study. The usability of CAPACITY (in terms of SUS) was rated as almost excellent, and UX (in terms of UEQ) as excellent; finally, the proposed technology, both from a software and a hardware perspective, seems to be highly accepted by the target population (in terms of TAM and ah-hoc questionnaires). Besides, adherence to using CAPACITY has been found optimal, which implies both that these superb results are correlated with maximizing the actual use of the proposed solution in a real environment and that data supporting the conclusions are based on reliable and solid information.

The main contribution of this paper is thus the demonstration that following an iterative UCD approach starting in a controlled laboratory environment to come up with a pre-validated interaction system, and later upscale it to a real uncontrolled environment is a valid strategy to maximize usability, UX, acceptance and actual adoption of a system. Furthermore, this research work contributes with a new experience to the scant number of RCTs studying how pre-frail and frail older persons interact with technology.

Findings support UCD as a key methodology. Involving potential end users and giving them voice during the design stage maximizes usability, UX, acceptance and usage. In this research, older persons were involved from the very beginning: first, older people's opinions were captured in a laboratory environment to later move towards clinical and home environments. Insights collected during this process enabled obtaining these excellent data within a RCT. Results indicate a potential high adoption in a wider deployment scenario (i.e., production phase). Some limitations must be taken into consideration when interpreting the results presented in this manuscript. First, the relatively small sample implies that findings need to be construed with prudence. Second, the external validity of the findings is not clear (i.e., whether the tested interaction system would obtain equivalent result in a population with different characteristics, such as culture, education, experience with technology, etc.); moreover, the assessment tools used to measure usability, UX, and acceptance, although they have been conceived to provide objective measurements, are highly dependent on the subjectivity of the respondents, so what is really measured is a perception on the different explored domains, given that humans are prone to bias while rating their experiences after interacting with a system. Third, no data related to the use of the technical assistance telephone line available for participants were collected, which has prevented integrating that information in the interpretation of the results. And, finally, no information on patient–physician communication through the platform was registered, limiting the extent of the presented usage analysis.

The CAPACITY technological ecosystem is constantly being improved, and new services added. From a service perspective, the current version of the solution incorporates functionalities to support a novel organizational model that interconnects all relevant people in the care process: the older person, the informal caregiver, and the primary and specialized care professionals. This evolved version of CAPACITY also integrates mechanisms (algorithms) to automatically detect functional decline and alert professionals and means to provide a multicomponent. Future work includes carrying out a new multicentric field experimentation (RCTs in Spain, Sweden, and Poland) with a higher sample size (ClinicalTrials.gov Identifier: NCT04592146) thus further assessment of the usability, UX and acceptance will be done, including extended work aimed at identifying ways of improving specific usability issues related to the individual answers to SUS, further exploring the relationship between usability and external factors (e.g., previous experience with technology, living conditions, etc.), and finding efficient ways to promote prompt responses to notifications. In addition, the home monitoring kit is in the process of being shifted towards a ubiquitous and transparent paradigm, that will probably maximize acceptance. These new devices will be based IoT technologies, easing their configuration, replacement, and scalability.

Supplementary Materials: The following are available online at https://www.mdpi.com/article/10.3390/s21196458/s1.

Author Contributions: L.R.-M. coordinated the research line. R.P.-R., E.V.-M., and X.F. conceptualized and materialized the technological approach. M.V.-A., R.P.-R., and L.R.-M. designed the experimental setup. R.P.-R. led and supervised the correct execution of the experiment, with support from M.V.-A., E.V.-M., C.M., and M.M.-R., and P.A.-S., R.P.-R., and E.V.-M. analyzed the results. All authors participated in writing the manuscript. All authors have read and agreed to the published version of the manuscript.

Funding: This research work was supported by the FACET (Integrated supportive services/products to promote FrAilty Care and wEll function. PGA 16003); and POSITIVE (maintaining and imPrOving the intrinSIc capaciTy involving primary care and caregIVErs. PGA 109091). Both projects funded by the EIT-Health.

Institutional Review Board Statement: The study was conducted according to the guidelines of the Declaration of Helsinki and approved by the Ethics Committee of Getafe University Hospital (protocol code 17/85 approved 14 December 2018).

Informed Consent Statement: Informed consent was obtained from all subjects involved in the study.

Data Availability Statement: No new data were created or analyzed in this study. Data sharing is not applicable to this article.

Acknowledgments: The authors would like to thank all other partners participating in the FACET (ABBOTT, ATOS, Genesis Biomed, GMV Quirónprevención and University of Aberystwyth) and in the POSITIVE (ATOS, Karolinska Institutet, KTH Royal Institute of Technology andMedical University of Lodz and IDNEO) projects.

Conflicts of Interest: The authors declare no conflict of interest. The funders had no role in the design of the study; in the collection, analyses, or interpretation of data; in the writing of the manuscript; or in the decision to publish the results.

References

1. WHO Integrated Care for Older People. Available online: https://www.who.int/publications/i/item/9789241550109 (accessed on 20 September 2021).
2. Fulop, T.; Larbi, A.; Witkowski, J.M.; McElhaney, J.; Loeb, M.; Mitnitski, A.; Pawelec, G. Aging, frailty and age-related diseases. *Biogerontology* **2010**, *11*, 547–563. [CrossRef]
3. Rodriguez-Mañas, L.; Fried, L.P. Frailty in the clinical scenario. *Lancet* **2015**, *385*, e7–e9. [CrossRef]
4. O'Caoimh, R.; Galluzzo, L.; Rodriguez-Laso, A.; Van Der Heyden, J.; Ranhoff, A.H.; Lamprini-Koula, M.; Ciutan, M.; López-Samaniego, L.; Carcaillon-Bentata, L.; Kennelly, S.; et al. Prevalence of frailty at population level in European ADVANTAGE Joint Action Member States: A systematic review and meta-analysis. *Ann. Ist. Super. Sanita* **2018**, *54*, 226–238. [CrossRef]
5. Rodriguez-Mañas, L.; Laosa, O.; Vellas, B.; Paolisso, G.; Topinkova, E.; Oliva-Moreno, J.; Bourdel-Marchasson, I.; Izquierdo, M.; Hood, K.; Zeyfang, A.; et al. Effectiveness of a multimodal intervention in functionally impaired older people with type 2 diabetes mellitus. *J. Cachex- Sarcopenia Muscle* **2019**, *10*, 721–733. [CrossRef]
6. Izquierdo, M.; Casas-Herrero, A.; Martínez-Velilla, N.; Alonso-Bouzón, C.; Rodríguez-Mañas, L. An example of cooperation for implementing programs associated with the promotion of exercise in the frail elderly. European Erasmus+ Vivifrail program. *Rev. Esp. Geriatr. Gerontol.* **2017**, *52*, 110–111. [CrossRef]
7. Izquierdo, M.; Vivifrail Investigators Group; Rodriguez-Mañas, L.; Sinclair, A.J. What is new in exercise regimes for frail older people—How does the Erasmus Vivifrail Project take us forward? *J. Nutr. Heal. Aging* **2016**, *20*, 736–737. [CrossRef]
8. Seldeen, K.L.; Lasky, M.G.; Leiker, M.M.M.; Pang, M.; Personius, P.K.E.; Troen, B.R. High Intensity Interval Training Improves Physical Performance and Frailty in Aged Mice. *J. Gerontol. Ser. A. Boil. Sci. Med. Sci.* **2017**, *73*, 429–437. [CrossRef]
9. Silva, R.B.; Aldoradin-Cabeza, H.; Eslick, G.D.; Phu, S.; Duque, G. The Effect of Physical Exercise on Frail Older Persons: A Systematic Review. *J. Frailty Aging* **2017**, *6*, 91–96. [CrossRef]
10. Makizako, H.; Shimada, H.; Doi, T.; Tsutumimoto, K.; Yoshida, D.; Suzuki, T. Effects of a community disability prevention program for frail older adults at 48-month follow up. *Geriatr. Gerontol. Int.* **2017**, *17*, 2347–2353. [CrossRef]
11. Dedeyne, L.; Deschodt, M.; Verschueren, S.; Tournoy, J.; Gielen, E. Effects of multi-domain interventions in (pre)frail elderly on frailty, functional, and cognitive status: A systematic review. *Clin. Interv. Aging* **2017**, *12*, 873–896. [CrossRef]
12. Yannakoulia, M.; Ntanasi, E.; Anastasiou, C.A.; Scarmeas, N. Frailty and nutrition: From epidemiological and clinical evidence to potential mechanisms. *Metabolism* **2017**, *68*, 64–76. [CrossRef]
13. Cruz-Jentoft, A.J.; Kiesswetter, E.; Drey, M.; Sieber, C.C. Nutrition, frailty, and sarcopenia. *Aging Clin. Exp. Res.* **2017**, *29*, 43–48. [CrossRef]
14. Herr, M.; Sirven, N.; Grondin, H.; Pichetti, S.; Sermet, C. Frailty, polypharmacy, and potentially inappropriate medications in old people: Findings in a representative sample of the French population. *Eur. J. Clin. Pharmacol.* **2017**, *73*, 1165–1172. [CrossRef]
15. Maclagan, L.C.; Maxwell, C.J.; Gandhi, S.; Guan, J.; Bell, C.M.; Hogan, D.B.; Daneman, N.; Gill, S.S.; Morris, A.M.; Jeffs, L.; et al. Frailty and Potentially Inappropriate Medication Use at Nursing Home Transition. *J. Am. Geriatr Soc.* **2017**, *7*, 2205–2212. [CrossRef]
16. Veronese, N.; Stubbs, B.; Noale, M.; Solmi, M.; Pilotto, A.; Vaona, A.; Demurtas, J.; Mueller, C.; Huntley, J.; Crepaldi, G.; et al. Polypharmacy Is Associated with Higher Frailty Risk in Older People: An 8-Year Longitudinal Cohort Study. *J. Am. Med. Dir. Assoc.* **2017**, *18*, 624–628. [CrossRef]

17. Suárez-Méndez, I.; Walter, S.; López-Sanz, D.; Pasquín, N.; Bernabé, R.; Gallo, E.C.; Valdés, M.; del Pozo, F.; Maestú, F.; Rodríguez-Mañas, L. Ongoing Oscillatory Electrophysiological Alterations in Frail Older Adults: A MEG Study. *Front. Aging Neurosci.* **2021**, *13*, 54. [CrossRef]
18. Kinney, J.M. Nutritional frailty, sarcopenia and falls in the elderly. *Curr. Opin. Clin. Nutr. Metab. Care* **2004**, *7*, 15–20. [CrossRef]
19. Binotto, M.A.; Lenardt, M.H.; Rodríguez-Martínez, M.D.C. Fragilidade física e velocidade da marcha em idosos da comunidade: Uma revisão sistemática. *Rev. Esc. Enferm. USP* **2018**, *52*, e03392. [CrossRef]
20. Carnemolla, P. Ageing in place and the internet of things—How smart home technologies, the built environment and caregiving intersect. *Vis. Eng.* **2018**, *6*, 7. [CrossRef]
21. WHO. Global Age-Friendly Cities: A Guide. Available online: https://www.who.int/ageing/publications/Global_age_friendly_cities_Guide_English.pdf (accessed on 20 September 2021).
22. Pilotto, A.; Boi, R.; Petermans, J. Technology in geriatrics. *Age Ageing* **2018**, *47*, 771–774. [CrossRef]
23. Alam, M.R.; Reaz, M.B.I.; Ali, M.A.M. A review of smart homes: Past, present, and future. *IEEE Trans. Syst. Man Cybern. Part. C Appl. Rev.* **2012**, *42*, 1190–1203. [CrossRef]
24. Turjamaa, R.; Pehkonen, A.; Kangasniemi, M. How smart homes are used to support older people: An integrative review. *Int. J. Older People Nurs.* **2019**, *14*, e12260. [CrossRef]
25. Majumder, S.; Aghayi, E.; Noferesti, M.; Memarzadeh-Tehran, H.; Mondal, T.; Pang, Z.; Deen, M.J. Smart Homes for Elderly Healthcare—Recent Advances and Research Challenges. *Sensors* **2017**, *17*, 2496. [CrossRef]
26. Kon, B.; Lam, A.; Chan, J. Evolution of Smart Homes for the Elderly. In Proceedings of the 26th International Conference on World Wide Web Companion, Perth, Australia, 3–7 April 2017. [CrossRef]
27. Kim, J.; Choi, H.S.; Wang, H.; Agoulmine, N.; Deen, M.J.; Hong, J.W.K. POSTECH's U-health smart home for elderly monitoring and support. In Proceedings of the 2010 IEEE International Symposium on "A World of Wireless, Mobile and Multimedia Networks", WoWMoM 2010—Digital Proceedings, Montreal, QC, Canada, 14–17 June 2010.
28. Mohler, M.J.; Wendel, C.S.; Taylor-Piliae, R.E.; Toosizadeh, N.; Najafi, B. Motor Performance and Physical Activity as Predictors of Prospective Falls in Community-Dwelling Older Adults by Frailty Level: Application of Wearable Technology. *Gerontology* **2016**, *62*, 654–664. [CrossRef]
29. Stavropoulos, T.G.; Papastergiou, A.; Mpaltadoros, L.; Nikolopoulos, S.; Kompatsiaris, I. IoT Wearable Sensors and Devices in Elderly Care: A Literature Review. *Sensors* **2020**, *20*, 2826. [CrossRef]
30. Changizi, M.; Kaveh, M.H. Effectiveness of the mHealth technology in improvement of healthy behaviors in an elderly population—A systematic review. *mHealth* **2017**, *3*, 51. [CrossRef]
31. Zaslavsky, O.; Thompson, H.; Demiris, G. The Role of Emerging Information Technologies in Frailty Assessment. *Res. Gerontol. Nurs.* **2012**, *5*, 216–228. [CrossRef]
32. Lee, C.; Coughlin, J.F. PERSPECTIVE: Older Adults' Adoption of Technology: An Integrated Approach to Identifying Determinants and Barriers. *J. Prod. Innov. Manag.* **2014**, *32*, 747–759. [CrossRef]
33. Merkel, S.; Kucharski, A. Participatory Design in Gerontechnology: A Systematic Literature Review. *Gerontol.* **2018**, *59*, e16–e25. [CrossRef]
34. Zhou, J.; Rau, P.-L.P.; Salvendy, G. Age-related difference in the use of mobile phones. *Univers. Access Inf. Soc.* **2013**, *13*, 401–413. [CrossRef]
35. Keränen, N.S.; Kangas, M.; Immonen, M.; Similä, H.; Enwald, H.; Korpelainen, R.; Jämsä, T. Use of Information and Communication Technologies Among Older People with and without Frailty: A Population-Based Survey. *J. Med. Internet Res.* **2017**, *19*, e29. [CrossRef]
36. Yazdani-Darki, M.; Rahemi, Z.; Adib-Hajbaghery, M.; Izadi, F.S. Older adults' barriers to use technology in daily life: A qualitative study. *Nurs. Midwifery Stud.* **2020**, *9*, 229.
37. Norman, D.A.; Draper, S.W. (Eds.) *User Centered System Design*; CRC Press: Boca Raton, FL, USA, 1986.
38. Duque, E.; Fonseca, G.; Vieira, H.; Gontijo, G.; Ishitani, L. A systematic literature review on user centered design and participatory design with older people. In Proceedings of the IHC 2019—Proceedings of the 18th Brazilian Symposium on Human Factors in Computing Systems, Vitória, Espírito Santo, 21–25 October 2019.
39. Wilkinson, C.; De Angeli, A. Applying user centred and participatory design approaches to commercial product development. *Des. Stud.* **2014**, *35*, 614–631. [CrossRef]
40. Najafi, B.; Armstrong, D.G.; Mohler, J. Novel Wearable Technology for Assessing Spontaneous Daily Physical Activity and Risk of Falling in Older Adults with Diabetes. *J. Diabetes Sci. Technol.* **2013**, *7*, 1147–1160. [CrossRef]
41. Gallucci, A.; Trimarchi, P.D.; Abbate, C.; Tuena, C.; Pedroli, E.; Lattanzio, F.; Stramba-Badiale, M.; Cesari, M.; Giunco, F. ICT technologies as new promising tools for the managing of frailty: A systematic review. *Aging Clin. Exp. Res.* **2020**, *33*, 1453–1464. [CrossRef]
42. VIVIFRAIL Project. Available online: https://ec.europa.eu/programmes/erasmus-plus/projects/eplus-project-details/#project/556988-EPP-1-2014-1-ES-SPO-SCP (accessed on 20 September 2021).
43. Pérez-Rodríguez, R.; Guevara-Guevara, T.; Moreno-Sánchez, P.A.; Villalba-Mora, E.; Valdés-Aragonés, M.; Oviedo-Briones, M.; Carnicero, J.A.; Rodríguez-Mañas, L. Monitoring and Intervention Technologies to Manage Diabetic Older Persons: The CAPACITY Case—A Pilot Study. *Front. Endocrinol.* **2020**, *11*, 300. [CrossRef]

44. Ferre, X.; Villalba-Mora, E.; Caballero-Mora, M.-A.; Sanchez, A.; Aguilera, W.; Garcia-Grossocordon, N.; Nuñez-Jimenez, L.; Rodríguez-Mañas, L.; Liu, Q.; Del Pozo-Guerrero, F. Gait Speed Measurement for Elderly Patients with Risk of Frailty. *Mob. Inf. Syst.* **2017**, *2017*, 1310345. [CrossRef]
45. Cobo, A.; Villalba-Mora, E.; Pérez-Rodríguez, R.; Ferre, X.; Escalante, W.; Moral, C.; Rodriguez-Mañas, L. Automatic and Real-Time Computation of the 30-Seconds Chair-Stand Test without Professional Supervision for Community-Dwelling Older Adults. *Sensors* **2020**, *20*, 5813. [CrossRef]
46. Fried, L.P.; Tangen, C.M.; Walston, J.D.; Newman, A.B.; Hirsch, C.; Gottdiener, J.S.; Seeman, T.; Tracy, R.P.; Kop, W.J.; Burke, G.L.; et al. Frailty in Older Adults: Evidence for a Phenotype. *J. Gerontol.* **2001**, *56*, M146–M157. [CrossRef]
47. Rubenstein, L.Z.; Harker, J.O.; Salvà, A.; Guigoz, Y.; Vellas, B. Screening for Undernutrition in Geriatric Practice: Developing the Short-Form Mini-Nutritional Assessment (MNA-SF). *J. Gerontol. Ser. A Boil. Sci. Med Sci.* **2001**, *56*, M366–M372. [CrossRef]
48. Barthel, D.; Mahoney, F. Baltimore City Medical Society Functional Evaluation: The Barthel Index. *Md. State Med. J.* **1965**, *14*, 56–61.
49. Morley, J.E.; Malmstrom, T.K.; Miller, D.K. A simple frailty questionnaire (FRAIL) predicts outcomes in middle aged African Americans. *J. Nutr. Health Aging* **2012**, *16*, 601–608. [CrossRef]
50. Pfeffer, R.I.; Kurosaki, T.T.; Harrah, C.H., Jr.; Chance, J.M.; Filos, S. Measurement of Functional Activities in older adults in the community. *J. Gerontol.* **1982**, *37*, 323–329. [CrossRef]
51. Villalba-Mora, E.; Ferre, X.; Pérez-Rodríguez, R.; Moral, C.; Valdés-Aragonés, M.; Sánchez-Sánchez, A.; Rodríguez-Mañas, L. Home Monitoring System for Comprehensive Geriatric Assessment in Patient's Dwelling: System Design and UX Evaluation. *Front. Digit. Heal.* **2021**, *3*, 40. [CrossRef]
52. International Organization for Standardization. ISO INTERNATIONAL STANDARD ISO 9241-11. Available online: https://www.iso.org/obp/ui/#iso:std:iso:9241:-11:ed-2:v1:en (accessed on 20 September 2021).
53. Brooke, J. SUS: A quick and dirty usability scale. *Usability Eval. Ind.* **1996**, *189*, 4–7. [CrossRef]
54. Brooke, J. SUS: A retrospective. *J. Usability Stud.* **2013**, *8*, 29–40.
55. Lewis, J.R.; Sauro, J. The factor structure of the system usability scale. In *International Conference on Human Centered Design*; Springer: Berlin/Heidelberg, Germany, 2009.
56. Tullis, T.S.; Stetson, J.N. A Comparison of Questionnaires for Assessing Website Usability ABSTRACT: Introduction. In Proceedings of the Usability Professionals Association (UPA) 2004 Conference, Minneapolis, MN, USA, 7–11 June 2004.
57. Schrepp, M.; Hinderks, A.; Thomaschewski, J. User Experience Questionnaire. *Mensch. Comput. 2017-Tag. Spiel. Einfach Interagieren* **2018**, *17*, 355.
58. Davis, F.D. Perceived Usefulness, Perceived Ease of Use, and User Acceptance of Information Technology. *MIS Q.* **1989**, *13*, 319. [CrossRef]
59. Sauro, J. Measuring Usability with The System Usability Scale (SUS). *Meas. Usability* **2011**, *2*, 1–11.
60. Lewis, J.R.; Sauro, J. Item Benchmarks for the System Usability Scale. *J. Usability Stud.* **2018**, *13*, 158–167.
61. Bangor, A.; Kortum, P.; Miller, J. Determining what individual SUS scores mean. *J. Usability Stud.* **2009**, *3*, 114–128.
62. Schrepp, D.M. User Experience Questionnaire Handbook. *Procedia Comput. Sci.* **2014**.
63. Rauschenberger, M.; Schrepp, M.; Olschner, S.; Thomaschewski, J.; Cota, M.P. Measurement of user experience: A Spanish Language Version of the User Experience Questionnaire (UEQ). In Proceedings of the 7th Iberian Conference on Information Systems and Technologies, Madrid, Spain, 20–23 June 2012; pp. 471–476. [CrossRef]
64. Schrepp, M.; Hinderks, A.; Thomaschewski, J. Construction of a Benchmark for the User Experience Questionnaire (UEQ). *Int. J. Interact. Multimedia Artif. Intell.* **2017**, *4*, 40. [CrossRef]
65. Cohen, J. Things I have learned (so far). *Am. Psychol.* **1990**, *45*, 1304–1312. [CrossRef]
66. Rosenthal, R. Parametric measures of effect size. In *the Handbook of Research Synthesis*; Russel Sage Foundation: New York, NY, USA, 1994; ISBN 0871542269.
67. Pérez-Rodríguez, R.; Moreno-Sánchez, P.A.; Valdés-Aragonés, M.; Oviedo-Briones, M.; Divan, S.; García-Grossocordón, N.; Rodríguez-Mañas, L. FriWalk robotic walker: Usability, acceptance and UX evaluation after a pilot study in a real environment. *Disabil. Rehabil.: Assist. Technol.* **2019**, *15*, 718–727. [CrossRef]
68. Weering, M.D.-V.; Jansen-Kosterink, S.; Frazer, S.; Vollenbroek-Hutten, M. User Experience, Actual Use, and Effectiveness of an Information Communication Technology-Supported Home Exercise Program for Pre-Frail Older Adults. *Front. Med.* **2017**, *4*, 208. [CrossRef]
69. Mehra, S.; Visser, B.; Cila, N.; Helder, J.V.D.; Engelbert, R.H.; Weijs, P.J.; Kröse, B.J. Supporting Older Adults in Exercising with a Tablet: A Usability Study. *JMIR Hum. Factors* **2019**, *6*, e11598. [CrossRef]
70. Adcock, M.; Sonder, F.; Schättin, A.; Gennaro, F.; De Bruin, E.D. A usability study of a multicomponent video game-based training for older adults. *Eur. Rev. Aging Phys. Act.* **2020**, *17*, 1–15. [CrossRef]
71. Cossu-Ergecer, F.; Dekker, M.; Van Beijnum, B.-J.F.; Tabak, M. Usability of a New eHealth Monitoring Technology That Reflects Health Care Needs for Older Adults with Cognitive Impairments and Their Informal and Formal Caregivers. *Computer* **2018**, *5*, 197–207. [CrossRef]
72. Hsieh, K.L.; Fanning, J.T.; Rogers, W.A.; Wood, T.A.; Sosnoff, J.J. A Fall Risk mHealth App for Older Adults: Development and Usability Study. *JMIR Aging* **2018**, *1*, e11569. [CrossRef]
73. Grindrod, K.A.; Li, M.; Gates, A. Evaluating User Perceptions of Mobile Medication Management Applications with Older Adults: A Usability Study. *JMIR mHealth uHealth* **2014**, *2*, e11. [CrossRef]

sensors

MDPI

Article

A CSI-Based Human Activity Recognition Using Deep Learning

Parisa Fard Moshiri [1], Reza Shahbazian [2], Mohammad Nabati [1] and Seyed Ali Ghorashi [3,*]

1 Cognitive Telecommunication Research Group, Department of Electrical Engineering, Shahid Beheshti University G. C., Tehran 1983969411, Iran; p.fardmoshiri@mail.sbu.ac.ir (P.F.M.); mo.nabati@mail.sbu.ac.ir (M.N.)
2 Electrical Engineering Research Group, Faculty of Technology and Engineering Research Center, Standard Research Institute, Alborz 31745-139, Iran; r.shahbazian@standard.ac.ir
3 Department of Computer Science & Digital Technologies, School of Architecture, Computing, and Engineering, University of East London, London E15 4LZ, UK
* Correspondence: s.a.ghorashi@uel.ac.uk

Abstract: The Internet of Things (IoT) has become quite popular due to advancements in Information and Communications technologies and has revolutionized the entire research area in Human Activity Recognition (HAR). For the HAR task, vision-based and sensor-based methods can present better data but at the cost of users' inconvenience and social constraints such as privacy issues. Due to the ubiquity of WiFi devices, the use of WiFi in intelligent daily activity monitoring for elderly persons has gained popularity in modern healthcare applications. Channel State Information (CSI) as one of the characteristics of WiFi signals, can be utilized to recognize different human activities. We have employed a Raspberry Pi 4 to collect CSI data for seven different human daily activities, and converted CSI data to images and then used these images as inputs of a 2D Convolutional Neural Network (CNN) classifier. Our experiments have shown that the proposed CSI-based HAR outperforms other competitor methods including 1D-CNN, Long Short-Term Memory (LSTM), and Bi-directional LSTM, and achieves an accuracy of around 95% for seven activities.

Keywords: activity recognition; Internet of Things; smart house; deep learning; channel state information

Citation: Fard Moshiri, P.; Shahbazian, R.; Nabati, M.; Ghorashi, S.A. A CSI-Based Human Activity Recognition Using Deep Learning. *Sensors* **2021**, *21*, 7225. https://doi.org/10.3390/s21217225

Academic Editor: Ivan Miguel Serrano Pires

Received: 6 September 2021
Accepted: 22 October 2021
Published: 30 October 2021

1. Introduction

The Internet of Things (IoT) is a dynamic global information network consisting of internet-connected devices [1]. Due to the recent advancements in communication systems and wireless technology over the last decade, IoT has become a vibrant research field [2]. The concept is straightforward; things or objects are connected to the internet and exchange data or information with each other over the network. Applications of IoT improve the quality of life [3]. As one of the main IoT applications, smart houses allow homeowners to monitor everything, including the health, especially for those with disabilities and elderly people, by exerting Human Activity Recognition (HAR) techniques [4]. Additionally, the joint task of HAR and indoor localization can be exerted in smart house automation [4]. A user's location can change how the IoT devices respond to identical gesture commands. For instance, users can use the "hand down" signal to reduce the temperature of the air conditioner, but they can also use the same gesture to lower the television in front of them [4]. HAR has emerged as one of the most prominent and influential research topics in several fields, including context awareness [5], fall detection [6], elderly monitoring [7], and age and gender estimation [8].

HAR techniques can be categorized into three groups: vision-based, sensor-based, and WiFi-based [7]. Existing sensor-based and vision-based methods for HAR tasks have achieved acceptable results. However, these methods still have limitations in terms of environmental requirements. Strictly speaking, camera-based recognition algorithms are susceptible to environmental factors such as background, lighting, occlusion, and social constraints such as privacy issues. Additionally, in sensor-based methods, people often object to these sensor modalities because they are bothersome or cumbersome. Although the underlying technology employed in these sensors is frequently inexpensive, IoT-connected versions of these sensors can be significantly more expensive due to added wireless hardware and branding. WiFi devices, which are less expensive and power-efficient than the aforementioned technologies, invariant to light, easier to implement, and have fewer privacy concerns than cameras, have recently attracted much interest in various applications [4].

The purpose of WiFi-based activity recognition is to distinguish the executed actions by analyzing the specific effects of each activity on the surrounding WiFi signals. In other words, the individual's movement affects the propagated signal from WiFi access points and can be used to recognize activities. WiFi signals can be described by two characteristics: Received Signal Strength (RSS) and Channel State Information (CSI) [4]. RSS is the estimated measure of received signals' power which has been mainly used in indoor positioning [9]. As RSS is not stable compared with CSI, it cannot properly capture dynamic changes in the signal while the activity is performed [10]. As a more informative specification of WiFi signals for HAR tasks, CSI has drawn more attention than RSS over recent years [10]. CSI can save physical layer information from each sub-carrier of the channel. When a person performs a particular activity between the transmitter and receiver, the reflected wireless signals from the body generate a unique pattern [11]. Furthermore, human body shapes, speed of performing an activity, environmental obstacles, and the path of performing an activity can cause different changes to received CSI signals. For instance, if a person walks in a straight line this activity has a different effect on CSI signal, comparing to the experiment that a person walks around a square path. Many WiFi devices use CSI to assess the quality of their connection internally. The device collects the experimental phase and strength of the signal at each antenna for each channel in the provided spectrum, allowing signal disruptions to be identified. The WiFi-based method takes advantage of the ubiquitous nature of radio frequency transmissions while also potentially allowing for developing a system that takes advantage of the existing WiFi infrastructure in smart houses [4].

Although business applications of HAR are in the beginning stages, many studies in this field introduce the issues that must be addressed before any practical action. One of the main issues is the specific hardware/software combination that is required to extract CSI data. After choosing the proper hardware, the collected CSI data can be further used as inputs of the Deep Learning (DL) algorithms for HAR task. The effects of each activity in characteristics of the collected CSI can be used in different DL algorithms to distinguish activities and finally classify them [11].

Since CSI is a time-series data with temporal dependency, Recurrent Neural Network (RNN) and its subsets have been exerted more than other DL algorithms in the HAR task. Long Short-Term Memory (LSTM) and RNN apply sequential processing to long-term information, meaning that these data pass through all cells in the network before reaching the present cell. RNNs structure cannot perform efficiently when we need to analyze long sequences, resulting in vanishing gradients. The vanishing gradient problem persists even when the switch gates and long memory in the LSTM network are maintained [11]. Furthermore, this module requires a significant amount of memory bandwidth due to the complexity of the sequential path and Multi-Layer Perceptron (MLP) layers in each cell. Despite the LSTMs proficiency for prediction and classification tasks in time series, they are incapable of learning terms with greater than 100 terms [12]. Additionally, LSTMs analyze the sequential data in one direction, meaning that only past CSI data will be considered [11]. Accordingly, they cannot distinguish between two similar activities, such as lie down and sit down, which have the same start position but different final positions.

In real-time activity monitoring, especially for elderly people, each activity's period and further information are essential. Therefore, we consider two approaches: 2D-CNN and Attention-based Bi-directional LSTM. Unlike RNNs and LSTMs, where long-term data is analyzed sequentially, convolutions analyze the data in parallel. Furthermore, the training time of LSTMs is slightly longer than the CNNs, and as a result, they require a greater memory bandwidth for processing. Less consumed time in training and lower computational complexity, along with mentioned problems, encouraged us to use 2D-CNN. Since 2D-CNN has high potential in image processing, we convert CSI data into RGB images. In order to generate RGB images, we made a pseudocolor plot from CSI matrices. Each element of the matrices is linearly mapped to the RGB colormap. Furthermore, we applied BLSTM on raw CSI data. HAR's performance can be improved by using attention-based BLSTM, which concentrates on regions of greater relevance and assigns them higher weights to improve performance. The main contributions of this research are as follows:

- We exploit Raspberry Pi for CSI data collection and offer a public CSI dataset for seven different activities including sit down, stand up, lie down, run, walk, fall and bend in an indoor environment using the Nexmon CSI tool [13]. Due to reflections induced by human activity, each subcarrier contains critical information that will increase HAR accuracy. The CSI matrices in our dataset are composed of 52 columns (available data subcarriers) and 600 up to 1100 rows depending on the period of each activity. The results demonstrate that this hardware is capable of providing tolerable data that is comparable to traditional technologies.
- We propose a new concept in improving the precision of HAR by converting CSI data into images using pseudocolor plots and feeding them into 2D-CNN. This method overcomes the mentioned limitations of LSTM and also the training time and computational complexity are less than those of other existing methods. We also exert a BLSTM network with an attention layer to address LSTM problems with future information. The results demonstrate that the conversion idea with 2D-CNN outperforms BLSTM in accuracy and consumed time.
- We also perform a deep evaluation by implementing two other algorithms, including 1D-CNN and LSTM, and compare our results with four different models used for HAR, including Random Forest (RF) [14], Hidden Markov Model (HMM) [14], DenseLSTM [15], ConvLSTM [16] and Fully Connected (FC) network [17]. We analyze the performance of our dataset and proposed DL algorithms.

The rest of this paper is organized as follows: Section 2 reviews HAR studies. In Section 3, we provide a brief explanation about CSI, required information on hardware, software and firmware. Furthermore, four used neural network's structures are discussed in this section. Additionally, we briefly discuss other datasets and their public accessibility. The main contributions of this research are summarized in Section 4. We discuss the device configuration to collect CSI, image generation from CSI, and feeding the data to the neural networks. In Section 5, measurement setups and experimental results are reported, and finally, conclusions are discussed in Section 6.

2. Related Works

HAR techniques can be divided into three groups: vision-based, sensor-based, and WiFi-based. Several image-based methods for HAR have been published in recent years, using datasets such as RGB (red, green, and blue [18]), depth [19], and skeleton images [20]. The RGB dataset may not be qualified and robust enough in this method when the video contains considerable sudden camera movements and cluttered background. To this end, Anitha et al. [21] propose a shot boundary detection method. In their proposed method, the features and edges of videos are extracted. The features were then extracted as images and subsequently merged with the video feature and fed into the classifier. The Kernel Principal Component Analysis (KCPA) technique is applied to locate image features and joint features. The preparation process is thus gradually proficient, making the independent vector analysis increasingly realistic for real-life applications. The human activity videos are classified by K-Nearest Neighbor, obtaining better results than other cutting edge activity methods. While capturing an image or video of an activity in RGB format, it generates many pixel values, making it more difficult to distinguish the subject from the surrounding background and resulting in computational complexity. The aforementioned obstacles and the view dependency, background, and light sensitivity impair RGB video-based HAR performance and persuade researchers to exert depth images and other formats of images or videos to improve HAR performance. Most of the methods introduced for HAR utilizing skeleton datasets are confined in various ways, including feature representation, complexity, and performance [22]. In [22], authors propose a 3D skeleton joint mapping technique that maps the skeleton joints into a spatio-temporal image by joining a line across the same joints in two adjacent frames, which is then used to recognize the person's activities. The 3D skeleton joint coordinates were mapped along the XY, YZ, and ZX planes to address the view dependency problem. They exploit transfer learning models including MobileNetV2, DenseNet121, and ResNet18 to extract features from images [22].

In sensor-based methods, wearable sensors capture activities, causing inconvenience and long-time monitoring unavailability [23]. In the past decade, smartphones have become more powerful with many built-in sensors, including the accelerator, gyroscope. The main impediments in using smartphones for HAR tasks are their higher noise ratio than wearable sensors and fast battery drain [23]. Several researchers have exerted Radio Frequency Identification (RFID) tags to recognize human activities [24]. Authors in [24] present a framework for HAR and activity prediction by using RFID tags. They utilize RFID tags to detect a high-level activity and object usage. Additionally, they employ weighted usage data and gain activity logs. Since human activities are time series data and the next activity is related to the current activity and previous ones, they use LSTM to predict activities with an accuracy of 78.3%. Although RFID tags are cheaper, RFID-based systems cannot achieve high accuracy in crowded environments. Additionally, as mentioned above, vision-based HAR needs cameras installation in the environment, which highly depends on the light source's consistency and is unable to pass through physical obstacles such as walls. Since indoor spaces such as smart houses, malls, and nursing homes are filled with wireless signals, WiFi-based systems have been exploited more than other approaches in recent years [25].

Due to the growing interest in sensor-less activity detection, the research and industry communities have joined on CSI analytics with the help of neural networks. Common CSI-based applications include a wide range of activity detection scenarios such as WiTraffic [26] to delicate activity recognition systems like Wifinger [27], breathtrack [28]. In [29], authors utilize CSI to sense distinct hand movements. They use predefined windows to monitor activity continuously. This method is time-consuming and yields lower accuracy. To overcome this problem, Wi-Chase [30] does not apply predetermined time windows. Due to detailed correlated information in different subcarriers, Wi-Chase also employs all available subcarriers, unlike Wi-Sleep [31] that uses only a subset of them. The extracted features were trained using machine learning algorithms, including KNN and Support Vector Machine (SVM) [30]. Although different WiFi-based HAR systems have been proposed, one of the major challenges has not been addressed properly. That is, WiFi signal changes are due to the various movement speeds and body types of people. Human activity is made up of many limb movements, such as lifting an arm or leg. The speed and scale of activity can naturally alter according to the scenario or period. Furthermore, physical traits such as body form and height are unique for each person. Therefore, human activity patterns can vary greatly amongst people. To address this problem, a WiFi-based HAR proposed in [15] incorporates synthesized activity data that reduces the influence of activity inconsistency such as varied motion speed. They collect CSI for 10 different activities including make phone calls, jumps, check wristwatch, lie down, walk, play guitar, fast walk, play piano, run, play basketball with Atheros AR9590 WiFi chipset. The combination of CSI spectrogram of overall subcarriers is fed into the network as image inputs. Then, four Dense layers are used to extract spatial features of activities. These features are entered to a convolutional layer. Then, a BLSTM is used to extract tempo features and a linear layer is applied to predict the activities. Three data synthesis methods are combined with eight types of transformation methods, including dropout, Gaussian noise, time-stretching, spectrum shifting, spectrum scaling, frequency filtering, sample mixture, and principal component coefficient. Dense LSTM with consistent accuracy of 90% is applied to efficiently optimize the system for the small-size dataset while keeping the model compact to minimize overfitting.

For multi-class classification based on extracted features such as HAR, a variety of machine learning algorithms such as RF, SVM, and HMM and also DL algorithms such as CNN, RNN, and LSTM can be applied. In [14], they apply RF, HMM, LSTM on their public dataset which have been collected with NIC 5300 with three antennas for six different activities including sit down, stand up, fall, walk, run, and bed. A 90-dimensional vector of CSI amplitude (3 antennas and 30 subcarriers) has been used as the input feature vector. They apply the PCA on the CSI amplitude for denoising, and Short-Time Fourier Transform (STFT) for feature extraction. First, they use RF with 100 trees for classification, which has unacceptable accuracy for bed, sit down and stand up activities. They also apply HMM on the extracted features obtained by STFT and DWT techniques. The accuracy is improved compared to RF, but with higher training time. Although HMM has obtained good results for walk and run activities, it cannot distinguish between stand up, sit down, and bed activities. They also apply LSTM on activities [14]. The LSTM extracts the features automatically and directly from raw CSI without any pre-processing. In other words in contrast to other methods, the LSTM approach does not need PCA and STFT, but it has more training time [14]. The accuracy of LSTM is reported over 75% for all activities in [15].

Since the static objects in an environment can also affect wireless signals and, respectively, HAR model, authors in [17] propose a deep neural network as baseline classifier based on the features for four simple activities, including standing up, sitting down, pushing and picking, performed in two different complex environments. More precisely, they propose a network with shared weight to make a similarity network for two different complex environments. They used one transmit antenna and two receive antennas and make four grayscale images from CSI amplitude and phase. In feature extraction stage, Gabor filter is applied on grayscale images to extract features. Gabor filter extracts spatial information of an image by convoluting the transformed image with a filter at specific wavelength λ and orientation θ [17]. For each gray-scale image, the final output is 5 (the number of λ) $\times$ 8 (number of θ) $\times$ 2 (mean and standard deviation) = 80, and a vector of dimensions 320 = 4 (number of grayscale images) $\times$ 80 are fed into the neural network as the input. They exert three FC hidden layers as the baseline network and two identical branches that share the same weight values as the similarity network. A pair of two random data are selected and fed into the two identical networks simultaneously and each one of them enters the fully connected network. If the two data belonged to the same category of activity, they are labeled as "similar", otherwise "non-similar". Their model obtains an accuracy of around 84% overall for the two different environment scenarios.

One of the main issues in Wifi-based HAR is the specific hardware/software combination for CSI data collection. In other words, Linux 802.11n CSI Tool is limited to older linux kernels versions and the required hardware cannot be found easily in market. Following the release of Nexmon CSI Tool [13], it is now possible to extract CSI from a BCM43455C0 wireless chipset, which is used in the Raspberry Pi 3B+ and 4B. As this is a recent release, Ref. [16] examines the performance of the Raspberry Pi 4 in CSI-based HAR. They collect CSI signals for different activities performed in normal life as listed: stand up, sit down, go-to-bed, cook, washing-dishes, brush-teeth, drink, pet a cat, sleeping, walk. They do not apply any denoising filter, as their results are acceptable comparing to other available datasets and also additional filtering may affect important information in data. They pack CSI vectors, collected by Raspberry Pi 4, into windows to train their classification model. As LSTMs and their extensions have been well-suited in HAR task, they use a deep convolutional variant of the LSTM model. They apply two 1D-convolutional layers along with four BLSTM, which have more training time and computational complexity. Their model achieves 92% accuracy which demonstrated the Raspberry Pi 4 capabilities for HAR in smart houses and it can be superseded the Linux 802.11 CSI Tool.

3. System Model

3.1. Preliminary

Transmitting a signal from the transmitter to the receiver, it is deflected, reflected, and scattered when it comes into contact with obstacles and objects. This results in multipath overlaid signals at the receiver when the signal encounters obstacles and objects [7]. Fine-grained CSI can be used to characterize this procedure. The Orthogonal Frequency-Division Multiplexing (OFDM) modulation is utilized in IEEE 802.11, and it distributes the available bandwidth across several orthogonal subcarriers [14]. Due to the limited bandwidth available, the fading that each subcarrier experiences are represented as flat fading [31]. Therefore, the small-scale fading aspect of the channel can be minimized by employing OFDM techniques. Narrow-band fading per subcarrier causes a considerable variation in the measured channel dynamics. The greatest advantage of employing CSI is that it can catch changes occurring at a single frequency and avoid averaging out changes across all WiFi bandwidth, unlike RSS.

Several subcarriers can be present in the physical link between each pair of transmitter and receiver antennas. As each subcarrier might serve many data streams, the CSI obtained from each subcarrier will be unique [14]. CSI can be represented as a channel matrix for t transmit and r receiving antennas, a given packet transmission n:

$$CSI_n = \begin{pmatrix} H_{1,1} & \cdots & H_{1,r} \\ \vdots & \ddots & \vdots \\ H_{t,1} & \cdots & H_{t,r} \end{pmatrix} \tag{1}$$

$H_{t,r}$ represents a vector that includes complex pairs for each subcarrier. Depending on the hardware we use and channel bandwidth, the number of available subcarriers is different [16]. Raspberry Pi 4 and Tp-link archer c20 paired over 5 GHz in 20 MHz bandwidth can access 56 data subcarriers. $H_{t,r}$ can be expressed as below for m data subcarrier in which h_m is a complex number, containing both amplitude and phase of the CSI:

$$H_{t,r} = [h_{t,r,1}, \ldots, h_{t,r,m}] \tag{2}$$

3.2. Hardware and Firmware

To the best of our knowledge, the specialized hardware/software combinations that is required to extract CSI data, are intel 5300 WiFi Network Interface Card (NIC) (Linux 802.11n CSI Tool) [32], Atheros AR9580, AR9590, AR9344, and QCA9558 (Atheros CSI tool) [33], Raspberry Pi (Nexmon CSI Tool) [13]. The intel 5300 NIC has been used for CSI collection since 2011 [32]. Although many researchers used 5300 NIC, such as [14], this hardware configuration has become less important over time since most laptops with this wireless card are not currently available in the market and third-party tools are required to collect CSI. More precisely, some type of Mini PCIe to PCI-Express Adapter with three antennas is required. Atheros CSI tool, as another 802.11n open-source experimental tool for CSI collection, allows extractions of physical layer wireless communication information, including CSI, RSS, the received payload packet, the timestamp, the data rate, etc. [33]. The ath9k open-source kernel driver supports Atheros 802.11n PCI or PCI-E chips; thus, this tool supports any sort of Atheros 802.11n WiFi chipsets. This tool was released in 2015 and there is more hardware with built-in Atheros 802.11n PCI or PCI-E chips rather than 5300 intel NIC, but more expensive.

The release of Nexmon CSI Tool [13] has enabled CSI extraction from Raspberry Pi 3B+ and 4B, Google Nexus 5, and some routers. One of the Nexmon tool benefits is that it permits several transmit-receive antenna configurations (up to 4 × 4 MIMO). Additionally, it includes customizable CSI collection filters that can extract relevant CSI from selected transmitters and the complete CSI data does not need to be suppressed. Although the Raspberry Pi utilizes a single transmit/receive antenna pairing, its price and prospective capabilities make it a suitable tool in WiFi-based healthcare monitoring in smart houses. Nexmon [13] provided a configuration option to assign a different interface to only the monitored frames after being configured on the host for monitoring on Raspberry Pi. This tool can use up to 80 MHz bandwidth and 242 subcarriers. There are three types of subcarriers in OFDM technology, including null subcarriers, pilot subcarriers, and data subcarriers. Null subcarriers (also called zero) are the unused subcarriers mainly employed as a guard against interference from adjacent channels. The pilot subcarriers do not convey modulated data; nevertheless, they are utilized for channel measurements and synchronization between the transmitter and receiver. Furthermore, pilot subcarriers broadcast using a predetermined data sequence and demonstrate an overhead for the channel. The remaining subcarriers from total subcarriers are called data subcarriers. These subcarriers will exploit the same modulation format as 802.11ac [34]. As mentioned in Table 1, we may have different numbers of subcarriers depending on the PHY standard and bandwidth.

Table 1. Subcarrier description for each PHY standard.

PHY Standards	Subcarriers Range	Pilot Subcarriers	Total/Data Subcarriers
802.11a/g	−26 to −1, +1 to +26	−21, −7, +7, +21	52/48
802.11n 802.11ac 20 MHz	−28 to −1, +1 to +28	−21, −7, +7, +21	56/52
802.11n 802.11ac 40 MHz	−58 to −2, +2 to +58	−53, −25, −11, +11, +25, +53	114/108
802.11ac 80 MHz	−122 to −2, +2 to +122	−103, −75, −39, −11, +11, +39, +75, +103	242/234

3.3. *Neural Network*

Once an activity is performed between transmitter and receiver, it will affect CSI characteristics. When a person performs a particular activity, the received CSI signals generates a unique pattern [7]. Recently, DL algorithms have been widely used to automatically learn features from the effects of activities on CSI. While having many layers in these algorithms offers improved classification skills, overfitting and performance deterioration become significant when implementing the neural network on a limited amount of dataset. Using traditional strategies such as weight decay, small batch size, and learning rate might not be enough to help avoid this problem. Accordingly, all of the pre-existed WiFi-based systems, such as those in Section 2, would require the implementation of dedicated numbers of particular neural layers to provide the desired performance. In this research, we present custom deep learning models that is best suited for situations with a small size dataset and has less computational complexity and consumed time compared to other methods.

3.3.1. CNN

CNN is a feed-forward neural network that excavates features from data with convolution operations. It contains several layers, including Convolution, Pooling, Dense and Flatten. This classification network requires less pre-processing rather than other classification techniques. Additionally, CNN can learn required filters or characteristics without the assistance of the user. CNNs use filters (also known as the kernel, feature detector) to extract features which are performed using the convolution function [35]. The initial Convolution Layer (ConvLayer) is designed to handle lower-level features, such as edges and color. When we employ several ConvLayers in the network topology, the network can achieve high recognition accuracy since it can also capture high-level features.

After each two 2D-ConLayer, we use the LeakyReLU activation function, an upgraded variant of ReLU (Rectified Linear Unit). According to the gradient in the negative direction, every value of inputs less than zero causes the gradient to be zero. Therefore, the neurons located in that region are deactivated and may suffer from the dying ReLU problem. In order to address this problem, instead of claiming that negative inputs values should be considered zero, a small linear component of S is defined. LeakyReLU can be formulated as $f(S) = \max(0.01 \times S, S)$, meaning that if the input is positive, the function returns S and if the input is negative, it returns $0.01 \times S$. This minor alteration causes a non-zero gradient for negative values; thus, we would not find any dead neurons in that location. Since the feature map output of ConvLayer specifies the specific position of features in the input, a slight movement in the location of the feature in the input data will create a significant difference in the feature map. To address this problem, we use the down-sampling strategy. A better and more widespread strategy is to utilize a pooling layer. After feature detection in ConvLayer, the Max pooling layer is applied to down-sampled feature maps and helps in extracting low-level features. After the first ConvLayer with Leaky ReLU activation function and max pooling, Batch Normalization (B.N.) is applied to stabilize the network during training and speed up training. B.N. makes the variable mean and standard deviation estimations more stable across mini-batches and, respectively, closer to 0 and 1. Dropout layers are applied between convolutional layers, decreasing overfitting while improving the network's generalization capability. The pooled features (the max pooling's output) should be flattened. Flattening involves concatenating the feature map matrix to create a single-column matrix. This matrix is passed through a dense layer where we get our predicted classes. The proposed 2D-CNN structure is depicted in Figure 1.

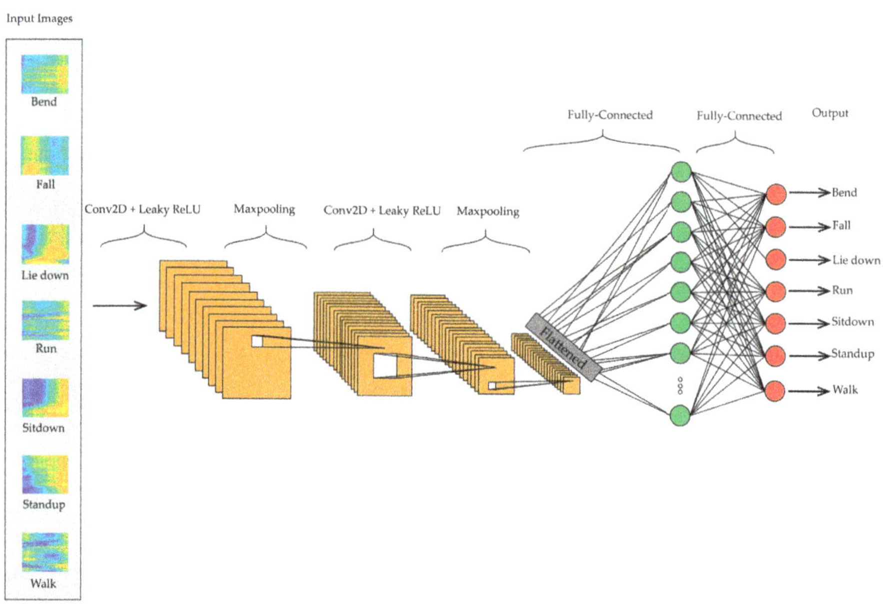

Figure 1. 2D-CNN structure used in this paper.

In addition to 2D-CNN exerted on converted RGB images, we also apply 1D-CNN to CSI data as depicted in Figure 2, which will convolve with moving along one dimension. Whether the input is 1D, 2D, or 3D, CNNs all have the same properties and use the same process. The crucial distinction is the dimensionality of the input data and the method in which the filter slides across it. The 1D-CNN has been trained to identify different activities based on sequential observations and map the internal features to different activities. It is particularly good at learning time-series data such as CSI, as it can leverage raw time series data and requires no domain expertise to hand-engineer input features. We use two ConvLayers with ReLU as an activation function. Same as 2D-CNN, after each ConvLayer, we apply max pooling layer, B.N., and dropout.

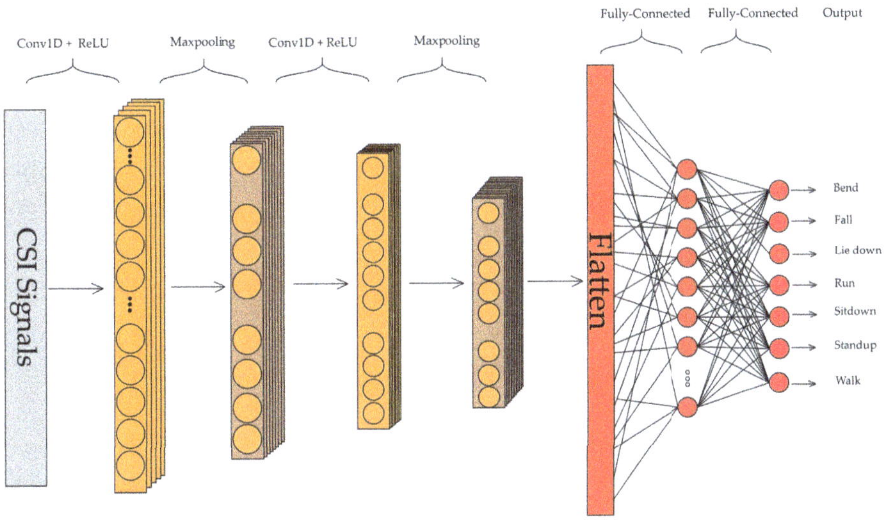

Figure 2. 1D-CNN structure used in this paper.

3.3.2. LSTM

RNN has been successfully applied to sequential modeling applications, such as language understanding [36] and HAR [37]. Nevertheless, when the learning sequence is long, the standard RNN frequently encounters the problem of the gradient vanishing and exploding. In order to address this issue, Hochreiter and Schmidhuber [38] designed a new RNN structure named the LSTM [38]. The LSTM network seeks to overcome gradient vanishing and exploding by utilizing memory cells with a few gates to retain essential information with long-term dependencies. The memory block comprises three gate sets. Each decides the block's state and produces an output, including forget gate, input gate, and output gate. The information to be eliminated from the unit is determined by the forget gate. The input gate handles which input values cause the memory state to be updated. The output gate determines the output of the block according to the input and the unit memory.

Since CSI signals are time-series and the LSTM can learn complicated and temporal dynamics, this network has obtained a remarkable performance for CSI-based HAR. In the HAR task, LSTM has two advantages. First, it can extract the features automatically without pre-processing. On top of that, it can hold temporal state information of the activity, resulting in better performance for similar activities such as lie down and sit down comparing to 1D-CNN, RF, and HMM. In this paper, we apply a simple LSTM with one hidden layer and 128 hidden units in which the feature vector is a 52-dimensional vector of CSI amplitudes. The proposed LSTM structure is depicted in Figure 3.

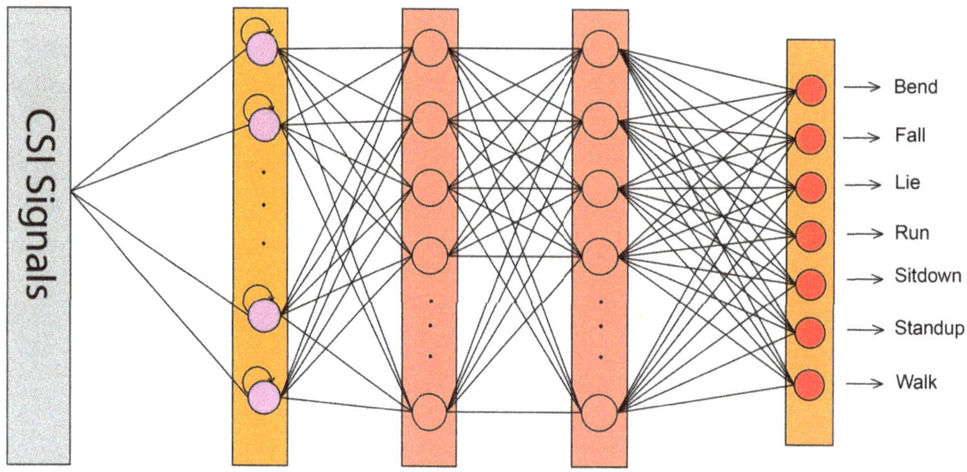

Input Signal LSTM Layer Dense Layer Dense Layer Prediction

Figure 3. LSTM structure used in this paper.

The traditional LSTM network only analyze the CSI data in one direction, meaning that the present hidden state only considers the past CSI information. Furthermore, future CSI information is also important for HAR. In this paper, an attention-based BLSTM is utilized to analyze past and future information and overcome long-term dependency. It contains a forward and backward layer for extracting information from the two directions. In other words, it's a two-layer LSTM sequence processing paradigm: one in which the input moves forward and the other in which the input moves backward. As the name suggests, attention is a technique that can allow input sequences of arbitrary length to pay attention to specified timesteps [11]. The concept is based on the studies about human vision systems, which indicate that humans consistently focus on a certain region of an image while identifying it and then altering their focus over time. It has been found to be effective in image recognition to have the machine focus on the region of interest while concealing the rest of the image at the same time for a recognition task. Due to the sequential features learned by the BLSTM network for WiFi-based HAR known to have high dimensions and feature contributions and time steps may vary from case to case, we seek to exploit the attention model to automatically learn features' significance and adjust feature weights based on activity recognition performance. In this paper, as depicted in Figure 4, a BLSTM with one attention layer with 400 units is used to learn the relative importance of features and timesteps and more important characteristics are given higher weights to obtain better performance.

The comparison between these four networks and five other networks in HAR researches, i.e., RF [14], HMM [14], DenseLSTM [15], ConvLSTM [16] and FC network [17], are discussed in Section 5. Note that, the proposed networks for our public dataset significantly outperforms other techniques in terms of accuracy, computational and structural complexity, and consumed time.

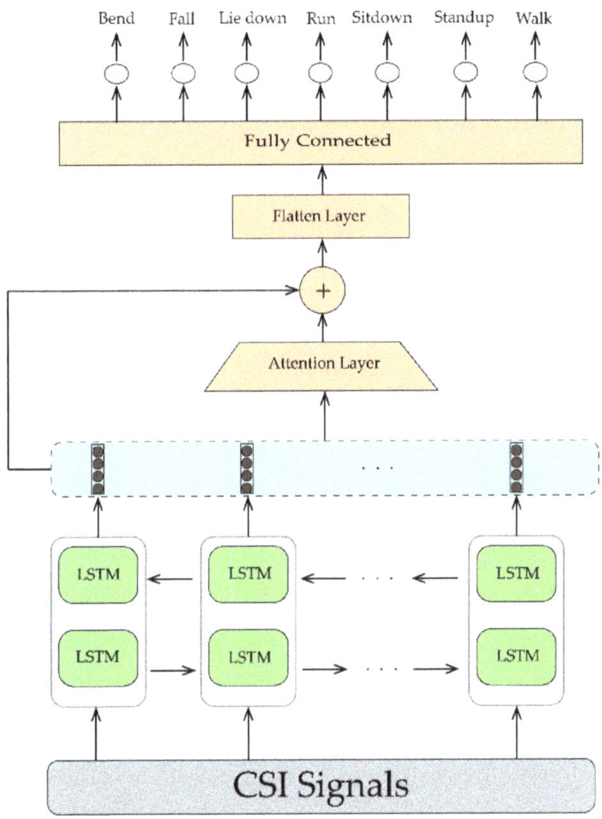

Figure 4. BLSTM structure used in this paper.

3.4. Human Activity Recognition Datasets

The amount of data we need for the HAR task depends on the complexity of the task and the chosen algorithm, hence there is no specific rule about the number of samples, needed to train a neural network and it is just a process of trial and error. For vision-based HAR task, [39] used 320 samples for 16 activities and [40] used 567 samples for 20 activities. We investigated the quantity of samples utilized in some CSI-based HAR researches. In ConvLSTM [16], they collected CSI data for 11 activities which were performed 100 times in a home environment (1100 samples). In [41], they collected 600 samples from 3 volunteers for 8 activities. In [30], they collected 720 samples of activities (12 volunteers × 20 samples × 3 activities). The authors in [42] collected 50 up to 100 samples for 4 actions (approximately 200 up to 400 samples). In [43], they collected 1400 samples from 25 volunteers. Authors in [44], collected 50 samples for 10 activities (500 samples). Siamak Yousefi et al. [14], as one of the most cited articles in WiFi-based human activity recognition, provided a public dataset for 6 different activities, performed by 6 users for 20 times (720 samples). According to other researches results, we asked 3 volunteers to perform 7 different activities 20 times, resulting in 420 samples. To the best of our knowledge, the WiFi-based researches data accessibility and number of samples are listed in Table 2. Furthermore, it should be mentioned that we plan to increase number of samples and perform activities in different scenarios.

Table 2. Number of samples and data accessibility in different CSI-based HAR researches.

Research	Number of Samples	Public Accessibility
[16]	1100	No
[41]	600	Yes
[30]	720	No
[42]	200–400	No
[43]	1400	No
[44]	500	No
[14]	720	Yes
Our Dataset	420	Yes

4. Proposed Method

Despite the numerous advantages that accessibility to CSI would provide to users, chip manufacturers continue to treat CSI as a private feature. Only a few devices that are still using the 802.11g and 802.11n technologies are capable of dumping CSI, and they do so with a number of restrictions. Additionally, the Linux 802.11n CSI Tool is only compatible with older Linux kernel versions, which can cause significant inconvenience. In IoT, wireless connectivity is critical for monitoring and control purposes such as HAR. When it comes to experimentation, the Raspberry Pi might be considered a cheap and available WiFi-enabled platform. We employ Nexmon Tool [13] and collect CSI data for seven daily human activities, including walk, run, fall, lie down, sit down, stand up, and bend. We use Raspberry Pi 4 and a Tp-link archer c20 as an Access Point (AP) in 20 MHz bandwidth on channel 36 in IEEE 802.11ac standard. As depicted in Figure 5, we use Personal Computer (PC) for traffic generation by pinging or watching a movie on the internet. The AP will reply with pong packets to the sent pings from the PC. The Pi is in monitor mode and will sniff through this connection and collect CSI for each sent-out pong packet. CSI is saved as a pcap file which can be analyzed in many software including MATLAB. CSI complex numbers are extracted and after removing null and pilot subcarriers, we export activity rows according to the period of each activity which has been detached depending on the video of activity performed by users and stopwatch. Due to reflections induced by human activity, each subcarrier for any given link experiences a variation [11]. Therefore, each subcarrier includes critical information that will increase recognition accuracy. A higher proportion of subcarriers boosts precise feature detection since it provides additional information and boosts identification of challenging features to analyze a subset of subcarriers. The CSI matrices have 52 columns (available data subcarriers) and 600 up to 1100 rows depending on the period of each activity. The dataset is available in GitHub https://github.com/parisafm/CSI-HAR-Dataset (accessed on 27 October 2021).

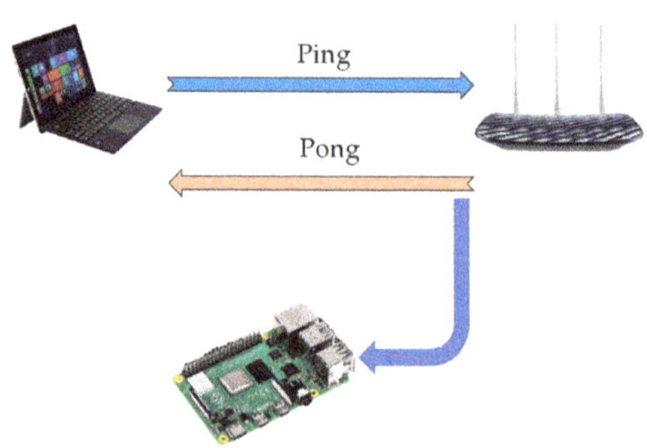

Figure 5. Configuration for CSI collection.

No data pre-processing is applied on the CSI amplitude since any additional filtering can result in losing important information and affect the system's performance. If the simulation results or generated images are disappointing, we can use a low pass filter for high-frequency reduction, as mentioned in [16]. In order to make RGB images, the data values must be normalized between 0 and 255 for all activities. We make a pseudocolor plot from matrices representing them as an array of colored faces in the x-y plane. In a pseudocolor plot, cells are arranged in a rectangular array with colors specified by the values in C as normalized CSI input matrices. MATLAB creates this plot by using four points near each corner of C to describe each cell. Each element of C is linearly mapped to the RGB colormap. The generated RGB images are resized to the desired size (64 × 64). Some of these images for each class of activities are depicted in Figure 6. Since the images are not noisy, we do not need to apply denoising filters and additional denoising technique may cause information lost.

These images and CSI data are then fed into neural networks. As CSI signals are typical time-series with temporal dependency, the future information in each step is crucial for HAR, and also LSTMs cannot effectively analyze more than 100 s term, we consider two methods. First, we convert CSI signals to RGB images using pseudocolor plot and feed them into 2D-CNN. By converting CSI to RGB images, the signal pattern for each activity can be seen in one look. Meaning that the pattern changes due to the human movements are depicted in image.

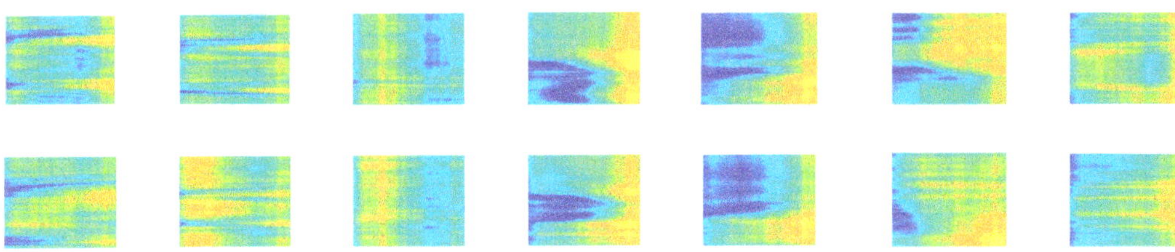

(a) (b) (c) (d) (e) (f) (g)

Figure 6. Generated RGB images: (**a**) walk; (**b**) run; (**c**) fall; (**d**) lie down; (**e**) sit down; (**f**) stand up; (**g**) bend.

Therefore, in contrast to LSTM that does not have any information about future steps, CNN can analyze the whole signals' alteration. Additionally, CNN process information parallelly, resulting in faster training than LSTMs with better accuracy. Another method to address LSTMs mentioned problems is to apply BLSTM on CSI data. BLSTM contains a forward and backward layer and can analyze both past and future information by extracting information from the two directions. Since the sequential features learned by the BLSTM network have high dimensions and feature contributions and timesteps may vary for each activity, we exploit the attention layer to learn the relative importance of features. Although BLSTM have high potential to recognize human activities, it needs a greater memory bandwidth for processing and thus it has more training time than the proposed 2D-CNN. Lower consumed time in training and less computational complexity, along with the ability to observe the whole pattern alteration in one look, make the novel image conversion idea and 2D-CNN implementation the best choice over other mentioned methods.

5. Evaluation

5.1. Measurement Setup

Buster lite 4.19.97 raspian and the main branch of nexmon-csi [45] were installed on the Raspberry Pi 4. Nexmon tool was configured as follows: Channel 36, bandwidth 20 MHz, Core 1, NSS mask 1, 4000 samples, 20 s. The AP's MAC address filter was set to make sure the Raspberry Pi will not connect to another AP on channel 36. The data collection was conducted from another device linked to the Pi over SSH to avoid interference, communicating over another 2.4 GHz network. The AP used is a Tp-link archer c20 wireless router operating a 5 GHz WiFi network on channel 36 at 20 MHz. A PC is paired with the AP to generate traffic by watching a video on the internet or pinging, for which the Pi can capture CSI. We put the Raspberry Pi in monitor mode and with the use of the sniffing method, we were able to collect CSI data. We collect 4000 samples at around 20 s which results in 200 Hz sample rate. Ap and Pi were both 1m above the ground to ensure an unobstructed signal path. They were 3 meters away from each other. The experimental environment is depicted in Figure 7. Each activity performed in the dataset was performed 20 times by three users of different ages. These activities are as listed: fall, stand up, sit down, lie down, run, walk, bend. CSI data were captured in the 20 s, in which an activity has been performed in the middle of this period. More precisely, users remain mostly stable at the start and the end of the capture. As the experiment was managed by the users, the length of time taken for the activity to begin and end may vary slightly, around 3 to 6 s (around 600 to 1100 rows of 4000 total rows). The activity period is extracted according to the video of the activity and stopwatch.

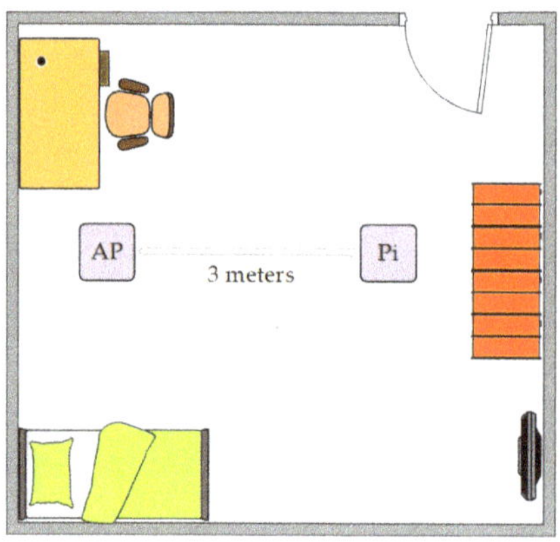

Figure 7. Experimental environment.

5.2. Simulations Results

The proposed deep learning architectures can discover more complex patterns in time series data, compared to hand-crafted features techniques such as RF [14] and HMM [14]. As shown in Figure 8, the ConvLSTM [16] model slightly outperforms the FC network in [17] and DenseLSTM [15]. Our proposed models have achieved better results compared with all of them without any extra data augmentations [15] and complex structure like ConvLSTM [16] and FC [17]. The detailed information about the mentioned methods are available in Section 2. The dataset was split into train and test in a 75% to 25% ratio. We implemented four neural networks on Keras for classification, which has been accelerated by Geforce RTX 2060. The raw CSI amplitude data is a 52-dimensional vector fed into 1D-CNN, LSTM, and attention-based BLSTM. In 1D-CNN model, we have two Conv1D with ReLU as an activation function and after each Conv1D layer, we added a MaxPooling layer. The LSTM network contains one LSTM hidden layer and 128 hidden units. For the BLSTM model, we used one BLSTM layer with 200 hidden nodes and one attention layer with 400 units. The converted RGB images were fed into 2D-CNN with 2xConv2D layer (with Leaky ReLU) and 2xMaxPooling layer (after each Conv2D). The structures of these networks are depicted in Figures 1–4.

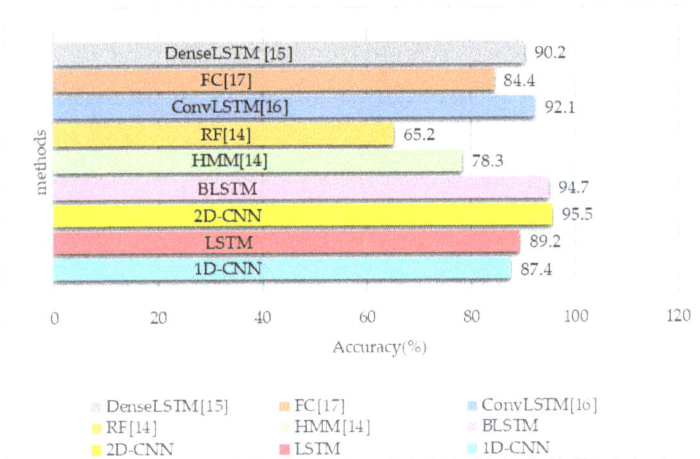

Figure 8. Accuracy of different methods implemented on the dataset.

CNN can detect simple patterns in data, which are subsequently utilized to create more complex patterns within higher layers. 1D-CNN is highly effective when features are derived from fixed-length parts of the dataset and the feature's position in the section is not crucial, including the analysis of time sequences data (such as gyroscope or accelerometer data or CSI). Since the LSTM network analyzes temporal dependencies in sequential data, it outperforms the 1D-CNN technique. As mentioned in Sections 1 and 2, LSTMs suffer from vanishing gradient and cannot access next step information. For activities like sit down and lie down which are different at last body movements, it is necessary to have knowledge about next step information. To address these problems, we converted CSI data into RGB images for each activity and used them as inputs for 2D-CNN, thus we can access all the information in past or next steps with one look at images. Additionally, we used BLSTM with attention layer to consider both past and next step information and automatically learn features' significance to assign higher weights based on HAR performance. The attention-based BLSTM approach and 2D-CNN have achieved the best performance for the recognition of all activities with an accuracy of around 95%. All of these comparisons are depicted in Figure 8.

Different activities have different CSI values, resulting in different recognition accuracy [7]. We use a confusion matrix (or error matrix) to describe the performance of our proposed classifiers for each activity in which the rows represent anticipated classes and the columns represent actual classes. The activities with more significant body movement, i.e., fall, walk, and run, have higher recognition accuracy (see Figure 9) since they have more influence on CSI characteristics. Furthermore, fall activity is crucial, particularly for elderly healthcare services. Our proposed 2D-CNN and BLSTM network have 98% and 96% accuracy for this activity, making these models efficient in elderly care systems. Another observation is that the action "Lie down" has a recognition accuracy similar to "Sit down" for most methods. The probable explanation is that these activities have a similar impact on CSI values since the start position is the same and the final positions are different. By applying attention-based BLSTM and 2D-CNN, the system is less confused between these two activities. As shown in Figure 9, the model is confused with these two activities around 3% in BLSTM and 2% in 2D-CNN which are acceptable when compared to LSTM with 8% and 1D-CNN with 9% confusion.

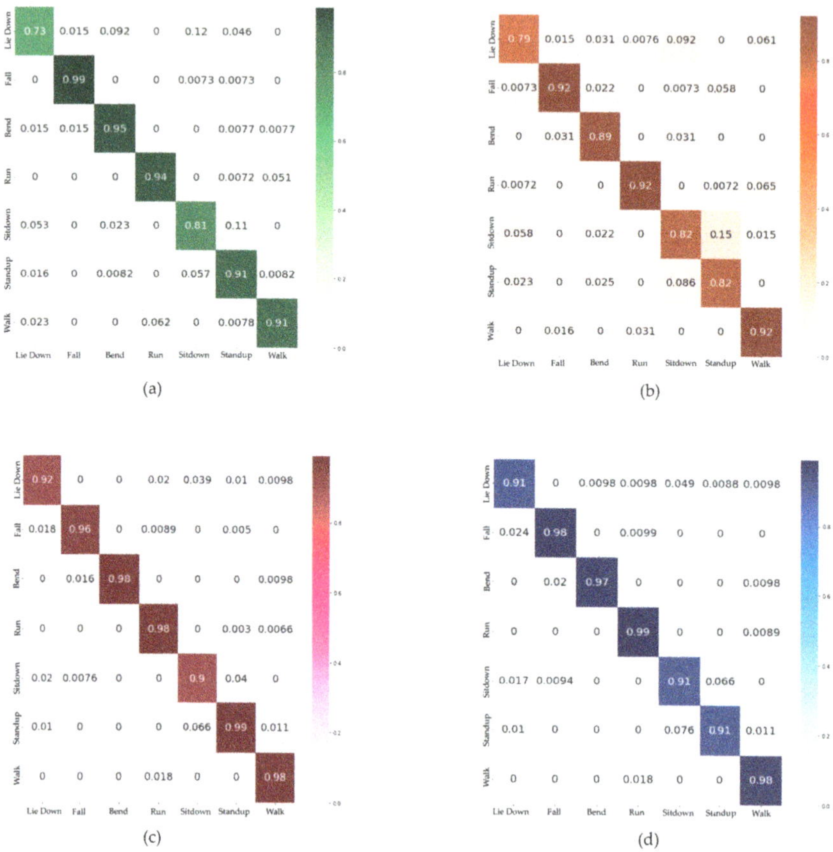

Figure 9. Confusion matrices of proposed methods: (**a**) LSTM; (**b**) 1D-CNN; (**c**) BLSTM; (**d**) 2D-CNN.

Consumed time is another critical performance evaluation indicator representing how much time the model spends training and testing. Table 3 compares the time consumption (milliseconds per step) of six DL approaches: ConvLSTM [16], DenseLSTM [15], LSTM, BLSTM, 1D-CNN, and 2D-CNN. We can observe that proposed 2D-CNN has the shortest time and highest accuracy (Figure 8) compared to the others, making 2D-CNN a better choice compared with BLSTM, ConvLSTM [16], and DenseLSTM [15] in a fraction of the time. More precisely, a long-term input is processed sequentially in LSTMs' gates, making them not hardware-friendly, as they require greater memory bandwidth to compute parameters, in addition to time-consuming simulations. In contrast, CNN extracts features by utilizing convolution operation, which is easier to compute and faster in training. Furthermore, the CNN accuracy rapidly improved while the BLSTM accuracy slowly improved in a longer training time.

Table 3. Consumed-time (milliseconds per step) comparison for different models.

Time	1D-CNN	LSTM	2D-CNN	BLSTM	ConvLSTM [16]	DenseLSTM [15]
Train	9	13	15	28	36	60
Test	3	6	7	12	19	41

6. Conclusions

Due to the ubiquity of WiFi devices, HAR based on wireless signals, including CSI, has witnessed more interest in smart house health monitoring systems. A few CSI datasets for the HAR task collected with 5300 NIC or Atheros PCI chips, are currently available. This paper presented a CSI dataset for indoor HAR using a Raspberry Pi, which is one of the most accessible embedded boards. In this work, we have designed four neural networks to conduct WiFi-based HAR with more than 87% accuracy for our dataset. We used a BLSTM network with an attention layer to address LSTM problems with future information. We also convert CSI data to images using pseudocolor plots and feeding them into 2D-CNN to overcome the mentioned limitations of LSTM. We showed that the idea of CSI conversion to images can obtain high accuracy of 95%, close to BLSTM, which is one of the most successful DL algorithms in time-sequential analysis. Additionally, as CNN processes different features parallelly, it is faster than other methods and less complex in computations. The strong performance of the proposed methods indicates that the data collected by Raspberry Pi can effectively be employed in smart house HAR. The proposed methods can boost elderly health monitoring systems since it meets the requirements for acceptable recognition accuracy and recognition speed for the most commonly performed actions in this task.

Nevertheless, we presented the first version of our public dataset and plan to improve it by investigating different environments and scenarios. In the future, we will study human-to-human interactions and the CSI changes in multiple user-multiple environments scenarios. Since different ages may perform activities differently, according to their physical ability, we collected CSI data from three different ages, including an adult, a middle-aged person, and an elderly person and try to study other ages, including child and teen. Additionally, we will investigate activities with different initial movements, such as standing + walking and running + walking.

Author Contributions: Conceptualization, P.F.M. and S.A.G.; methodology, P.F.M. and S.A.G.; software, P.F.M.; validation, P.F.M. and M.N.; formal analysis, P.F.M. and R.S.; investigation, P.F.M.; resources, S.A.G. and P.F.M.; data curation, P.F.M.; writing—original draft preparation, P.F.M.; writing—review and editing, S.A.G., R.S. and M.N.; visualization, P.F.M.; supervision, S.A.G. and R.S.; project administration, S.A.G. and R.S.; funding acquisition, S.A.G. All authors have read and agreed to the published version of the manuscript.

Funding: This research received no external funding.

Informed Consent Statement: Informed consent was obtained from all subjects involved in the study.

Data Availability Statement: The data presented in this study are available in GitHub: https://github.com/parisafm/CSI-HAR-Dataset (accessed on 27 October 2021).

Conflicts of Interest: The authors declare no conflict of interest.

References

1. Hassan, Q.F. *Internet of Things A to Z: Technologies and Applications*, 1st ed.; Wiley: Hoboken, NJ, USA, 2018; pp. 5–45; ISBN 978-1-119-45674-2.
2. Dey, N.; Hassanien, A.E.; Bhatt, C.; Ashour, A.S.; Satapathy, S.C. *Internet of Things and Big Data Analytics toward Next-Generation Intelligence*, 1st ed.; Springer: New York, NY, USA, 2018; Volume 30, pp. 199–243; ISBN 978-3-319-86864-6.
3. Perera, C.; Liu, C.H.; Jayawardena, S. The Emerging Internet of Things Marketplace from an Industrial Perspective: A Survey. *IEEE Trans. Emerg. Top. Comput.* **2015**, *3*, 585–598. [CrossRef]
4. Wang, F.; Feng, J.; Zhao, Y.; Zhang, X.; Zhang, S.; Han, J. Joint Activity Recognition and Indoor Localization with WiFi Fingerprints. *IEEE Access* **2019**, *7*, 80058–80068. [CrossRef]
5. Vlachostergiou, A.; Stratogiannis, G.; Caridakis, G.; Siolas, G.; Mylonas, P. *Smart Home Context Awareness Based on Smart and Innovative Cities*; Association for Computing Machinery: New York, NY, USA, 2015; ISBN 9781450335805.
6. Palipana, S.; Rojas, D.; Agrawal, P.; Pesch, D. FallDeFi: Ubiquitous Fall Detection using Commodity WiFi Devices. In Proceedings of the ACM on Interactive, Mobile, Wearable and Ubiquitous Technologies, Singapore, 8–12 October 2018; Volume 1, pp. 1–25. [CrossRef]

7. Moshiri, P.F.; Navidan, H.; Shahbazian, R.; Ghorashi, S.A.; Windridge, D. Using GAN to Enhance the Accuracy of Indoor Human Activity Recognition. In Proceedings of the 10th Conference on Information and Knowledge Technology, Tehran, Iran, 31 December 2019–2 January 2020.
8. Ahad, M.A.R.; Ngo, T.T.; Antar, A.D.; Ahmed, M.; Hossain, T.; Muramatsu, D.; Makihara, Y.; Inoue, S.; Yagi, Y. Wearable Sensor-Based Gait Analysis for Age and Gender Estimation. *Sensors* **2020**, *20*, 2424. [CrossRef]
9. Nabati, M.; Ghorashi, S.A.; Shahbazian, R. Joint Coordinate Optimization in Fingerprint-Based Indoor Positioning. *IEEE Commun. Lett.* **2021**, *25*, 1192–1195. [CrossRef]
10. Zhang, W.; Zhou, S.; Yang, L.; Ou, L.; Xiao, Z. WiFiMap+: High-Level Indoor Semantic Inference with WiFi Human Activity and Environment. *IEEE Trans. Veh. Technol.* **2019**, *68*, 7890–7903. [CrossRef]
11. Chen, Z.; Zhang, L.; Jiang, C.; Cao, Z.; Cui, W. WiFi CSI based passive Human Activity Recognition Using Attention Based BLSTM. *IEEE Trans. Mob. Comput.* **2019**, *18*, 2714–2724. [CrossRef]
12. Elbayad, M.; Besacier, L.; Verbeek, J. Pervasive attention: 2d Convolutional Neural Networks for Sequence-to-Sequence Prediction. *arXiv* **2018**, arXiv:1808.03867.
13. Gringoli, F.; Schulz, M.; Link, J.; Hollick, M. Free Your CSI: A Channel State Information Extraction Platform For Modern Wi-Fi Chipsets. In Proceedings of the 13th International Workshop on Wireless Network Testbeds, Experimental Evaluation & Characterization, New York, NY, USA, 4 October 2019; pp. 21–28. [CrossRef]
14. Yousefi, S.; Narui, H.; Dayal, S.; Ermon, S.; Valaee, S. A survey on behavior recognition using WiFi channel state information. *IEEE Commun. Mag.* **2017**, *55*, 98–104. [CrossRef]
15. Zhang, J.; Fuxiang, W.; Wei, B.; Zhang, Q.; Huang, H.; Shah, S.W.; Cheng, J. Data Augmentation and Dense-LSTM for Human Activity Recognition Using WiFi Signal. *IEEE Internet Things J.* **2021**, *8*, 4628–4641. [CrossRef]
16. Forbes, G.; Massie, S.; Craw, S. Wifi-based human activity recognition using Raspberry Pi. In Proceedings of the IEEE 32nd International Conference on Tools with Artificial Intelligence, Baltimore, MD, USA, 9–11 November 2020; pp. 722–730. [CrossRef]
17. Zhou, N.; Sun, W.; Liang, M. Human Activity Recognition based on WiFi Signal Using Deep Neural Network. In Proceedings of the IEEE 8th International Conference on Smart City and Informatization, Guangzhou, China, 11 February 2020; pp. 26–30. [CrossRef]
18. Mahjoub, A.B.; Atri, M. Human action recognition using RGB data. In Proceedings of the 11th International Design & Test Symposium, Hammamet, Tunisia, 18–20 December 2016. [CrossRef]
19. Zhang, B.; Wang, L.; Wang, Z.; Qiao, Y.; Wang, H. Real-Time Action Recognition with Deeply Transferred Motion Vector CNNs. *IEEE Trans. Image Process.* **2018**, *27*, 2326–2339. [CrossRef] [PubMed]
20. Agahian, S.; Farhood, N.; Cemal, K. Improving bag-of-poses with semi-temporal pose descriptors for skeleton-based action recognition. *Vis. Comput.* **2019**, *35*, 591–607. [CrossRef]
21. Anitha, U.; Narmadha, R.; Sumanth, D.; Kumar, D. Robust Human Action Recognition System via Image Processing. *Procedia Comput. Sci.* **2020**, *167*, 870–877. [CrossRef]
22. Tasnim, N.; Islam, M.K.; Baek, J.-H. Deep Learning Based Human Activity Recognition Using Spatio-Temporal Image Formation of Skeleton Joints. *Appl. Sci.* **2021**, *11*, 2675. [CrossRef]
23. Rustam, F.; Reshi, A.A.; Ashraf, I.; Mehmood, A.; Ullah, S.; Khan, D.M.; Choi, G.S. Sensor-Based Human Activity Recognition Using Deep Stacked Multilayered Perceptron Model. *IEEE Access* **2020**, *8*, 218898–218910. [CrossRef]
24. Du, Y.; Lim, Y.; Tan, Y. A Novel Human Activity Recognition and Prediction in Smart Home Based on Interaction. *Sensors* **2019**, *19*, 4474. [CrossRef] [PubMed]
25. Nabati, M.; Navidan, H.; Shahbazian, R.; Ghorashi, S.A.; Windridge, D. Using Synthetic Data to Enhance the Accuracy of Fingerprint-Based Localization: A Deep Learning Approach. *IEEE Sens. Lett.* **2020**, *4*, 1–4. [CrossRef]
26. Won, M.; Zhang, S.; Son, S.H. WiTraffic: Low-Cost and Non-Intrusive Traffic Monitoring System Using WiFi. In Proceedings of the 26th International Conference on Computer Communication and Networks, Vancouver, BC, Canada, 18 September 2017; pp. 1–9. [CrossRef]
27. Tan, S.; Yang, J. WiFinger: Leveraging commodity WiFi for fine-grained finger gesture recognition. In Proceedings of the 17th ACM International Symposium on Mobile Ad Hoc Networking and Computing, New York, NY, USA, 5 July 2016; pp. 201–210. [CrossRef]
28. Zhang, D.; Hu, Y.; Chen, Y.; Zeng, B. BreathTrack: Tracking indoor human breath status via commodity WiFi. *IEEE Internet Things J.* **2019**, *6*, 3899–3911. [CrossRef]
29. Zeng, Y.; Pathak, P.H.; Xu, C.; Mohapatra, P. Your AP knows how you move: fine-grained device motion recognition through WiFi. In Proceedings of the 1st ACM Workshop on Hot Topics in Wireless, New York, NY, USA, 11 September 2014; pp. 49–54. [CrossRef]
30. Arshad, S.; Feng, C.; Liu, Y.; Hu, Y.; Yu, R.; Zhou, S.; Li, H. Wi-chase: A WiFi based human activity recognition system for sensorless environments. In Proceedings of the IEEE 18th International Symposium on A World of Wireless, Mobile and Multimedia Networks, Macau, China, 13 July 2017; pp. 1–6. [CrossRef]
31. Liu, X.; Cao, J.; Tang, S.; Wen, J. Wi-sleep: Contactless sleep monitoring via WiFi signals. *IEEE Real-Time Syst. Symp.* **2014**, 346–355. [CrossRef]
32. Halperin, D.; Hu, W.; Sheth, A.; Wetherall, D. Tool Release: Gathering 802.11n Traces with Channel State Information. *ACM SIGCOMM Comput. Commun. Rev.* **2011**, *41*, 53. [CrossRef]

33. Xie, Y.; Li, Z.; Li, M. Precise Power Delay Profiling with Commodity WiFi. *IEEE Trans. Mob. Comput.* **2015**, *18*, 53–64. [CrossRef]
34. Gast, M.S. *802.11ac: A Survival Guide*; O'Reilly Media, Inc.: Sepastopol, CA, USA, 2013; pp. 11–20; ISBN 9781449343149.
35. Li, Z.; Yang, W.; Peng, S.; Liu, F. A Survey of Convolutional Neural Networks: Analysis, Applications, and Prospects. *IEEE Trans. Neural. Netw. Learn. Syst.* **2021**, 1–21. [CrossRef]
36. Peng, B.; Yao, K. Recurrent Neural Networks with External Memory for Language Understanding. *arXiv* **2015**, arXiv:1506.00195.
37. Wang, L.; Liu, R. Human Activity Recognition Based on Wearable Sensor Using Hierarchical Deep LSTM Networks. *Circuits Syst. Signal Process.* **2020**, *39*, 837–856. [CrossRef]
38. Hochreiter, S.; Schmidhuber, J. Long Short-Term Memory. *Neural Comput.* **1997**, *9*, 1735–1780. [CrossRef] [PubMed]
39. Wu, Y. Mining action let ensemble for action recognition with depth cameras. In Proceedings of IEEE Conference on Computer Vision and Pattern Recognition, Providence, RI, USA, 16–21 June 2012; pp. 1290–1297.
40. Li, W.; Zhang, Z.; Liu, Z. Action recognition based on a bag of 3D points. In Proceedings of the IEEE Computer Society Conference on Computer Vision and Pattern Recognition—Workshops, San Francisco, CA, USA, 13–18 June 2010; pp. 9–14.
41. Yang, J.; Liu, Y.; Liu, Z.; Wu, Y.; Li, T.; Yang, Y. A Framework for Human Activity Recognition Based on WiFi CSI Signal Enhancement. *Int. J. Antennas Propag.* **2021**, *2021*, 6654752. [CrossRef]
42. Ding, X.; Jiang, T.; Zhong, Y.; Wu, S.; Yang, J.; Xue, W. Improving WiFi-based Human Activity Recognition with Adaptive Initial State via One-shot Learning. In Proceedings of the IEEE Wireless Communications and Networking Conference, Nanjing, China, 29 March–1 April 2021; pp. 1–6. [CrossRef]
43. Wang, W.; Liu, A.X.; Shahzad, M.; Ling, K.; Lu, S. Understanding and modeling of wifi signal based human activity recognition. In Proceedings of the 21st Annual International Conference on Mobile Computing and Networking, New York, NY, USA, 7 September 2015; pp. 65–76.
44. Zhang, Y.; Wang, X.; Wang, Y.; Chen, H. Human Activity Recognition Across Scenes and Categories Based on CSI. *IEEE Trans. Mob. Comput.* **2020**, *1*. [CrossRef]
45. Schulz, M.; Wegemer, D.; Hollick, M. Nexmon: The C-Based Firmware Patching Framework. 2017. Available online: https://nexmon.org (accessed on 27 October 2021).

sensors

MDPI

Article

Evaluations of Deep Learning Approaches for Glaucoma Screening Using Retinal Images from Mobile Device

Alexandre Neto [1,2], José Camara [2,3] and António Cunha [1,2,*]

1 Escola de Ciências de Tecnologia, University of Trás-os-Montes and Alto Douro, Quinta de Prados, 5001-801 Vila Real, Portugal; alexandre.h.neto@inesctec.pt
2 INESC TEC—Institute for Systems and Computer Engineering, Technology and Science, 4200-465 Porto, Portugal; jcrcamara@hotmail.com
3 Departamento de Ciências e Tecnologia, University Aberta, 1250-100 Lisboa, Portugal
* Correspondence: acunha@utad.pt; Tel.: +351-931-636-373

Abstract: Glaucoma is a silent disease that leads to vision loss or irreversible blindness. Current deep learning methods can help glaucoma screening by extending it to larger populations using retinal images. Low-cost lenses attached to mobile devices can increase the frequency of screening and alert patients earlier for a more thorough evaluation. This work explored and compared the performance of classification and segmentation methods for glaucoma screening with retinal images acquired by both retinography and mobile devices. The goal was to verify the results of these methods and see if similar results could be achieved using images captured by mobile devices. The used classification methods were the Xception, ResNet152 V2 and the Inception ResNet V2 models. The models' activation maps were produced and analysed to support glaucoma classifier predictions. In clinical practice, glaucoma assessment is commonly based on the cup-to-disc ratio (CDR) criterion, a frequent indicator used by specialists. For this reason, additionally, the U-Net architecture was used with the Inception ResNet V2 and Inception V3 models as the backbone to segment and estimate CDR. For both tasks, the performance of the models reached close to that of state-of-the-art methods, and the classification method applied to a low-quality private dataset illustrates the advantage of using cheaper lenses.

Keywords: deep learning; glaucoma screening; retinal images; segmentation; classification

Citation: Neto, A.; Camara, J.; Cunha, A. Evaluations of Deep Learning Approaches for Glaucoma Screening Using Retinal Images from Mobile Device. *Sensors* **2022**, *22*, 1449. https://doi.org/10.3390/s22041449

Academic Editor: Ivan Miguel Serrano Pires

Received: 29 December 2021
Accepted: 10 February 2022
Published: 14 February 2022

1. Introduction

Glaucoma is one of the main causes of vision loss, mainly due to increased fluid pressure and improper drainage of fluid in the eye. In 2013, it was estimated that 64.3 million people aged 40–80 years were diagnosed with glaucoma worldwide. This disease is expected to reach nearly 76 million by 2020 and 111.8 million by 2040. The prevalence of glaucoma is 2.5% for people of all ages and 4.8% for those above 75 years of age [1]. Glaucoma is an asymptomatic condition, and patients do not require medical assistance until a late stage, making the diagnosis frequently too late to prevent blindness. Population-level surveys suggest that only 10–50% of people with glaucoma are aware that they have the disease. As early diagnosis and treatment of the condition can prevent vision loss, glaucoma screening has been tested in numerous studies worldwide [2]. An ophthalmologist can directly examine the eye with an ophthalmoscope or can examine a fundus image capture with a fundus camera, as can be seen in Figure 1. The examination of these fundus images is important because the ophthalmologist can record indicators and parameters related to cupping to detect glaucoma, such as disc diameter, the thickness of the neuroretinal rim (decreasing in the order inferior (I) > superior (S) > nasal (N) > temporal (T) (ISNT rule)), peripapillary atrophy, notching and cup-to-disc ratio (CDR), with this last indicator being the most used measurement by specialists [3–5].

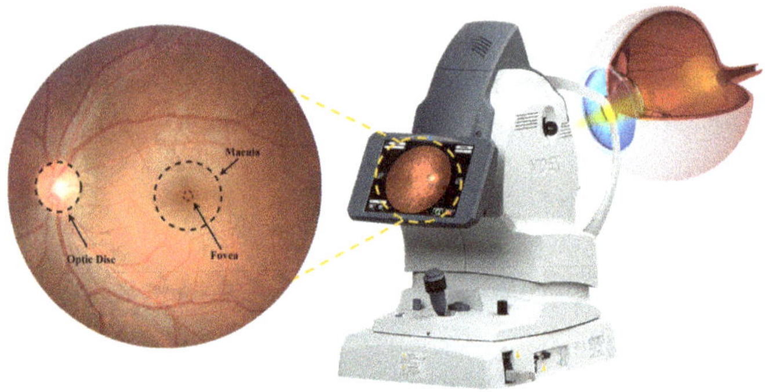

Figure 1. Representation of capturing an image of the interior surface of the eye (retina).

Usually, glaucoma is diagnosed on the basis of the patient's medical history, measures of intraocular pressure (IOP), a visual field loss test and manual evaluation of the optic disc (OD) using ophthalmoscopy to examine the shape and colour of the optic nerve. The examination of the OD is important since glaucoma begins to form a cavity and develops an abnormal depression/excavation at the front of the nerve head, called the optic cup, which, in advanced stages, facilitates the progression of glaucoma, blocking the OD (Figure 2) [6–8].

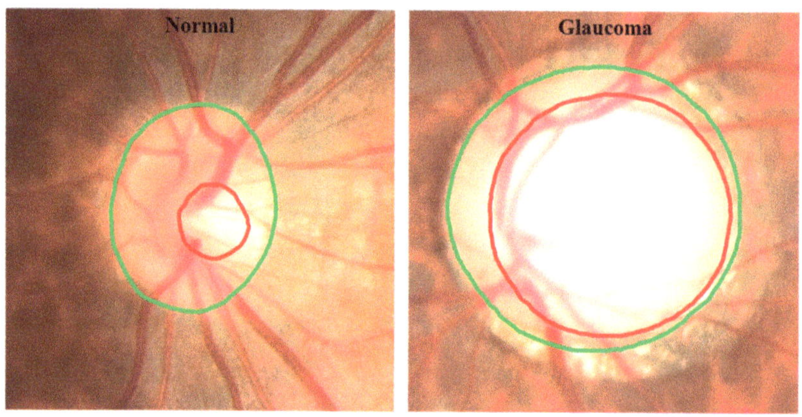

Figure 2. Retinal image from the normal and glaucomatous eye. **Green line:** OD boundary; **red line:** cup boundary.

After gathering the retinal images, they must be inspected and analysed to look for indicators of ophthalmologic pathologies. These diagnostic systems offer the potential to be used on a large scale for the early diagnosis and treatment of glaucoma. However, they require subjective evaluation by qualified experts, and it is time-consuming and costly to inspect each retinal image manually. In this regard, deep learning (DL) algorithms help in the automatic screening of glaucoma and assist ophthalmologists in achieving higher accuracy in glaucoma screening, especially in repetitive tasks [3,6,9].

The main objective is to develop a system that allows the screening of low-resolution retinal images captured by a low-cost lens attached to a smartphone. To accomplish this, secondary objectives must be achieved. In this study, state-of-the-art DL methods were explored, tested and applied to high-resolution public databases and then applied to a private database containing low-quality images captured through a low-cost lens attached to a mobile device. These classification methods provide activation maps that allow the model's decision to be analysed and discussed. Segmentation methods were applied as well, using the CDR to classify images after OD and cup segmentation. These segmentation methods can help an ophthalmologist in a subjective and difficult task, enabling more consistent results that are similar to a clinician's segmentations. For this purpose, state-of-the-art works on classification and segmentation methods for glaucoma screening were reviewed.

2. Literature Review

DL techniques have yielded good results in research on glaucoma screening due to the development of technologies to detect, diagnose and treat glaucoma. The main approach is to conduct screening through computer-aided diagnosis (CAD) systems that use DL to learn and train models through previously labelled available data, identifying patterns and making decisions with minimal human intervention [10]. This section surveys key works with methods using automatic classification models and classification methods using segmentation models.

2.1. Classification Methods

The use of classification methods for screening glaucoma lesions in retinal images is another well-established approach. An overview of the best methods recently published is provided in the following.

The study of Gómez-Valverde [2] used the VGG19, GoogLeNet (also known as Inception V1), ResNet50 and DENet models. With these models, Valverde compared the performance between transfer learning and training from scratch. To confirm the performance of VGG19, 10-fold cross-validation (CV) was applied. Valverde used three different databases: RIM-ONE and DRISHTI-GS (public) and Esperanza (private dataset). In the RIM-ONE database, the images classified as suspect were considered to be glaucomatous for the study. The best result was obtained with the VGG19 model using transfer learning.

Diaz-Pinto [11] applied five different ImageNet pre-trained models (VGG16, VGG19, InceptionV3, ResNet50 and Xception) for glaucoma classification and used a 10-fold CV strategy to validate the results. Five databases were used for this work: ACRIMA, HRF, DRISHTI-GS, RIM-ONE and Sjchoi86-HRF. The images were cropped around the OD using a bounding box of 1.5 times the OD radius. All models passed the AUC threshold of 0.96, indicating excellent results.

Serener et al. [12] selected the ResNet50 and GoogLeNet models and trained them with two public databases: a database from Kim's Eye Hospital (total of 1542 images, including 786 photos from normal patients and 756 from glaucoma patients) and RIM-ONE r3. The database from Kim's Eye Hospital was used to train the two models, and for the performance evaluation, the models were tested with the RIM-ONE r3 database. With GoogLeNet, Serener obtained better results for early-stage glaucoma than for the advanced glaucoma stage.

The work performed by Norouzifard [13] used two DL models, namely, VGG19 and Inception ResNet V2. These two models were pre-trained and then fine-tuned. For this work, two databases were used: one from the University of California Los Angeles (UCLA) and another publicly available one called high-resolution fundus (HRF). From the UCLA database, they randomly selected 70% of the images for training, 25% for validation and the remaining 5% for testing. To solidify the work, the models were then re-tested with the HRF database. The Inception ResNet V2 model with the UCLA database obtained a specificity and sensitivity above 0.9, even when re-tested with the HRF database.

The study by Sreng [5] was performed in two stages: first, DeepLabv3+ detected and extracted the OD from the entire image, and then three types of convolutional neural networks (CNNs) were used to identify images in the segmented OD region as glaucomatous or normal. After the image was cropped around the OD, 11 ImageNet pre-trained models 23re used: AlexNet, GoogleNet, InceptionV3, Xception, ResNet-50, SqueezeNet, ShuffleNet, MobileNet, DenseNet, InceptionResNet and NasNet-Large. This method was trained with five public databases: REFUGE, ACRIMA, ORIGA, RIM-ONE and DRISHTI-GS. The results showed that DenseNet with the ACRIMA database had the best performance, followed by MoblieNet with the REFUGE database.

2.2. Segmentation Methods

Several methods have been published in the literature on segmenting the OD and the cup disc, mostly using adaptations of U-Net. The following presents an overview of the best methods recently published.

Al-Bander [14] proposed a method with a DenseNet incorporated with an FCN with U-shaped architecture. Al-Bander's approach involved the use of five databases of colour fundus images: ORIGA, DRIONS-DB, DRISHTI-GS, ONHSD and RIM-ONE. For the pre-process, only the green channel of the colour images was considered since the other colour channels contain less useful information. The images were then cropped to isolate the ROI. For OD segmentation, the model achieved better Dice and intersection-over-union (IoU) results with the DRISHTI-GS database compared to RIM-ONE, and the same results were obtained for cup segmentation but with lower values of Dice and IoU compared to OD segmentation.

In the work of Singh [15], a conditional generative adversarial network (cGAN) model was proposed to segment the OD. The cGAN is composed of a generator and a discriminator and can learn statistically invariant features, such as the colour and texture of an input image, and segment the region of interest. For this method, skip connections were used for concatenating the feature maps of a convolutional layer with those resulting from the corresponding deconvolutional layer. To train and evaluate the model, the DRISHTI-GS and RIM-ONE databases were used, with the size of the images reduced to 256×256 and the value of each pixel normalised between 0 and 1. For OD segmentation, the model for both databases achieved values above 0.9 for accuracy, Dice and IoU.

Qin [16] proposed neural network constructs utilising the FCN and inception building blocks in GoogleNet. The FCN is the main body of the deep neural network architecture, and to this method, they added several convolution kernels for feature extraction after deconvolution based on the Inception structure in GoogLeNet. Qin's experiments used two databases: REFUGE and one from the Second Affiliated Hospital of Zhejiang University School of Medicine. For this technique, the authors used a fully automatic method using the Hough circle transform that recognises and cuts the image to obtain an image of the ROI. In the segmentation of the OD and the cup, the model obtained values above 0.9 for Dice and the IoU.

In the work by Yu and others [17], a modified U-Net with a pre-trained ResNet-34 model was developed. This work comprised two steps: first, one single-label modified U-Net model was applied to segment an ROI around the OD, and then after this, the cropped image was used in a multi-label model whose objective was to segment the OD and cup simultaneously. In Yu's study, the RIGA database was used to train and evaluate the CNN, but then to achieve robust performance, the model trained on RIGA was applied on the DRISHTI-GS and RIM-ONE r3 databases. All of the database images were pre-processed with contrast enhancement, followed by resizing to 512x512 dimensions. In this method, the segmentation of the OD and the cup produced better results with DRISHTI-GS than RIM-ONE r3.

Cup-to-Disc Ratio

Glaucoma progression is assessed based on the ratio between OD and cup measurements. The cup-to-disc ratio (CDR) is a clinical method that compares the ratio of the cup to disc, which is currently determined manually, limiting its potential in mass screening. Manual segmentation is dependent on the experience and expertise of the ophthalmologist, so it ends up being subjective and differing between observers [18]. The CDR is commonly used in clinics to classify glaucoma, and specific patterns of change in the region of the OD and cup are used as evidence of glaucoma or glaucoma progression along with other clinical tests, such as intraocular pressure and visual field acuity [18,19].

Accurate segmentation of the OD and cup is essential to a reliable CDR measurement, and reliance on manual effort restricts the deployment of CDR for mass screening, which is fundamental in the detection of early glaucoma for effective medical intervention [18]. Machine learning approaches automatically segment the OD and cup regions and then measure the CDR or extract features that may help to determine whether or not the images contain glaucoma, as can be seen in Figure 3. A higher CDR indicates a higher risk of glaucoma [5,20].

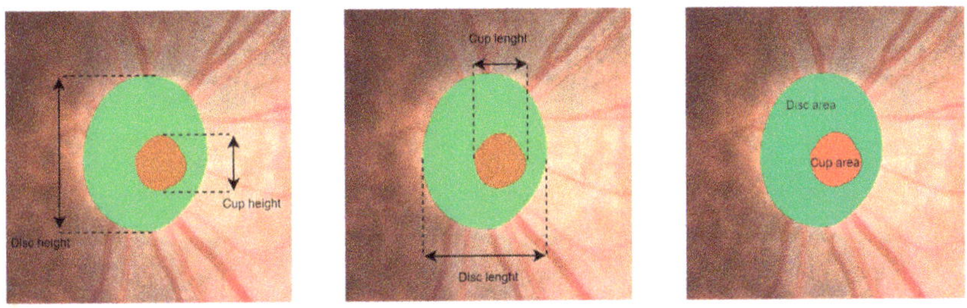

Figure 3. CDR: (**a**) VCDR; (**b**) HCDR; (**c**) ACDR.

Different parameters can be measured for the CDR to determine the cupping and assess the eye for the presence of glaucoma, such as the horizontal cup diameter to the horizontal OD diameter, the vertical cup disc diameter to the vertical OD diameter and the area of the cup to the area of the OD [19]. If the vertical CDR (VCDR) and horizontal CDR (HCDR) are more than 0.5, the eye is considered to be at risk of abnormality; otherwise, it is considered a normal eye [21]. VCDR and HCDR equations are presented in Equations (1) and (2):

$$\text{VCDR} = \frac{V_{\text{cup}}}{V_{\text{disc}}}; \tag{1}$$

$$\text{HCDR} = \frac{H_{\text{cup}}}{H_{\text{disc}}}. \tag{2}$$

Alternatively, considering the criteria by Diaz [21], the assessment can be performed through the area CDR (ACDR) using a threshold of 0.3, as presented in Equation (3):

$$\text{ACDR} = \frac{A_{\text{cup}}}{A_{\text{disc}}}. \tag{3}$$

Diaz [21] presented an automatic algorithm that uses several colour spaces and the stochastic watershed transformation to segment the cup and then obtains handcrafted features, such as the VCDR, HCDR and ACDR. Diaz's method was evaluated on 53 images, obtaining a specificity and sensitivity of 0.81 and 0.87.

After segmentation, Al-Bander [14] calculated the VCDR with varying thresholds and compared the results with an expert's glaucoma diagnosis, achieving an AUC of 0.74, very close to the 0.79 achieved using ground-truth segmentation. After that, the same approach was used, but with HCDR achieving an AUC of 0.78, close to the 0.77 achieved by the expert's annotation and higher than the results obtained with the VCDR.

3. Materials and Methods

The model pipeline in this work is illustrated in Figure 4. In the first task (Task 1: Data preparation), data pre-processing and organisation processes are described. In the second task (Task 2: Glaucoma screening), different glaucoma classification methods are explained based on classification models alongside the respective activation maps and based on OD and cup segmentation models for CDR calculation. The models and hyper-parameters used for each approach are described in the model setups. In the third and last task (Task 3: Evaluation), the models are evaluated based on each approach's glaucoma classification.

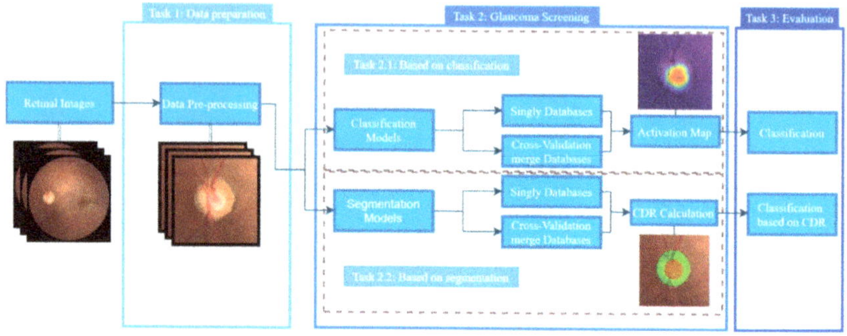

Figure 4. The model pipeline for glaucoma screening.

3.1. Data Preparation

Three public databases were used: RIM-ONE r3, DRISHTI-GS and REFUGE. The RIM-ONE r3 database has a balanced proportion between normal and glaucomatous samples, with 85 healthy images and 74 glaucomatous images with a resolution of 2144 × 1424 pixels. The images in this database vary significantly the quality of the illumination and contrast: some are low-light images, making it difficult to identify the OD and cup, and others have good illumination and contrast, helping to identify the retinal components. DRISHTI-GS has a larger representation of glaucoma samples (70 images) than healthy samples (31 images), and the images have a resolution of 2896 × 1944 pixels. Compared to RIM-ONE r3, DRISHTI-GS images have more homogeneous illumination and contrast, which helps to identify and segment the OD and the cup. The REFUGE database is composed of 400 images with a resolution of 2124 × 2056 pixels, but we only had access to the validation set, which has a lower representation of glaucoma samples compared to healthy samples (40 glaucomatous images and 360 normal images).

For each dataset, retina images were divided into a training set (70%), validation set (15%) and test set (15%). The models were trained with each database separately for the segmentation and classification approach. The respective OD and cup masks are available in all of these databases. In the RIM-ONE r3 database, the images classified as suspect were considered glaucomatous, as was also the case in the work by Gómez-Valverde [2]. Since we had little data, the three databases were merged into a larger database (called K-Fold CVDB, standing for K-Fold Cross-Validation DataBase) to perform K-fold cross-validation (CV). The K-Fold CVDB was divided into 5 similar folds for the cross-validation, and one set was left out to test and validate each model and verify the robustness after the training as the final step. The data organisation process is explained in Figure 5.

All images used to train and test the different models were normalised and centralised in the OD and then cropped to focus the CNNs on the ROI. The cropped images have 512 × 512 resolution and did not suffer from changes in illumination or contrast. Augmentation processes were applied to the databases to avoid overfitting the model, such as rotations (range = 0.2), zooms (range = 0.05), shifts (width and heigh shift range = 0.05) and horizontal flips.

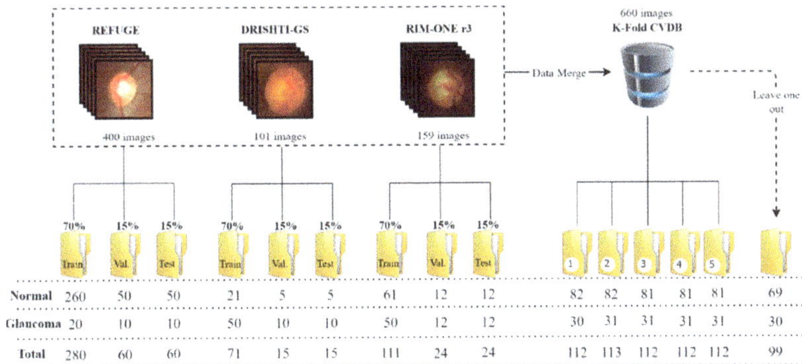

Figure 5. Data organisation for training the models.

3.2. Glaucoma Screening

For both approaches, different models were trained with each database separately, and then CV was performed (more precisely, leave-one-out K-fold CV). For this step, the data were partitioned into K equal-sized subsets. K-1 subsets were used to train the CNN, and the remaining set was used for testing. Additionally, the leave-one-out dataset was used for testing the model at the end, giving a more thorough evaluation of each model's performance since these data were not used to train or test any model. All models were fine-tuned either for image classification or for OD and cup segmentation. Fine-tuning is a procedure based on transfer learning to optimise and minimise the error through the weight initialisation of the convolutional layers using pre-trained CNN weights with the same architecture. The exception is the layer whose number of nodes depends on the number of classes. After the weight initialisation, in the last fully connected layer, the network can be fine-tuned, starting with tuning only the last layer and then tuning the remaining layers, incrementally including more layers in the update process until achieving the desired performance. The early layers learn low-level features, and the late layers learn high-level features specific to the problem in the study [22,23]. For all of the classification and segmentation models used to detect glaucoma, ImageNet pre-trained weights were used. All models selected were based on the best results reported in the reviewed literature.

3.2.1. Classification Methods

Classification used the same principles as segmentation, using pre-trained models with good results inspired by state-of-the-art works. These models were trained with transfer learning using ImageNet weights. First, the four additional layers were pre-trained, freezing the remaining layers before the new ones, and after that, the models were fine-tuned, unfreezing the first layers and training all layers present in the models. We selected the Xcpetion (C1), ResNet 152 V2 (C2) and Inception ResNet V2 (C3) models.

Xception is an extension of the Inception architecture and stands for Extreme Inception. It replaces the standard Inception modules with depthwise separable convolutions called "separable convolution" in frameworks such as TensorFlow and Keras. In the Inception module, filters of different sizes and dimensions are concatenated into a single new filter, acting as a "multi-level feature extractor" by computing 1×1, 3×3 and 5×5 convolutions within the same module of the network. Based on these modules, a more complex and deeper architecture compared to all previous CNN architectures was developed [24]. Depthwise convolution is a spatial convolution performed independently over each channel, followed by a pointwise convolution, i.e., a 1×1 convolution. This architecture's premise is that cross-channel correlations and spatial correlations are sufficiently decoupled to be mapped separately [25].

ResNet is a deep residual network developed with the idea that identifying shortcut connections allows for increasing the depth of convolutional networks while avoiding the gradient degradation problem. These shortcut connections help gradients flow easily in the backpropagation step, which leads to increased accuracy during the training phase. ResNet is composed of 4 blocks with a lot of convolutional blocks inside. Each convolutional operation has the same format in all versions of ResNet (50, 101 and 152), with the only difference being in the number of subsequent convolutional blocks. This deep residual network exploits residual blocks to overcome gradient gradation [23,26].

Inspired by the performance of ResNet, hybrids of Inception and ResNet models were developed. They are two sub-versions of Inception ResNet, i.e., V1 and V2. Inception ResNet V1 has a computational cost similar to that of Inception V3 and Inception ResNet V2 and is similar to Inception v4, with the only difference being in the hyper-parameter settings. They introduce residual connections that use the output of the inception module's convolution operation as the input of the next module. Therefore, the input and output after convolution must have the same dimensions. To increase depth after convolution, 1x1 convolutions were used after the original convolutions [24].

To train all of these models, images and their respective labels (normal or glaucoma) were used as inputs, and the probability of being one of the classes, normal or glaucoma, was the output.

3.2.2. Segmentation Methods

The availability of a huge dataset such as ImageNet with a high capacity to train the model led to a large variety of pre-trained models for the feature encoder in a CNN. The encoder in a U-Net model is a stack of convolution layers combined with activation functions and pooling layers that can adopt the architecture that is frequently employed for feature extraction with pre-trained models. For the segmentation approach in glaucoma screening, the pre-trained models selected were Inception ResNet V2 and Inception V3 (for simplification, called S1 and S2, respectively). These pre-trained models are used as feature encoders in modified U-Net and use the retina image as input and the respective masks of the OD and cup for training. As output prediction is given, a mask of the OD or cup segmentation is also then used to measure the indicators of glaucoma presence, such as CDR. The predicted mask applies morphological processes to remove holes and anomalies of the prediction if they are present.

3.2.3. Model Setups

Segmentation models: The models trained for segmentation were pre-trained for 20 epochs and fine-tuned for 100 epochs with a batch size of 2 for the validation and training sets. The encoder weights were frozen for the pre-training step, and for fine-tuning, the encoder layers were unfrozen; the model was trained again to update all weights. The learning rate started at 10^{-4} with Adam optimiser, and binary cross-entropy was used as the loss function. To prevent the learning rate from stalling on the plateau, the callback reduces the learning rate on the plateau by a factor of 0.90 and only saves the best training weights.

Classification models: The classification model was pre-trained for 20 epochs and fine-tuned for 200 epochs with a batch size of 2 for validation and training sets. The learning rate started at 10^{-4} with Adam optimiser, and binary cross-entropy was used as the loss function; to prevent the learning rate from stalling on the plateau, the callback reduces the learning rate on the plateau by a factor of 0.90. All of these models are available in TensorFlow Core and were loaded. The classification layer (last layer/dense layer) was removed, and after that, 4 new layers were added: a global average pooling 2d layer, a dropout layer (dropout = 0.5), a batch normalisation layer and, finally, a dense layer with 2 outputs with SoftMax as the activation function (2 outputs for 2 classes, glaucoma and normal).

3.3. Model Evaluation

The metrics for the evaluation of the segmentation model were the intersection over union (IoU) and the Dice coefficient.

The IoU metric measures the accuracy of an object detector applied to a particular database. It measures the common area between the predicted (P) and expected (E) regions, divided by the total area of the two regions, as presented in Equation (4):

$$\text{IoU} = \frac{\text{Area}(\text{P} \cap \text{E})}{\text{Area}(\text{P} \cup \text{E})} \tag{4}$$

The Dice coefficient is a statistic used to gauge the similarity between two samples (in this case, between predicted and reference (Ref) segmentation). TP is true positives, FP is false positives and FN is false negatives, as can be seen in Equation (5):

$$\text{Dice} = \frac{2\text{TP}}{2\text{TP} + \text{FP} + \text{FN}} \tag{5}$$

The CDR equations are described in the previous section in Equations (1)–(3). For the evaluation of the classification models, other metrics were used. The accuracy (Acc) (6) is the fraction of correct predictions by the model.

$$\text{Accuracy} = \frac{\text{TP} + \text{TN}}{\text{TP} + \text{TN} + \text{FP} + \text{FN}} \tag{6}$$

where TP is true positives, TN is true negatives, FP is false positives and FN is false negatives. Sensitivity (Sen) (7) measures the proportion of positives that are correctly identified, and specificity (Sep) (8) measures the proportion of negatives that are correctly identified.

$$\text{Sen} = \frac{\text{TP}}{\text{TP} + \text{FN}} \tag{7}$$

$$\text{Sep} = \frac{\text{TN}}{\text{TN} + \text{FP}} \tag{8}$$

The F1-score (F1) (9) indicates the balance between precision and recall, where precision is the number of TP divided by the number of all positives, and recall is the number of TP divided by the number of all samples that should have been identified as positive. The F1-score is the harmonic mean of the two.

$$\text{F1 Score} = \frac{2\text{TP}}{2\text{TP} + \text{FP} + \text{FN}} \tag{9}$$

4. Results and Discussion

The results are organised in the same way that the methodology is presented in the workflow. For both methods, glaucoma screening was performed by training the models with each database separately and with merged data with the K-Fold CVDB. The results are discussed and compared with the results published by the scientific community. In the end, both methods are compared to assess their capability for glaucoma screening and determine how much they can contribute to supporting this important and challenging task.

4.1. Glaucoma Screening Based on Classification Methods

First, for the classification approach, each database was used to train each model separately to determine which model has the best performance and which database has the best quality to produce better model results. The challenge of this methodology is that separately training the models on each database decreases the amount of data that the models learn since there are fewer data to train and validate the training. The results are presented in Table 1.

Table 1. Results for the models trained separately on each database.

Database	Model	Acc	Sen	Spe	AUC	F1
RIM-ONE	C1	0.83	0.79	0.90	0.84	0.85
	C2	0.63	0.64	0.60	0.75	0.67
	C3	0.67	0.79	0.50	0.70	0.73
DRISHTI-GS	C1	0.67	0.67	0.67	0.80	0.71
	C2	0.60	0.89	0.17	0.72	0.73
	C3	0.73	0.78	0.67	0.79	0.79
REFUGE	C1	0.97	0.83	0.98	0.98	0.83
	C2	0.85	0.67	0.87	0.74	0.47
	C3	0.95	0.83	0.96	0.99	0.77

Overall, the database that showed better results was REFUGE for the C1 and C3 models, with an AUC close to one and with high sensitivity and specificity close to the results presented in Table 2. The C1 and C3 models outperformed those reported by Sreng [5], who also used the REFUGE database for pre-trained networks with a transfer learning model ensemble.

Table 2. Results of state-of-the-art glaucoma classification methods.

Author	Model	Database	Sen	Spe	AUC
Gómez-Valverde [2]	VGG19	RIM-ONE, DRISHTI-GS and Esperanza	0.84	0.89	0.92
	GoogLeNet		0.89	0.83	0.93
	ResNet50		0.84	0.89	0.92
	DENet		0.81	0.88	0.91
Diaz-Pinto [11]	VGG16	ACRIMA, HRF, DRISHTI-GS, RIM-ONE and Sjchoi86-HRF	0.91	0.88	0.96
	VGG19		0.92	0.88	0.96
	Inception V3		0.92	0.88	0.97
	ResNet50		0.90	0.89	0.96
	Xception		0.93	0.86	0.96
Norouzifard [13]	Inception ResNet V2	UCLA and HRF	0.91	0.93	-
Sreng [5]	Pre-train CNNs as feature extractors models ensemble	RIM-ONE	-	-	0.99
		DRISTHI-GS	-	-	0.92
		REFUGE	-	-	0.94
		ORIGA	-	-	0.85
		ACRIMA	-	-	0.99

The RIM-ONE and DRISHTI-GS databases showed lower AUC values than REFUGE and the results reported by Sreng [5], who also evaluated each database separately. Better results with a significant difference suggest that the quality of the samples in the REFUGE database is superior to that of the others, with homogeneity in the contrast, illumination and resolution of all samples, in contrast to RIM-ONE and DRISHTI-GS, whose images have heterogeneous quality with a lot of variations in the same factors.

As mentioned in the methodology, the K-fold CV technique was used to train the models with K-1 folds, and then the other test set was used for testing. The K-Fold CVDB was divided into five folds, with an extra set left out to test at the end (leave-one-out method) in all iterations of each model. Of these five folds, four were used to train the model, and the other was used to evaluate it, changing the test fold and training folds in each iteration. The results of the classification in each fold are presented in Table 3.

Table 3. Results for the models with K-fold CV for each test set of each fold with the mean results of the models for the 5 folds and the standard deviation.

Model	Acc	Sen	Spe	AUC	F1-Score
C1	0.91 (±0.04)	0.86 (±0.10)	0.93 (±0.04)	0.96 (±0.02)	0.84 (±0.07)
C2	0.90 (±0.02)	0.81 (±0.08)	0.93 (±0.02)	0.94 (±0.02)	0.81 (±0.05)
C3	0.88 (±0.03)	0.77 (±0.10)	0.92 (±0.04)	0.94 (±0.02)	0.78 (±0.05)

Compared to the results presented above, when the models were trained with each database separately, the K-fold CV technique showed an immediate enhancement and direct correlation between the amount of data for training and the quality of classification. All of the models showed similar results, with slightly better performance for the C1 model. These models outperformed most of the state-of-the-art works mentioned previously, with some exceptions (Diaz-Pinto [11] and some results of Sreng [5]) owing to fewer data for training. To evaluate the robustness of the models, they were tested with the test set omitted from training, and the results are shown in Table 4.

Table 4. Results for the models in K-fold CV for the leave-one-out test set with the respective mean and standard deviation for the 5 folds of each model.

Model	Acc	Sen	Spe	AUC	F1-Score
C1	0.84 (±0.01)	0.78 (±0.03)	0.87 (±0.01)	0.91 (±0.01)	0.75 (±0.02)
C2	0.82 (±0.03)	0.69 (±0.10)	0.87 (±0.02)	0.87 (±0.03)	0.69 (±0.07)
C3	0.80 (±0.02)	0.65 (±0.10)	0.86 (±0.04)	0.86 (±0.04)	0.66 (±0.05)

The results of the classification of the leave-one-out set decreased compared to those discussed above. Nevertheless, most of the models yielded better results than most of the state-of-the-art works, with the same exceptions as those noted previously. The most significant decrease in the results was in the sensitivity, showing a lack of representation and a high rate of false positives for glaucoma samples. The evaluation with the leave-one-out dataset demonstrated that the best model of the three used is C1, as mentioned previously, with the smallest decrease in every metric among all of the models.

The classification models can be a "black-box": extremely hard to explain and hard for non-experts to understand. Explainable artificial intelligence (AI) approaches are methods and techniques that can explain to humans why the DL models arrived at a specific decision. Explainable AI can create transparency, interpretability and explainability as a foundation for the output of the neural networks. For a visual interpretation of the output to supplement the results of the classification models, activation maps (Figure 6) were created that show the regions of the input images that cause the CNNs to classify the samples as glaucomatous or normal, thus helping clinicians to understand the reason for the output classification.

Gradient-weighted glass activation mapping (Grad-CAM) uses the gradients of any target concept flowing into the final convolutional layer to produce a coarse localization map highlighting the important regions in the image for predicting the concept. These heatmaps can reveal some important indicators or factors for the classification. The used models focussed more on the centre of the OD, where the cupping zone is responsible for and highly correlated with glaucoma cases. The larger the cup area, the more suspicious, and the more probable that the patient has a case of glaucoma. This type of indicator can help ophthalmologists to make a better and more reliable decision, with one of these indicators being the CDR. To calculate this ratio, first, the OD and cup must be segmented, and the trustworthiness of the screening depends on how well they are segmented. The segmentation procedure is time-consuming and inconsistent when performed manually, so to facilitate a more consistent segmentation, we present models for segmentation with a consequent glaucoma classification based on CDR calculation.

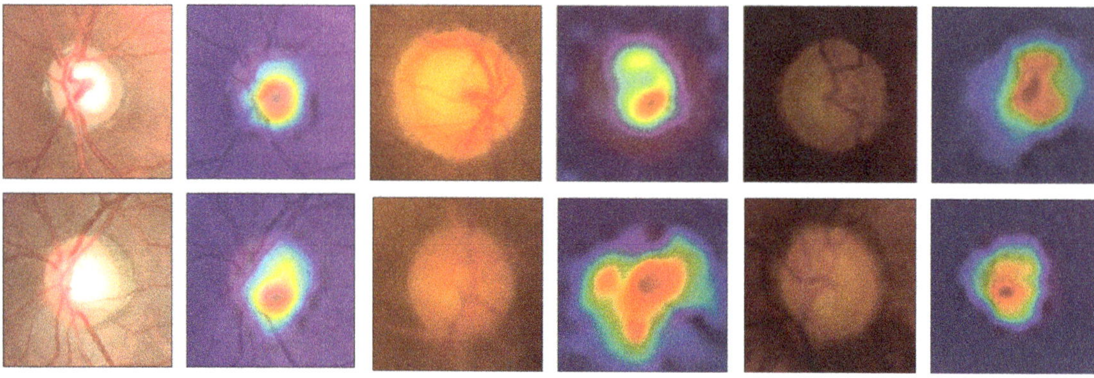

Figure 6. Activation maps of the classification models of the K-fold CV (left, original; right, heatmap). The importance is indicated by the emphasized shades in the following order, from the most important to the least important: red, orange, yellow, green and blue.

4.2. Glaucoma Screening Based on Segmentation Methods

The OD and cup were segmented by two different CNNs, and then the different CDRs were calculated; glaucoma was then classified based on the CDR model. This requires the reference (Ref) masks of each database with annotations of the segmentation made by clinicians, and these were available in the databases selected for this work. Finally, the segmentation and glaucoma screening were compared with the reference masks using the same criteria of the glaucoma classification based on CDR values. To perform the segmentation, two different models, S1 and S2, were used. First, the OD segmentation results are presented, followed by the cup segmentation results, and finally, the glaucoma classification based on the CDR calculation with segmentation masks is provided.

4.2.1. OD Segmentation

The procedure in the segmentation methods is the same as the one presented for the classification approach, with the segmentation in each database performed separately and with the K-Fold CVDB. For the K-fold CV, the means of IoU and Dice of the five folds in each model were obtained. The final mask is the intersection/agreement of at least four masks of the five iterations of each model to compute the final CDRs. The results for OD segmentation are presented in Table 5.

At first view, the results in every dataset segmentation are very similar to every compared state-of-the-art method, with a slight but non-significant difference that does not change the outcome of the CDR calculation. This can be explained by the fact that the segmentation of OD is an easy task because of the visible contrast and outline of the OD and the retina, which facilitate identification and segmentation by the neural network. The K-fold CV showed decreases in the IoU and Dice in both models compared to the other results since they represent the mean of five iterations in each model. This can affect the final results, with divergence in the agreement of OD segmentation. However, this difference was not significant enough to jeopardise the CDR calculation, at least in most of the samples. The two models had similar results, with a slightly better performance for S1. After OD segmentation, the procedure was repeated but with different CNN models, this time training the model to segment the cup.

Table 5. Results for OD segmentation for each model and the K-Fold CVDB. For comparison, the results from the literature review are presented as well. S1 is Inception ResNet V2, and S2 is Inception V3.

Database	Model	IoU	Dice
RIM-ONE	S1	0.91 (±0.08)	0.95 (±0.05)
	S2	0.89 (±0.05)	0.94 (±0.03)
	Al-Bander [14]	0.83	0.90
	Singh [15]	0.93	0.98
	Yu [17]	0.93	0.96
DRISHTI-GS	S1	0.94 (±0.02)	0.97 (±0.01)
	S2	0.93 (±0.02)	0.96 (±0.01)
	Al-Bander [14]	0.90	0.95
	Singh [15]	0.96	0.97
	Yu [17]	0.95	0.97
REFUGE	S1	0.93 (±0.03)	0.96 (±0.02)
	S2	0.92 (±0.03)	0.96 (±0.02)
	Qin [16]	0.92	0.97
K-Fold CVDB	S1	0.83 (±0.01)	0.91 (±0.01)
	S2	0.81 (±0.01)	0.89 (±0.004)

4.2.2. Cup Segmentation

For cup segmentation, the same models were used, but this time, the network was trained to localise and segment the excavation region inside the OD. Contrary to the previous task, cup segmentation is much harder since there is not a high contrast between the exaction zone and the OD (at least not as high as the contrast between the OD and the retina). The results from the two models are presented in Table 6 with the same structure as the one presented for OD segmentation.

Table 6. Results for cup segmentation for each model and the K-Fold CVDB. For comparison, the results from the literature review are represented as well.

Database	Model	IoU	Dice
RIM-ONE	S1	0.70 (±0.13)	0.82 (±0.09)
	S2	0.68 (±0.15)	0.80 (±0.12)
	Al-Bander [14]	0.56	0.69
	Yu [17]	0.74	0.84
DRISHTI-GS	S1	0.76 (±0.22)	0.84 (±0.19)
	S2	0.74 (±0.20)	0.83 (±0.19)
	Al-Bander [14]	0.70	0.82
	Yu [17]	0.80	0.89
REFUGE	S1	0.81 (±0.07)	0.90 (±0.04)
	S2	0.80 (±0.06)	0.89 (±0.04)
	Qin [16]	0.90	0.92
K-Fold CVDB	S1	0.64 (±0.03)	0.77 (±0.02)
	S2	0.60 (±0.02)	0.74 (±0.02)

Overall, the results achieved the same baseline as the state-of-the-art methods. When directly compared on the RIM-ONE database, the S1 model had better results than S2 and had better Dice than Al-Bander [14], and IoU and Dice were only worse compared to Yu's [17] work. With DRISHTI-GS, the two models had better IoU and Dice than Al-Bander [14] and a slight difference in IoU and Dice compared to the remaining works, with an overall better performance observed for the S1 model. In the REFUGE database, the results from our models and Qin's [16] work are very similar, with a minor difference in the Dice, and as observed in the segmentation of the other databases, the S1 model had better results as well.

In the K-fold CV, both models had a major decrease in performance compared to the other works and the performance of the same models using each database separately. As in the previous verification, the S1 model continued to produce better results. Compared to OD segmentation, the IoU and Dice were much lower, which is a consequence of these coefficients being too sensitive to small errors when the segmented object is small and not sensitive enough to large errors when the segmented object is larger.

The results of OD and cup segmentation were used to calculate the CDRs to use as an indicator of glaucoma presence. Reference segmentation by clinicians was used as the ground truth but is not an absolute truth since the segmentation process can be subjective, and the results can differ between clinicians. Thus, the segmentation predicted by the CNNs can sometimes cause the misclassification of the images but can be considered another opinion, especially in cup segmentation since the perimeter of the cup is not as delimitated and visible as the OD. After the segmentation of the OD and cup, the CDRs were calculated to obtain the glaucoma classification.

4.2.3. Glaucoma Screening Based on Estimated CDR

The segmentation masks of the OD and cup of both models were computed and used to calculate the ratio between them. In this work, all CDRs were calculated, including the vertical and horizontal CDRs and the ratio between the areas of the OD and cup. For the VCDR and HCDR, the criteria used were CDR < 0.5 for normal and CDR ≥ 0.5 for glaucomatous, and the ACDR was normal if <0.3 and glaucomatous if ≥ 0.3, as described in Diaz's work [21]. The same criteria were used for the Ref masks to allow a direct comparison between the results of our models and the segmentation performed by ophthalmologists to gauge the reliability of segmentation by the S1 and S2 models. The results are expressed in Table 7.

Table 7. Results of glaucoma classification with CDR calculations for S1, S2 and Ref masks.

		S1				S2				Ref			
		Sen	**Spe**	**F1**	**AUC**	**Sen**	**Spe**	**F1**	**AUC**	**Sen**	**Spe**	**F1**	**AUC**
RIM-ONE	**VCDR**	0.71	0.80	0.77	0.90	0.79	0.90	0.85	0.92	0.75	1.00	0.86	0.84
	HCDR	0.71	0.70	0.74	0.78	0.64	0.90	0.75	0.86	0.75	0.86	0.80	0.80
	ACDR	0.71	0.90	0.80	0.90	0.64	1.00	0.78	0.92	0.50	1.00	0.67	0.82
DRISHTI-GS	**VCDR**	1.00	0.17	0.78	1.00	1.00	0.50	0.86	1.00	1.00	0.67	0.90	0.94
	HCDR	1.00	0.20	0.82	0.96	1.00	0.33	0.82	0.93	1.00	0.67	0.90	0.94
	ACDR	1.00	0.50	0.86	1.00	1.00	0.50	0.86	0.98	1.00	0.67	0.90	0.94
REFUGE	**VCDR**	0.80	0.67	0.30	0.87	0.80	0.73	0.33	0.87	1.00	0.72	0.33	0.87
	HCDR	0.80	0.36	0.18	0.78	0.80	0.44	0.20	0.78	1.00	0.43	0.20	0.80
	ACDR	0.60	0.82	0.33	0.84	0.80	0.85	0.47	0.84	1.00	0.79	0.40	0.86
K-Fold CVDB	**VCDR**	0.90	0.68	0.68	0.87	0.90	0.70	0.69	0.87	1.00	0.60	0.91	0.88
	HCDR	0.83	0.30	0.49	0.78	0.80	0.46	0.53	0.73	1.00	0.40	0.87	0.83
	ACDR	0.80	0.78	0.70	0.84	0.70	0.84	0.68	0.83	0.90	0.60	0.86	0.86

The results for both models were similar to the results using the Ref masks, which indicates that they produced similar segmentation results or at least provided similar CDRs. Overall, the results from CDRs based on the Ref masks were better than the results from the two models, but the difference between the models' classification and the classification in the Ref masks, in a lot of cases, was not significant.

With RIM-ONE, the Ref had a better F1-score for the VCDR and HCDR, but the difference in the F1-scores between the two models was very small. S2 achieved better sensitivity and specificity, but this difference was also small, which may indicate that the masks were very close to each other or had similar forms that led to the computation of similar CDR values. DRISHTI-GS and the K-Fold CVDB were the two datasets with the worst results for both models in comparison with the Ref results, showing a greater difference, but the AUC indicated that the difference was not that large. The results from

REFUGE were better for the S2 model compared to S1 and Ref for sensitivity, specificity and F1-score, but in all models and the Ref, the values were very low, which may suggest that, in this case, the CDRs are not a sufficient indicator to produce a classification of glaucoma or normal; thus, for a better decision, complementary information is needed to support the final call. All of the ROC curves from the different databases for all CDRs of the models and Ref masks are presented in Figure 7.

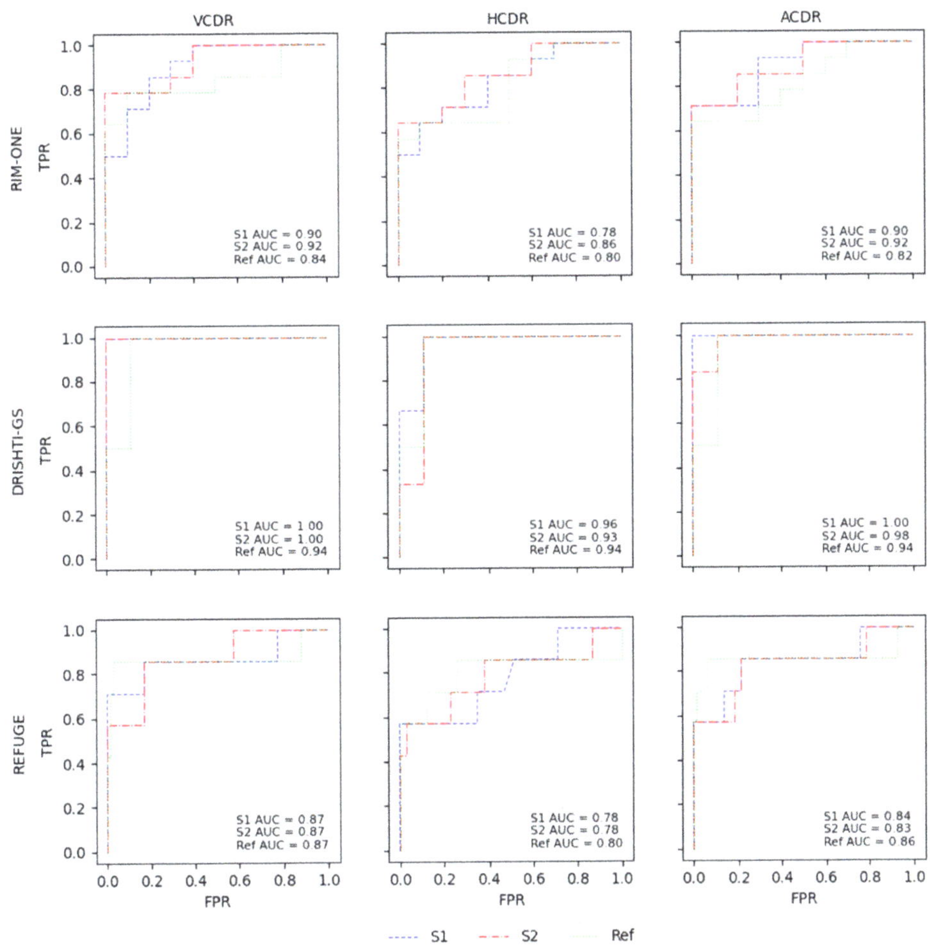

Figure 7. ROC curves for the glaucoma classification through CDR calculations for each database separately with the S1 and S2 models and the respective CDR calculations with the Ref masks.

Of all CDRs, the VCDR and ACDR had the best results. The HCDR was the worst result in the two models and the Ref, and the model with the overall best results was S2. This is also shown in the ROC curves, with the models and the Ref having very similar results in all AUCs for the glaucoma classification based on the different CDRs. The difference between the AUCs of the models and the Ref was not significant and was generally very small, with the Ref showing slightly better performance than the S2 model. This can reinforce the notion that the masks originating from the S1 and S2 models are very close to the Ref masks or compute similar CDRs that lead to a similar glaucoma classification based on CDRs. In the work by Diaz [21], the model obtained specificity of 0.81 and sensitivity of 0.87, and Al-Bander [14] achieved an AUC of 0.74 using the

VCDR and 0.78 using the HCDR. The majority of our results surpass the results of the state-of-the-art glaucoma classification methods based on CDRs.

For the K-fold CV, the results of the ROC curves for both models were very similar to those of the Ref and had close AUC values, except for the HCDR. As mentioned previously, the HCDR was the CDR that differed the most, as can be seen in Figure 8.

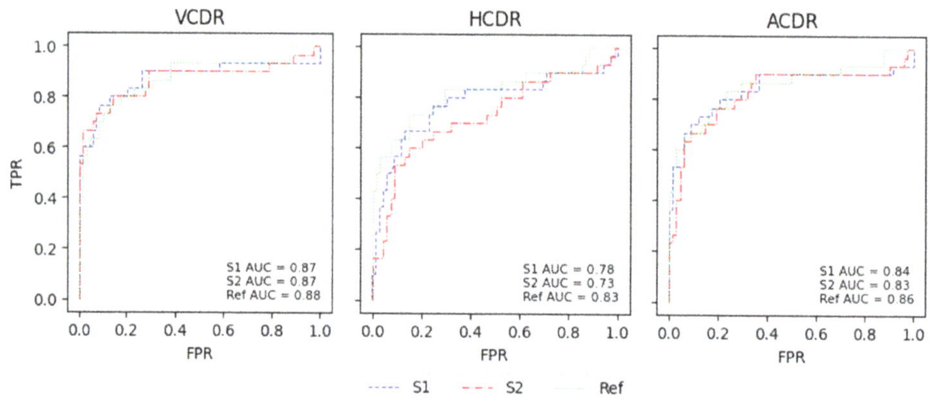

Figure 8. ROC curves for the glaucoma classification through CDR calculation for K-Fold CVDB with the S1 and S2 models and with the respective CDR calculations with the Ref masks.

For visual comparison, in the following images, the masks and outlines of both models are drawn in red, and those of the Ref masks are drawn in green. The intersection between the masks predicted by the models and the Ref masks is indicated by the combination of green and red (true positive). The green area represents a false negative since there is no intersection between the masks, and the red area represents a false positive since the model's prediction does not correspond to the same result as the Ref mask.

In Figure 9, the CDRs values for higher Dice cases are extremely close to the Ref CDR values. In the K-fold, the resulting masks are the intersection of at least four agreements in the model of the different folds. The predicted masks of the OD and cup are very close to the Ref masks, reflecting high IoU and Dice values. In the lower Dice cases, the CDRs significantly differ compared to Ref CDRs, but despite this, there is complete agreement in the final decision for the classification based on CDRs since all apply the same threshold values, although they differ more than the higher Dice cases.

The two models used achieved state-of-the-art results for the segmentation, and the outcome was similar to the glaucoma classification based on the CDR with the Ref masks, indicating that these types of models can mitigate these labour-intensive and subjective tasks, that is, the segmentation of the OD and cup, providing a more consistent final result. To complement the CDR indicator, additional examination must be performed to make the final diagnosis of the patient using, for example, IOP values, anamnesis data and medical records. Another problem is the thin margin in the threshold CDRs, potentially resulting in an arbitrary classification; to resolve this obstacle, more diagnosis classes can be added based on CDRs, such as a suspicious case of glaucoma in the samples for which the CDR value barely passes or reaches the threshold.

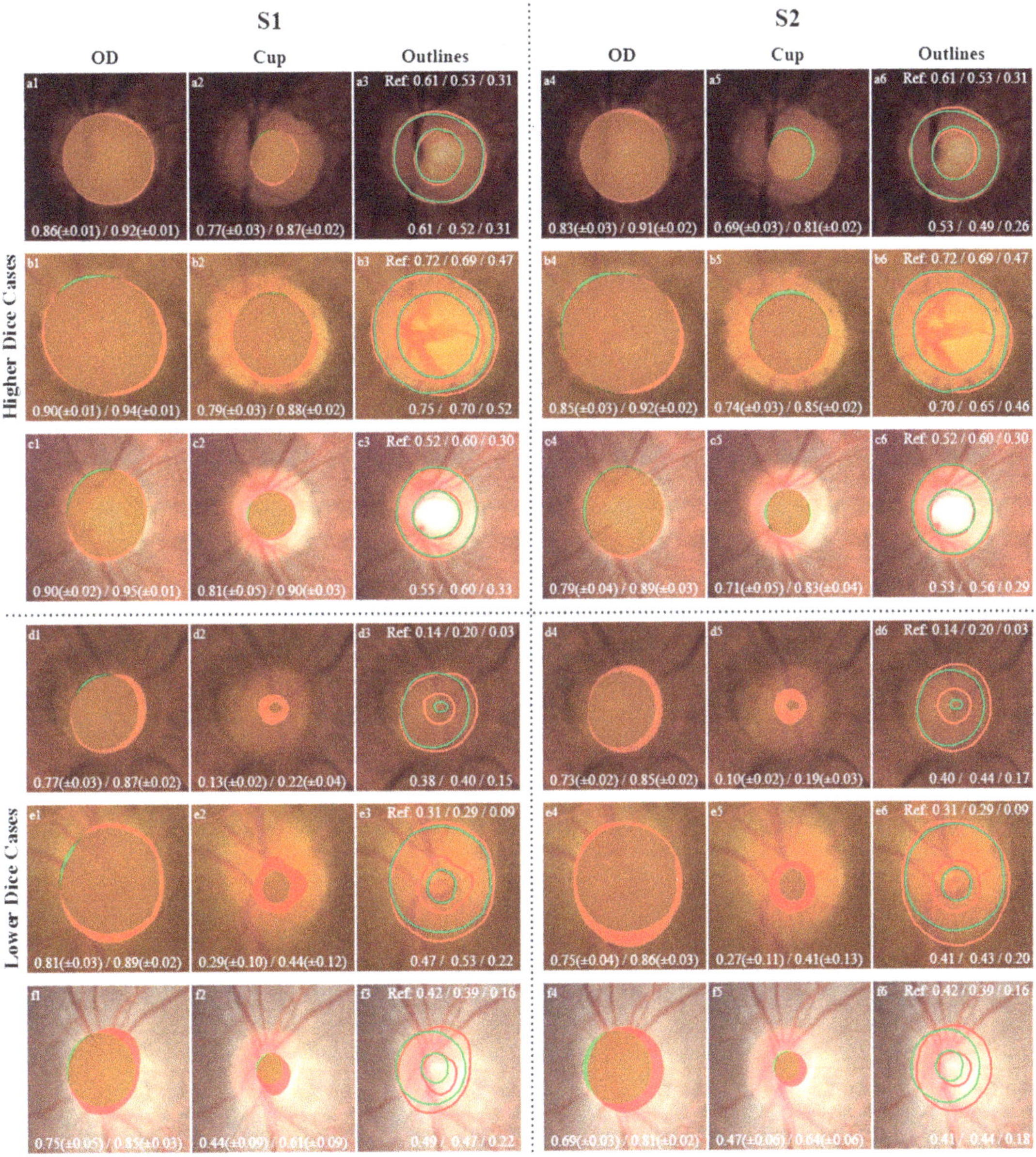

Figure 9. Higher Dice cases and lower Dice cases for each model for OD and cup segmentation with the final outlines of the OD and cup for the K-Fold CVDB. Green represents the masks and outlines of the Ref, and red represents the masks and outlines of the predictions from the models. The lower right corner shows the mean IoU/Dice for OD and cup segmentation for every fold in each model with the respective standard deviation; the "Outlines" columns at the top are the Ref values of the CDRs (VCDR/HCDR/ACDR), and the bottom of the images show the results from CDRs of the models in the same order as that described for the Ref.

4.3. Classification Methods on a Private Dataset

For the glaucoma screening based on DL methods, only classification models were applied on the private dataset, since this did not have ground-truth masks for the application of segmentation techniques. For the classification, the same K-fold CV approach was applied to a private dataset of D-EYE (Portable Retinal Imaging System) images with lower resolution. The goal was to see if applying the classification methods to images acquired by mobile devices could achieve similar results to those obtained using high-resolution images captured by clinical equipment. This would impart some portability to the glaucoma screening process, expanding it to more people and preventing glaucoma cases. This dataset was approved by the Ethical Committee of the Universidade Aberta of Lisbon and by the Health Ministerium of Brazil following the dispatch of the information DW/2018 of 02-21-2019 provided by the Brazilian Research Ethics Committee. This dataset consists of D-EYE images collected between October 2018 and March 2020 from patients aged above 40, either treated or untreated for glaucoma; subjects accepted the research protocols and allowed the use of data for studies on applications of automatised methods of glaucoma screening.

The images were obtained using a lens of D-EYE coupled to the camera of an iPhone 6S, which allows photographing the patient's optical papilla through 75 and 90 diopters and recording a short video that is stored in .mp4 format and collected in an environment with low light. From the videos, images that had 1080×1920 pixels of resolution and underwent the same pre-processing treatment as described for the glaucoma screening with the public databases were selected, and the OD was cropped and centred, obtaining dimensions of 512×512. From the database, a total of 347 images were selected, of which 293 were classified as normal, and 54 were classified as glaucomatous.

For the classification, since it is a small database, K-fold CV was applied with a leave-one-out set to classification neural networks. Each database was divided into five folds, in which each fold had 49 normal samples and 9 glaucomatous samples. The leave-one-out set had 48 normal samples and 9 glaucomatous ones and was used to validate the models after training. To train the CNNs, the same pre-trained classification models were used, namely, C1, C2 and C3. The results are presented in Table 8.

Table 8. Results for the models with K-fold CV for each test set of each fold with the mean of results of the models for the 5 folds and the standard deviation.

Model	Acc	Sen	Spe	AUC	F1-Score
C1	0.87 (±0.03)	0.36 (±0.11)	0.96 (±0.03)	0.82 (±0.05)	0.45 (±0.10)
C2	0.86 (±0.02)	0.16 (±0.11)	0.99 (±0.02)	0.84 (±0.06)	0.24 (±0.16)
C3	0.87 (±0.05)	0.47 (±0.15)	0.94 (±0.07)	0.87 (±0.05)	0.52 (±0.14)

The models obtained high AUC values and specificity but low sensitivity and F1-score results, showing that they had difficulty classifying the glaucomatous samples since the database lacks sufficient representation of glaucoma samples, and most of these images were classified based on the clinical record, family history and IOP values. The glaucoma images do not show consistent patterns that indicate glaucoma incidence directly in the image. To validate the model's performance, each one was tested with the leave-one-out dataset, and the results are presented in Table 9.

In Figure 10, image Figure 10a illustrates a case of glaucoma with cataract opacity that worsens the overall quality of the image. Nevertheless, the C3 model predicted the sample correctly. The activation map points to a peripheric region with the presence of vases, but the spot that indicates the incidence of glaucoma is located in the centre of the excavation zone. Other good examples of a poor focus on the region of interest are images Figure 10b,c, where the output prediction was correctly classified, but the activation map points to an excentric zone instead of focussing on the cup area.

Table 9. Results for the models in K-fold CV for the leave-one-out test set with the respective mean and standard deviation for the 5 folds of each model for the private dataset.

Model	Acc	Sen	Spe	AUC	F1-Score
C1	0.85 (±0.02)	0.36 (±0.16)	0.95 (±0.01)	0.79 (±0.04)	0.42 (±0.14)
C2	0.85 (±0.04)	0.29 (±0.05)	0.95 (±0.05)	0.77 (±0.22)	0.38 (±0.07)
C3	0.85 (±0.04)	0.49 (±0.22)	0.92 (±0.08)	0.80 (±0.07)	0.49 (±0.10)

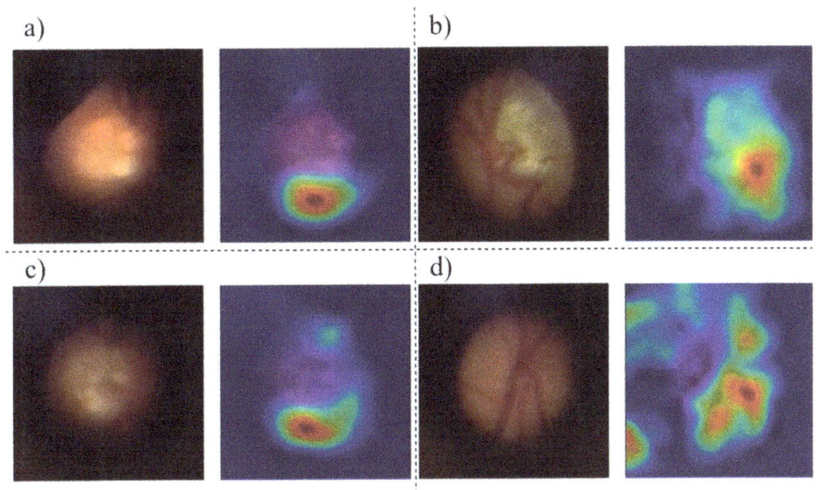

Figure 10. Activation maps for D-EYE images of the CV models. The importance is indicated by the emphasized shades in the following order, from the most important to the least important: red, orange, yellow, green, blue. All images report glaucoma cases (**a**–**d**).

Figure 10c shows slightly better recognition of the centre zone, but the models focus on the veins and not on the optic cup. The CNNs provide additional information for diagnosis by an ophthalmologist but, in most cases, have an excentric focus (on veins) instead of focussing on the centre area (cupping of the OD).

The CNNs can help clinicians to expand glaucoma screening and accelerate the early screening of glaucoma. This database lacks visual representation of glaucoma samples since most of the glaucoma images collected do not have visual signs of glaucoma, and the diagnoses were made based on other indications, such as IOP, clinical records and family history, as mentioned previously. To improve the results, a more balanced database is needed with more glaucoma samples with visual patterns and indicators that evidence the presence of glaucoma. Another way to obtain better results would be to use other types of clinical data in neural networks to complement the image data.

5. Conclusions

Glaucoma disease has a high incidence around the world, the main cause of which is the lack of tools for and accessibility to early screening to prevent the evolution of the disease. Since it is a disorder that is usually asymptomatic, it is frequently detected in the late stage; by this time, medical treatment cannot reverse the injuries and vision loss but can only prevent the spread of glaucoma. Glaucoma screening is carried out in clinical centres by specialised clinicians with expensive tools. Mass screening is time-consuming and, most of the time, subjective, especially in the early stage, depending on the expertise of the ophthalmologist. For this reason, different approaches using CNNs can help to expand mass glaucoma screening, save time and money and help medical staff to perform more reliable screening with more consistent decisions, speeding up the process and relieving hard and repetitive work.

All classification models achieved results similar to those of state-of-the-art methods, with the Xception model showing an overall better performance. The CNN models for classification, unlike the CDR and segmentation method, are "black-boxes"; they do not provide a visual representation of their decisions. Thus, in this work, the model's activation maps are presented to provide visual interpretations and analyse the model's classification, thus helping medical experts to understand the CNN's decision. A careful analysis reveals that, in this case, the CNNs focus on the centre of the OD in the cup, reinforcing the significance of the cupping area as an indicator of glaucoma presence. The OD and cup can provide the CDR, which is usually used as an indicator for glaucoma screening. In other cases, the activation maps focus on peripheral veins that, in most cases, do not correlate with the incidence of glaucoma.

Since the ratio between the OD and cup is the most used indicator in the ophthalmology field, segmentation methods were applied to classify the samples after classification based on CDRs. The segmentation of the cup is more difficult than segmenting the OD, which does not usually have a well-defined boundary to help in the segmentation, making it difficult for clinicians to perform this task. For this reason, the CNNs have proved to be helpful in facilitating a subjective and hard task that highly depends on the experience of the ophthalmologist. The CDRs computed through the segmented masks were very close to the Ref CDRs, reinforcing that the CNNs can conduct an evaluation similar to that performed by a clinician. The model that produced the best results overall for these tasks was Inception V3 as the backbone of U-Net, with slightly better performance for the different CDRs. A way to improve classification based on CDR calculations is to use an additional class instead of binary classification, providing an extra margin for the threshold.

The classification methods were applied to a private database with images collected through a lens attached to a mobile device, and the results are promising since this lens is cheaper and can expand the accessibility and accelerate mass glaucoma screening. The model with the best results in the private database was Inception ResNet V2, which had higher sensitivity compared to the remaining models. The Xception model achieved similar AUC results but had a lower sensitivity compared to Inception ResNet V2. The classification results of images in this private database are promising but did not achieve the sensitivity of the models trained with public databases. The model's classification can facilitate mass screening with images collected by lenses attached to mobile devices, serving as an extra opinion and providing activation maps to explain the model's decision. These new approaches of collecting retinal images with posterior CNN classification models can accelerate and contribute to mass screening, mostly in remote areas, helping to redirect people to medical centres to prevent glaucoma as early as possible.

Author Contributions: Conceptualisation, A.N. and J.C.; methodology A.N. and J.C.; validation, A.C.; formal analysis, A.N. and J.C.; investigation, A.N. and J.C.; writing—original draft preparation, A.N. and J.C.; writing—review and editing, A.N., J.C. and A.C.; supervision, A.C.; project administration, A.C.; funding acquisition, A.C. All authors have read and agreed to the published version of the manuscript.

Funding: This work is financed by National Funds through the Portuguese funding agency, FCT—Fundação para a Ciência e a Tecnologia, within project LA/P/0063/2020.

Institutional Review Board Statement: Not applicable.

Informed Consent Statement: Not applicable.

Data Availability Statement: Not applicable.

Conflicts of Interest: The authors declare no conflict of interest.

References

1. Maheshwari, S.; Pachori, R.B.; Acharya, U.R. Automated Diagnosis of Glaucoma Using Empirical Wavelet Transform and Correntropy Features Extracted from Fundus Images. *IEEE J. Biomed. Health Inform.* **2017**, *21*, 803–813. [CrossRef] [PubMed]
2. Gómez-Valverde, J.J.; Antón, A.; Fatti, G.; Liefers, B.; Herranz, A.; Santos, A.; Sánchez, C.I.; Ledesma-Carbayo, M.J. Automatic glaucoma classification using color fundus images based on convolutional neural networks and transfer learning. *Biomed. Opt. Express* **2019**, *10*, 892–913. [CrossRef] [PubMed]
3. Zhang, Z.; Srivastava, R.; Liu, H.; Chen, X.; Duan, L.; Wong, D.W.K.; Kwoh, C.K.; Wong, T.Y.; Liu, J. A survey on computer aided diagnosis for ocular diseases. *BMC Med. Inform. Decis. Mak.* **2014**, *14*, 80. [CrossRef] [PubMed]
4. Kanse, S.S.; Yadav, D.M. Retinal Fundus Image for Glaucoma Detection: A Review and Study. *J. Intell. Syst.* **2019**, *28*, 43–56. [CrossRef]
5. Sreng, S.; Maneerat, N.; Hamamoto, K.; Win, K.Y. Deep learning for optic disc segmentation and glaucoma diagnosis on retinal images. *Appl. Sci.* **2020**, *10*, 4916. [CrossRef]
6. Claro, M.L.; Veras, R.; Santos, L.; Frazão, M.; Filho, A.C.; Leite, D. Métodos computacionais para segmentação do disco óptico em imagens de retina: Uma revisão. *Rev. Bras. Comput. Apl.* **2018**, *10*, 29–43. [CrossRef]
7. Thakur, N.; Juneja, M. Survey on segmentation and classification approaches of optic cup and optic disc for diagnosis of glaucoma. *Biomed. Signal Process. Control* **2018**, *42*, 162–189. [CrossRef]
8. Bajwa, M.N.; Malik, M.I.; Siddiqui, S.A.; Dengel, A.; Shafait, F.; Neumeier, W.; Ahmed, S. Correction to: Two-stage framework for optic disc localization and glaucoma classification in retinal fundus images using deep learning. *BMC Med. Inform. Decis. Mak.* **2019**, *19*, 136. [CrossRef] [PubMed]
9. Hagiwara, Y.; Koh, J.E.W.; Tan, J.H.; Bhandary, S.V.; Laude, A.; Ciaccio, E.J.; Tong, L.; Acharya, U.R. Computer-aided diagnosis of glaucoma using fundus images: A review. *Comput. Methods Programs Biomed.* **2018**, *165*, 1–12. [CrossRef] [PubMed]
10. Barros, D.M.S.; Moura, J.C.C.; Freire, C.R.; Taleb, A.C.; Valentim, R.A.M.; Morais, P.S.G. Machine learning applied to retinal image processing for glaucoma detection: Review and perspective. *Biomed. Eng. Online* **2020**, *19*, 20. [CrossRef] [PubMed]
11. Diaz-Pinto, A.; Morales, S.; Naranjo, V.; Köhler, T.; Mossi, J.M.; Navea, A. CNNs for automatic glaucoma assessment using fundus images: An extensive validation. *Biomed. Eng. Online* **2019**, *18*, 29. [CrossRef] [PubMed]
12. Serener, A.; Serte, S. Transfer learning for early and advanced glaucoma detection with convolutional neural networks. In Proceedings of the 2019 Medical Technologies Congress (TIPTEKNO), Izmir, Turkey, 3–5 October 2019; pp. 1–4. [CrossRef]
13. Norouzifard, M.; Nemati, A.; Gholamhosseini, H.; Klette, R.; Nouri-Mahdavi, K.; Yousefi, S. Automated glaucoma diagnosis using deep and transfer learning: Proposal of a system for clinical testing. In Proceedings of the 2018 International Conference on Image and Vision Computing New Zealand (IVCNZ), Auckland, New Zealand, 19–21 November 2018. [CrossRef]
14. Al-Bander, B.; Williams, B.M.; Al-Nuaimy, W.; Al-Taee, M.A.; Pratt, H.; Zheng, Y. Dense fully convolutional segmentation of the optic disc and cup in colour fundus for glaucoma diagnosis. *Symmetry* **2018**, *10*, 87. [CrossRef]
15. Singh, V.K.; Rashwan, H.A.; Akram, F.; Pandey, N.; Sarker, M.M.K.; Saleh, A.; Abdulwahab, S.; Maaroof, N.; Barrena, J.T.; Romani, S.; et al. Retinal Optic Disc Segmentation Using Conditional Generative Adversarial Network. *Front. Artif. Intell. Appl.* **2018**, *308*, 373–380. [CrossRef]
16. Qin, Y.; Hawbani, A. A novel segmentation method for optic disc and optic cup based on deformable U-net. In Proceedings of the 2019 2nd International Conference on Artificial Intelligence and Big Data (ICAIBD), Chengdu, China, 25–28 May 2019; pp. 394–399. [CrossRef]
17. Yu, S.; Xiao, D.; Frost, S.; Kanagasingam, Y. Robust optic disc and cup segmentation with deep learning for glaucoma detection. *Comput. Med. Imaging Graph.* **2019**, *74*, 61–71. [CrossRef] [PubMed]
18. Wong, D.W.K.; Liu, J.; Lim, J.H.; Jia, X.; Yin, F.; Li, H.; Wong, T.Y. Level-set based automatic cup-to-disc ratio determination using retinal fundus images in argali. In Proceedings of the 2008 30th Annual International Conference of the IEEE Engineering in Medicine and Biology Society, Vancouver, BC, Canada, 20–25 August 2008; pp. 2266–2269. [CrossRef]
19. Nath, M.K.; Dandapat, S. Techniques of Glaucoma Detection from Color Fundus Images: A Review. *Int. J. Image Graph. Signal Process.* **2012**, *4*, 44–51. [CrossRef]
20. Cheng, J.; Liu, J.; Xu, Y.; Yin, F.; Wong, D.W.K.; Tan, N.M.; Tao, D.; Cheng, C.Y.; Aung, T.; Wong, T.Y. Superpixel classification based optic disc and optic cup segmentation for glaucoma screening. *IEEE Trans. Med. Imaging* **2013**, *32*, 1019–1032. [CrossRef] [PubMed]
21. Diaz, A.; Morales, S.; Naranjo, V.; Alcoceryz, P.; Lanzagortayz, A. Glaucoma diagnosis by means of optic cup feature analysis in color fundus images. In Proceedings of the 2016 24th European Signal Processing Conference (EUSIPCO), Budapest, Hungary, 29 August–2 September 2016; pp. 2055–2059. [CrossRef]
22. Tajbakhsh, N.; Shin, J.Y.; Gurudu, S.R.; Hurst, R.T.; Kendall, C.B.; Gotway, M.B.; Liang, J. Convolutional Neural Networks for Medical Image Analysis: Full Training or Fine Tuning? *IEEE Trans. Med. Imaging* **2017**, *35*, 1299–1312. [CrossRef] [PubMed]
23. Pashaei, M.; Kamangir, H.; Starek, M.J.; Tissot, P. Review and evaluation of deep learning architectures for efficient land cover mapping with UAS hyper-spatial imagery: A case study over a wetland. *Remote Sens.* **2020**, *12*, 959. [CrossRef]
24. Szegedy, C.; Ioffe, S.; Vanhoucke, V.; Alemi, A.A. Inception-v4, inception-ResNet and the impact of residual connections on learning. In Proceedings of the Thirty-first AAAI Conference on Artificial Intelligence, San Francisco, CA, USA, 4–9 February 2017; pp. 4278–4284.

25. Chollet, F. Xception: Deep learning with depthwise separable convolutions. In Proceedings of the IEEE Conference on Computer Vision and Pattern Recognition 2017, Honolulu, HI, USA, 21–26 July 2017; pp. 1800–1807. [CrossRef]
26. Carranza-García, M.; Torres-Mateo, J.; Lara-Benítez, P.; García-Gutiérrez, J. On the performance of one-stage and two-stage object detectors in autonomous vehicles using camera data. *Remote Sens.* **2021**, *13*, 89. [CrossRef]

sensors MDPI

Article

Multi-Device Nutrition Control

Carlos A. S. Cunha *,† and Rui P. Duarte †

CISeD—Research Centre in Digital Services, Polytechnic Institute of Viseu, 3504-510 Viseu, Portugal; pduarte@estgv.ipv.pt
* Correspondence: cacunha@estgv.ipv.pt
† Current address: Department of Informatics, School of Technology and Management, Campus de Repeses, 3504-510 Viseu, Portugal.

Abstract: Precision nutrition is a popular eHealth topic among several groups, such as athletes, people with dementia, rare diseases, diabetes, and overweight. Its implementation demands tight nutrition control, starting with nutritionists who build up food plans for specific groups or individuals. Each person then follows the food plan by preparing meals and logging all food and water intake. However, the discipline demanded to follow food plans and log food intake results in high dropout rates. This article presents the concepts, requirements, and architecture of a solution that assists the nutritionist in building up and revising food plans and the user following them. It does so by minimizing human–computer interaction by integrating the nutritionist and user systems and introducing off-the-shelf IoT devices in the system, such as temperature sensors, smartwatches, smartphones, and smart bottles. An interaction time analysis using the keystroke-level model provides a baseline for comparison in future work addressing both the use of machine learning and IoT devices to reduce the interaction effort of users.

Keywords: precision nutrition; food plans; IoT; machine learning; food logging

Citation: Cunha, C.A.S.; Duarte, R.P. Multi-Device Nutrition Control. *Sensors* **2022**, *22*, 2617. https://doi.org/10.3390/s22072617

Academic Editor: Ivan Miguel Serrano Pires

Received: 25 February 2022
Accepted: 21 March 2022
Published: 29 March 2022

1. Introduction

Disease caused by inappropriate diets is responsible for 11 million deaths and hundreds of millions of disability-adjusted life years [1]. The use of technology to support health (eHealth) opens an expansive landscape of opportunities. The emergence of a large set of smart devices capable of facilitating physiological data recording and other forms of recording the health status has potentiated many new eHealth applications. Mobile phones and smartwatches are among the devices with the most potential because of their ubiquity and sensor capabilities installed [2–4].

The importance of nutrition to health is unquestionable. However, the specificity of nutritional requirements for a person demands personalized nutrition control. Nutritional requirements lean on body parameters, genetic and epigenetic makeup, daily routines, and history of disease or allergies. Thus, health professionals (e.g., doctors and nutritionists) must intervene to keep food plans adequate for the target person. Nonetheless, the biggest challenge is not elaborating the food plan but instead is the follow-up. That includes keeping the food plan always present to the user, replacing unavailable or undesired foods, adjusting food quantities to exceptional energy consumption, and using logged food intake data to readjust future food plan revisions. Food intake logging, in particular, benefits from automation since it is time-consuming, and the discipline demanded by its operationalization leads to high dropout rates of food plan execution.

State-of-the-art approaches for automation food intake logging exploit the recognition of food and quantities in images [5,6] taken using the phone camera and unconventional intrusive devices to detect swallowing patterns associated with calories intake [7]. Notwithstanding the innovation inherent to these approaches, they suffer from measurement errors summing to the error introduced by food tables to quantify nutrients. Plus, these solutions still require some interaction (e.g., opening the application and taking pictures). A more realistic solution to reduce human interaction costs is integrating the nutritionist and user systems and resorting to off-the-shelf smart devices.

Smart devices are essential tools to enable the ubiquity of food plans by allowing their visualization anywhere. Plus, they act as a data-gathering mechanism for logging macronutrients, micronutrients, and hydration levels. These data feed into a nutritional model that can support the nutritionist (or other health professionals) adjusting the next food plan iteration.

This article presents the requirements and concepts of a solution covering the food plan life-cycle from its creation by the nutritionist to its visualization, adaptation, and logging of food intake by the person. It also discusses the system architecture and design by focusing on

- Devices for food plan creation, visualization, and food logging (smartphones, smartwatches, and smart bottles).
- Devices for capturing relevant data for food plan adaption (e.g., energy consumption).
- Data integration mechanism.

The rest of this article is organized as follows. Section 2 presents the related work. Section 3 defines the problem addressed in this article and enumerates the requirements of a possible solution. Section 4 presents the concepts and formulas used in food plan creation. Section 5 describes the system architecture and implementation. Sections 6 and 7 describe the scenarios where the system will be tested. Finally, Section 9 presents the conclusions.

2. Related Work

This paper addresses a multidisciplinary problem connecting several research areas, such as precision nutrition, Internet of Things (IoT), web technologies, and machine learning.

Precision nutrition is an eHealth research area that depends on the person's characteristics to deliver nutritional advice [8]. One prominent research topic in this area is when advice is supported by machine learning models created from several sources of data—e.g., dietary intake (content and time), personal, genetics, nutrigenomics, activity tracking, metabolomics, and anthropometric. Food intake monitoring, in particular, provides a fundamental source of data to machine learning algorithms for creating adequate diet models. However, traditional food logging systems are intrusive, forcing users to change their routines. Hence, user interaction with the system makes this activity one of the main contributors to food plan execution dropouts.

Several approaches for automatic food intake logging have been proposed. Wearables are devices with high potential in healthcare [9], since they could automate the process of food intake logging. The results of their exploratory use in nutrition to reduce the burden of manual food intake logging are presented in [7]. The authors explored using a smart necklace that monitors vibrations in the neck and a throat microphone to classify eaten food into three food categories. The resultant models trained with data produced by these wearables revealed higher accuracy for the microphone when compared to the vibrations sensor. Notwithstanding the potential of wearables for automatic logging of food intake, they are still in their infancy, requiring development to reduce intrusiveness and achieve close to perfect accuracy.

Visual-based dietary assessment approaches represent another type of appealing solution that resorts to pictures to determine the intake of food nutrients. Lo et al. [5] explores deep learning view synthesis for the dietary assessment using images from any viewing angle and position. An unsupervised segmentation method identifies the food item, and a 3D image reconstruction estimates the portion size of food items. Despite the high accuracy of the approach, the results depend on depth images with separable and straightforward objects, notwithstanding typical dishes that may overlap several food items. Another work estimates food energy based on images using the generative adversarial network (GAN) architecture [6]. It resorts to a training-based system, which contrasts with approaches based on predefined geometric models which bound the evolution of models to food with known shapes. The authors' approach provides visualization of how food energy estimation is spatially distributed across the image, enabling spatial error evaluation.

While visual food inference represents a promising research topic for automatic logging of food intake, its accuracy is still unacceptable for most applications. An alternative method for food logging is using speech-to-text conversion to reduce the user's interaction effort required to introduce nutrient information in the software application. Speech2Health [10] allows recording of food intake through natural language. A user-acceptance study using Speech2Health has shown several advantages of a speech-based approach over text-based or image-based food intake recording. Nevertheless, even minor errors resulting from identifying food names and portion sizes from voice excerpts are unacceptable for generic use. Privacy represents another issue that speech-to-text introduces in public environments.

Most related work addresses the problem of automatic food intake monitoring. Instead of explicitly addressing that problem, we devised a holistic approach that depends on food plans created by nutritionists and followed by target users. By confirming meals or logging changes, these users produce data for feeding the feedback loop that approximates the food plan progressively to the actual user's needs. The availability of a baseline plan and the use of intelligent devices to record hydration, temperature, and energy expenditure reduce user interaction effort. Additionally, machine learning is applied to user preferences modeling, helping nutritionists choose the best food for the plan.

3. Problem Definition

Nutrition is a topic that has received more attention in the last decades due to its potential for benefiting from advances in technology. The ubiquity of smartphones and the emergence of wearable devices has created the opportunity to gather data automatically and support the user in deciding the best food to eat at each meal.

Many smartphone apps provide features to log intake meals and present nutritional statistics. However, choosing the best food plan for an individual requires a professional analysis that considers their physical condition (e.g., fat mass, lean mass, and weight), clinical condition, and goals. Discarding the health practitioner from the process may lead to inadequate food plans and be dangerous for individuals with health issues. Fortunately, it is possible to use technology to reduce the manual effort needed to manage the food plan life-cycle. The problems solved by a holistic solution span over the nutritionist and user (person following the food plan) domains.

We specified the requirements for the user and nutritionist domains with the support of several experts, such as nutritionists and doctors from a private hospital. We scheduled several meetings with these experts in two different phases: (1) requirement analysis, with the support of high-definition interface prototypes, and (2) deliverable analysis, where we tested software increments within a limited group of people by creating appointments, food plans, and performing food logging. Appendices A.1 and A.2 in Appendix A present the use cases for each of these domains.

3.1. Nutritionist Domain

We identified the following requirements for the nutritionist domain:

1. The nutritionist creates an appointment with the person's data and all the parameters needed to obtain the nutrients required to build the food plan. The system should calculate the energy expenditure.
2. The nutritionist creates the food plan aligned with the nutrition goals obtained from the appointment. The system should suggest food according to user preferences and goals.

Nutritionists gather several types of data in the course of the appointment, which allows determining the person energy expenditure (Section 4) and other metrics and goals that can further support decisions during food-plan-making.

Energy expenditure is the core metric for devising the food plan. It provides the calories further distributed between macronutrients (i.e., proteins, carbs, and lipids) as follows:

$$\text{energy} = \alpha * \text{protein} + \gamma * \text{carbs} + \beta * \text{lipids} \quad (1)$$

After providing the data required to calculate the energy expenditure to the system, the nutritionist defines values for α, γ, and β. These values represent the contribution ratio of each macronutrient to the energy expenditure, which is fixed to a specific day and distributed between meals.

The energy expenditure and its distribution between macronutrients and meals are dependent on the person's profile. For example, athletes have an increased demand for energy compared with sedentary people, and distribution of nutrients needs to be adapted to specific days (e.g., carbohydrate intake before and after exercise to help restore suboptimal glycogen reserves).

Fiber, water, and micronutrients are essential food plan elements unrelated to energy expenditure. The nutritionist adjusts the quantity of each nutrient to the person's goals and condition. For instance, during demanding physical activity, the person may need drinks with added sodium to replace electrolyte losses. On the other side, a person with the risk of high blood pressure would benefit from lowering sodium intake.

Food plan creation is time-consuming because it involves the combination of different types of food adequate to the person. That combination should fulfill the target energy expenditure and its distribution between macronutrients, and approximate the micronutrients specified for the food plan. As for selecting alternative food when the user follows the plan (user domain), the user preferences model also supports the nutritionist in choosing the food to be added to the plan. Here, the contribution of each food to the goals established for energy, macronutrients, and micronutrients represents a crucial input for the classifier.

The nutritionist needs to revise the food plan to adjust the energy and nutrients to the user goals, respecting the subsequent appointments. For example, suppose the user goal is not to reduce fat mass but increase muscle instead. In that case, the total energy intake specified for the plan must be reduced and, consequently, the proportion of macronutrients contributing to that energy. Since energy expenditure occupies the top of the energy breakdown hierarchy, it will drive food plan adaption according to data gathered during previous food plan executions. Smart devices may improve the accuracy of energy expenditure in further food plan revisions. The *physical activity energy expenditure* (Section 4.2) represents one component of energy expenditure that can be easily captured with acceptable accuracy by smartwatches (or fit bands), alone or combined with heart rate straps. These data combined with food and water intake logs—registered through the system interface or obtained through intelligent bottles—provide elements required to tune the successive food plan revisions.

3.2. User Domain

We identified the following requirements for the user domain:

1. The person accesses the meals defined in the food plan for the current day or specific event using the mobile phone or smartwatch.
2. The person confirms the ingestion of the meal as it is in the food plan.
3. The person searches for alternatives to the current meal with equivalent nutrition characteristics aligned with their preferences model.
4. The person logs other food eaten not present in the food plan.
5. The smart bottle logs water ingestion with respect to a specific period.
6. The smartwatch logs the calories spent by the person during the day with respect to physical activity.
7. All logs are either associated with a specific day or to an event (e.g., sports practice).

Nutritionists must design food plans aligned with user conditions and preferences. Further, users demand ubiquitous food plan visualization and logging mechanisms with small interaction costs. While interaction efforts depend heavily on user interface design, off-the-shelf IoT devices can be valuable tools to reduce human interaction with the system. These devices may be balanced with efficient user interfaces to reduce food plan execution abandonment.

As food plans are fixed to days of the week, repeating for several weeks, users may often lack some ingredients when executing the plan. Hence, the system may suggest alternative food according to the nutritional equivalence and user preferences—using historical data for similar meals, days of the week, months, or even weather contexts.

3.3. Automation Limits

The number of interactions with the system and the individual interaction cost determine the total user interaction effort. Logging of meals intake as in the food plan requires a small interaction effort since the only input is the user confirmation in either the smartphone or smartwatch. Sometimes that happens in batches (e.g., by the end of the day), resulting in low interaction costs and a small number of interactions (one per meal), as presented in Table 1. In this scenario, the user domain can benefit from integrating the food plan built by the nutritionist with the smartphone application that allows its visualization and confirmation of intake meals.

Water intake logging demands a higher number of user interactions when compared with meal confirmation. The user may take a sip of water dozens or hundreds of times a day to be hydrated. Consequently, water intake logging is more complex unless they stick to a standard behavior, such as drinking from the same bottle and logging the bottle storage capacity when they finish. However, even that standard method has flaws because the user may never finish the last bottle refill during the day or replace it with new water. Smart bottles may potentially reduce the number of user interactions for water intake logging since all the logged water intake is sent to the cloud service and made accessible to our system without user interaction.

While the previous scenarios offer an automation opportunity, some actions are difficult to automate, such as logging food not registered in the food plan. As shown in Table 1, notwithstanding the small number of interactions during the day, the interaction cost of individual actions is high—justified mainly by the search for additional food and the introduction of respective quantities. In addition, their automation is complex, and the closest state-of-the-art approaches rely on machine learning to identify food in pictures taken using the phone. However, these approaches are still far from one hundred percent of accuracy, which leads to large errors summed from

- Errors resultant from the identification of food objects;
- Errors inherent to values presented in food nutrient composition tables;
- Food quantification errors, introduced either by visual approximation or predicted from the picture.

Reduction of interaction costs with respect to activities with low automation potential needs to be handled at the interface design level. The user application interface should be optimized to reduce the effort of food-searching for the *changing meal* and *add extra food* actions.

Table 1. Interaction effort of main actions for each device.

Action	Smartphone		Smartwatch		Smart Bottle	
	Interaction Cost	Number of Interactions	Interaction Cost	Number of Interactions	Interaction Cost	Number of Interactions
Meal confirmation	low	low	low	low	n/a	n/a
Changing meal	high	low	n/a	n/a	n/a	n/a
Add extra food	high	low	n/a	n/a	n/a	n/a
Water logging	low	high	low	high	none	none

4. Energy Expenditure

The user energy expenditure drives the creation of food plans. Adequate diets approximate intaken calories to the total energy expenditure, which includes the resting energy expenditure (REE), physical activity energy expenditure (PEE), and thermic effect of food (TEF).

CB (caloric balance) in the human body approximates the CC (caloric consumption) to the sum of PEE, REE, and TEF.

$$CB = CC - PEE - REE - TEF \quad (2)$$

This section presents the calculation of PEE and REE. Notwithstanding the low contribution of the TEF (between 3% and 10%) to the total energy expenditure (TEE), it may have an impact on obesity. However, we do not handle it in this article due to its high measurement complexity [11] created by dependency on several other variables (e.g., measurement duration) [12].

4.1. Resting Energy Expenditure

REE is considered equivalent to the basal metabolic rate (BMR). BMR is the minimum number of calories required for basic functions at rest. On the other side, RMR is the number of calories our body burns while at rest. Despite both definitions slightly differing, the Harris–Benedict equation [13,14] can approximate REE or other equivalent equations presented in Table 2 for calculation of BMR.

4.2. Physical Activity Energy Expenditure

PEE calculation involves converting metabolic equivalents of activities to calories expended per minute (cal/min), based on body weight and the varying exercise intensities. The physical activity level (PAL) is an inexpensive and accurate method for calculation of PEE, based on the average values of 24 h of TEE and REE, as follows:

$$PAL = TEE/REE \quad (3)$$

The effect of gender does not interfere with PAL calculation because the BMR absorbs the gender difference in energy needs accentuated by the heavier weight of men.

A table that associates physical intensity lifestyles to PAL values (Table 3) can simplify PAL calculation. In that context, TEE is the result of multiplying REE by the PAL value associated with the person's lifestyle category [15].

Another method for PAL calculation combines the time allocated to habitual activities and the energy cost of those activities (Table 4). In this case, PAL represents an energy requirement expressed as a multiple of 24-hour physical activity ratio (PAR). Here, PAR is a factor of BMR (PAR is 1 when there is no energy requirement above REE). Intuitively, the energy cost (PAR) is multiplied by the activity time to obtain PAL [15,16].

4.3. Distribution of Nutrients

The TEE estimate represents the total calories in the food selected for the food plan. TEE is then broken down into macronutrients complemented with micronutrients.

Macronutrients are typically specified in grams per kilo of body weight; such is the case of protein, carbohydrates, and fat (lipids). The exception is fibers that are specified in total grams. Water is frequently classified also as being a macronutrient [17]. However, water and fiber have zero calories, unlike protein, fat, and carbs. Notwithstanding that fibers do not usually count as calories in food plans, one type of fiber, named soluble fiber [18], may be absorbed by the organism and thus provide the body with calories.

Compared with macronutrients, the number of micronutrients is vast, and for that reason, nutritionists only select a few to be used as control metrics during food plan creation. From the conversation with several nutritionists, we have chosen iron, calcium, sodium, and magnesium, because of their transversality over several population groups. However, the selection of micronutrients depends always on the target population group (e.g., elderly, young people, and athletes).

Table 2. Metrics provided by the user (rows) and calculated by the application (columns).

		Body Composition			Basal Metabolic Rate				Obesity		
		Muscle Mass (Lee)	**Fat-Free Mass**	**AEC Rate**	**Harris-Benedict [19]**	**Mifflin-St Jeor [20]**	**Katch-McArdle [21]**	**Cunningham [22]**	**Body Mass Index [23]**	**Evans 3SKF [24]**	**Withers [25]**
body composition	weight		✓		✓	✓	(muscle mass)	(muscle mass)	✓		
	height	✓			✓	✓			✓		
	fat mass		✓								
	skeletal muscle										
	bone mass										
	body cell mass										
	bone mineral content										
	intracellular water			✓							
	extracellular water			✓							
	gender	✓			✓	✓				✓	
	age	✓			✓	✓					
	race	✓								✓	
girths	tight	✓									
	calf	✓									
	relaxed biceps										
	contracted biceps										
	waist										
	gluteus										
	chest										
	crural										

Table 2. *Cont.*

		Body Composition			Basal Metabolic Rate				Obesity		
		Muscle Mass (Lee)	**Fat-Free Mass**	**AEC Rate**	**Harris-Benedict [19]**	**Mifflin-St Jeor [20]**	**Katch-McArdle [21]**	**Cunningham [22]**	**Body Mass Index [23]**	**Evans 3SKF [24]**	**Withers [25]**
lean mass segments	left/right arm										
	trunk										
	left/right leg										
fat mass segments	left/right arm										
	trunk										
	left/right leg										
skinfold	corrected upper arm	✓									
	calf										✓
	biceps										✓
	triceps									✓	✓
	supraspinal										✓
	subscapular										✓
	chest										
	axila										
	iliac crest										
	abdomen									✓	✓
	thigh									✓	✓
level of fat	visceral fat										

Table 3. Classification of lifestyles according to physical intensity (PAL values).

Category	PAL
Sedentary or light activity lifestyle	1.40–1.69
Active or moderately active lifestyle	1.70–1.99
Vigorous or vigorously active lifestyle	2.00–2.40

Table 4. Total energy expenditure for a population group.

Activities	Time Allocation	PAR	Time × PAR	Mean PAL
Sleeping	6	1.0	6.0	
Personal Care (dressing, showering)	2	2.3	4.6	
Eating	2	1.5	3.0	
Walking without a load	2	3.2	6.4	
Sitting	4	1.5	6.0	
Cooking	2	2.1	4.2	
Household work	2	2.8	5.6	
Light leisure activities	2	1.4	2.8	
Driving car	2	2.0	4.0	
Total	**24**		**42.6**	**42.6/24 = 1.8**

5. Architecture and Implementation

This section presents the architecture and implementation of the solution proposed in this article, divided between two front-ends: *nutritionist front-end* and *user front-end*.

5.1. Nutritionist Front-End

The nutritionist front-end (Figure 1) implements two important concepts: *appointment* and *food plan*.

The appointment is the concept responsible for managing the energy expenditure—and its distribution throughout macronutrients—and micronutrients, as presented in Section 4. Moreover, to support user monitoring between appointments, it should present all historical data entailing previous food plans and energy distribution by day of the week, event, and meal type.

Monitoring of physical conditions frequently resorts to the person's goals, specified in terms of:

- Weight.
- Body fat.
- Visceral fat.
- Fat-free mass.
- Muscle mass.
- Body mass index.
- Exercise performance.

Control and analysis of generic user goals depend on the previous metrics, although specific people groups may require other specific metrics; such is the case of groups with specific diseases that require the control of specific body parameters.

Other important appointment data required for food plan making include the following:

- Bowel function.
- Sleep quality, and wake up and sleeping times.

- Person's race.
- Food likes and dislikes.
- Night shifts.
- Job.
- Lifestyle.
- Clinical conditions.
- Current water intake.

Nutritionists rely on the appointment of data for food plan creation. While adding new meals and foods to the food plan, the nutritionist can balance the food calories with target energy and nutrients. They can also visualize other relevant information gathered during the elaboration of appointments.

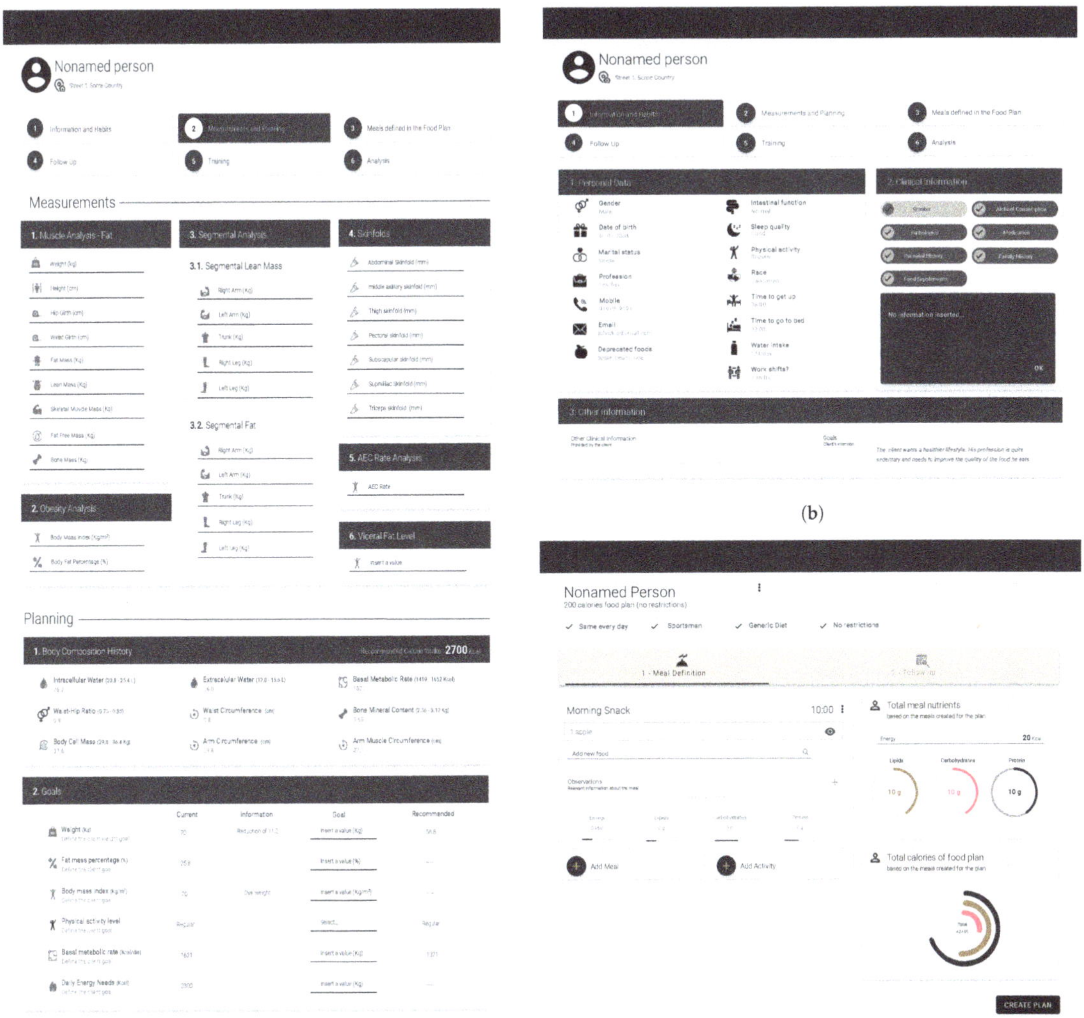

(b)

(a) (c)

Figure 1. Nutritionist front-end. (**a**) Appointment. (**b**) Client details. (**c**) Food plan.

5.2. User Front-End

The user front-end (Figure 2a) uses the food plan as the basis for preparing meals, searching for alternative foods, monitoring consumption of water and calories during the day, and food logging. Food is presented on the plate (Figure 2b)—useful for elderly, people with vision impairment, or those that may find it difficult using mobile/smartphones with mobile devices—and in the list format.

Daily statistics (Figure 2c) are valuable assets for monitoring calories, macronutrients, micronutrients, and hydration during the day. These values are paired with target values defined by the nutritionist in the food plan.

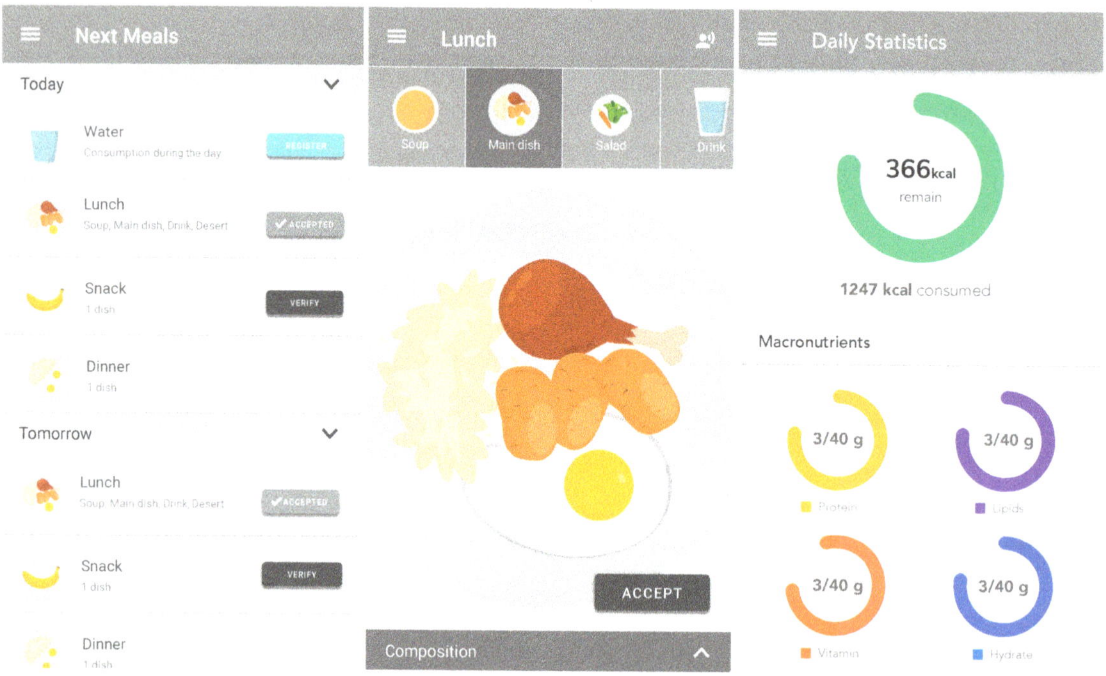

(a) (b) (c)

Figure 2. User front-end. (**a**) Daily meals. (**b**) Meal visualization. (**c**) Daily statistics.

Notwithstanding the small screen sizes of smartwatches, they are practical for presenting meals (Figure 3a), sending notifications, and logging food intake. They also present statistics regarding nutrients intake (Figure 3b) and hydration (Figure 3c).

5.3. Architecture

Figure 4 presents the solution architecture composed of four different interfaces. The nutritionist interacts with the system to create appointments and food plans using a web application. On the other side, the user visualizes the current food plan or logs food ingestion using a mobile phone or smartwatch.

5.3.1. Web Applications

The mobile application is delivered as a PWA (progressive web application). PWAs represent a new class of applications alternative to traditional mobile phone apps, with several advantages over them. Instead of being developed to a specific platform (e.g., iOS or Android), they are built as a web application that can work offline and be installed on any smartphone. A previous study reported PWAs 157 times smaller than React Native-based interpreted apps and 43 times smaller than Ionic hybrid apps [26]. The Twitter PWA consumes less than 3% of the device storage space as compared to Twitter for Android [27], and the Ola PWA is 300 times smaller than their Android app [28]. Additionally, they are cross-platform, although current implementations may require adaptation between some browsers.

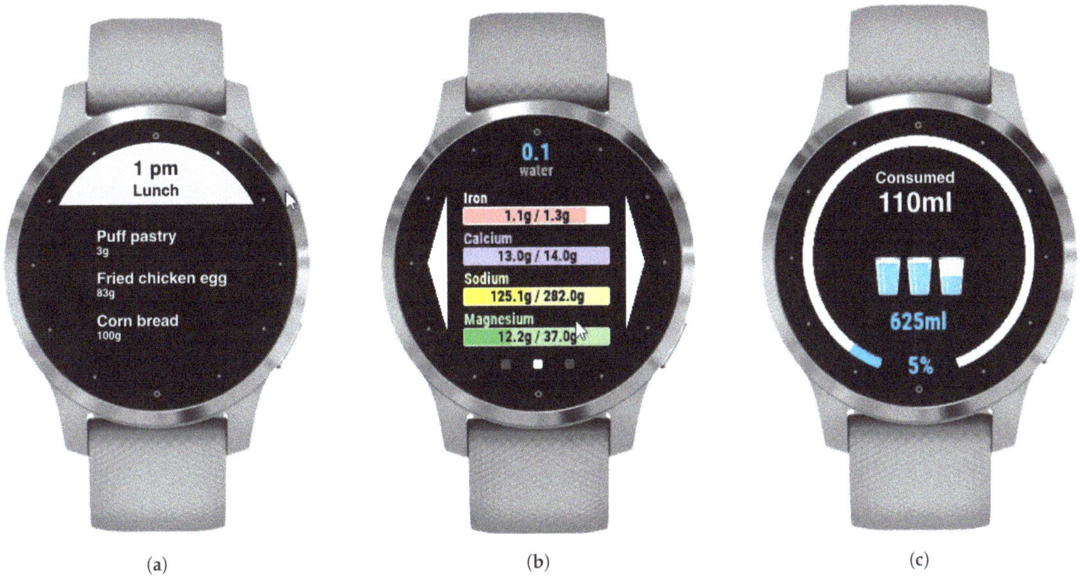

Figure 3. Smartwatch. (**a**) Food plan visualization and logging. (**b**) Daily control of nutrients. (**c**) Daily control of water.

Both applications with respect to the user and nutritionist front-ends were developed in LitElement [29], a base class to create lightweight web components. Design of the user front-end for smartphones embraces the PWA principles [30] (e.g., web application installability, and offline usage).

5.3.2. Smart Bottle

Water consumption is logged either by the user—using the smartphone or smartwatch—or automatically by a smart bottle. We tested several smart bottles and decided on the Hidratespark [31], justified by its mature API and good construction and usability of the bottle. Plus, it can be easily integrated with Fitbit [32], which is used as a gateway to retrieve data to the user's back-end.

Water intake goals defined in the food plan are adjusted according to the environment temperature. Temperature sensors provide the inputs to make that adjustment according to the rules stated in the food plan.

5.3.3. Smartwatch

As explained in Section 4, determining the energy expenditure of one person is one of the main challenges in the creation of a food plan. Modern smartwatches provide a good approximation of energy consumption during physical activity. They provide valuable information to be used by food plan revision activities, enabling correction of energy expenditure values predicted by traditional methods during follow-up appointments (Section 4.2). Pedometers and heartbeat monitors incorporated in devices provide a good approximation of calories burned data [33].

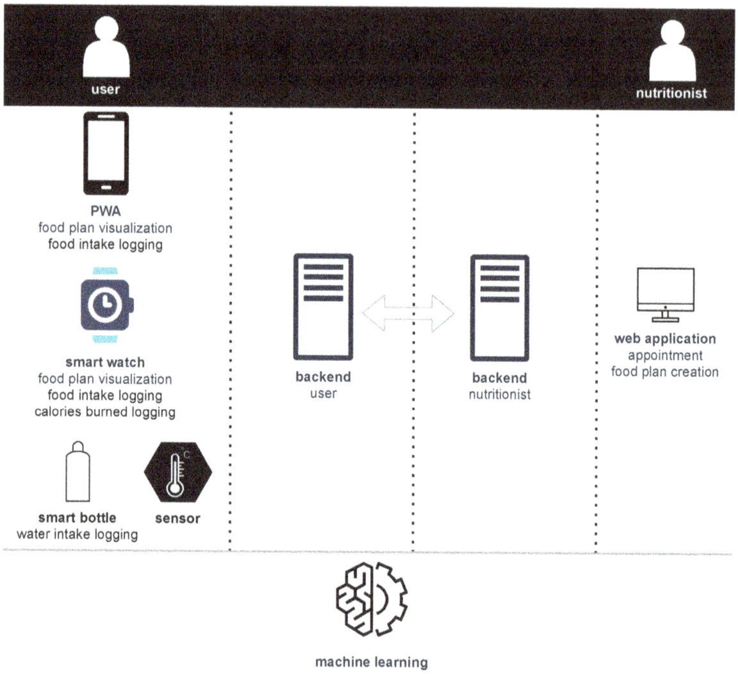

Figure 4. Architecture.

5.3.4. Preference Learning

Exploring machine learning techniques on logged data makes it possible to help nutritionists model user food preferences. These techniques build up a recommendation system [34], based on food preference models, that supports the selection of food during food plan creation. That system will also allow proposing food alternatives to the person following the plan. That may occur when the food is unavailable, or the person prefers other equivalent food.

Reinforcement learning seems an adequate tool for applying preference learning to food recommendation [35]. Starting without knowledge, the agent helps the nutritionist to choose the food and quantity for the food plan without breaking the constraints imposed by the goals established for macronutrients and micronutrients. The agent accuracy improves with the feedback received from the nutritionist and the intake of food logged by the user. The same agent can help the user choose equivalent food and quantities when executing the plan based on learned preferences and goals of nutrients.

5.4. Security

Security is a complex and wideband problem. It spans the human-related processes and the system level (e.g., network and application). Human misconduct is in the origin of several security threats in eHealth systems [36]. Training people and auditing security procedures is a natural way of reducing the risk of threats occurrence. Coordination between developers, users, organizations, and government regulators represents another security flaw source [37].

In this work, we handle security at the system design level. E-Health systems contain data that are sensitive to confidentiality, integrity, and availability threats [38]. There are different types of data sensitiveness. Personal data are the most critical data under management; thus, ensuring the confidentiality of these data is of the utmost importance. Hence, we segregate the user data in the application and provide one feature to remove these data anytime without compromising their food plan while an anonymous entity. The latter offers less security risk when unrelated to the person.

The design of the nutritionist application allows deletion of the user's personal data without compromising the food plan management features, as long as an *ID* can identify the user. The segregation of functionality and data between the user and nutritionist applications offers an additional protective barrier. The user application uses an application token to communicate with the nutritionist application, and the former does not store or handle personal data—an ID identifies the user.

As much as personal data, authentication credentials are sensitive data demanding theft protection. The HTTPS already ensures protocol-level privacy in the communication channel. Plus, the front-end encrypts passwords before transmitting them to the back-end, and they are then handled and stored in an encrypted form.

Feature-oriented access control constrains the access to features available on each web page. There are three profile types: nutritionists, administrators, and users.

Risk management models, such as the one presented in [39], may complement our system design. Additionally, other protection schemes against complex attacks [40] are orthogonal to our system and may also be used.

6. Case Study: Alzheimer's

Alzheimer's disease is a progressive loss of mental function, characterized by degeneration of brain tissue, including loss of nerve cells, accumulation of an abnormal protein, and development of neurofibrillary braids [41]. Alzheimer's patients become dependent on others, even for the most basic tasks. Controlling feeding and hydrating for an Alzheimer's patient is thus a crucial activity performed by the person who supports their daily routine, called the informal caregiver (IC).

Conditions of malnutrition, super nutrition, and dehydration are common in people with diseases causing dementia. The loss of autonomy also manifests itself in their inability to demonstrate food needs. Therefore, it is fundamental to support the nutritionist in the preparation and follow-up of a food plan aligned with the patient's needs. Food plan monitoring is undoubtedly a process that demands much discipline from the IC and the ability to deal with possible circumstantial adaptations, such as replacing foods prescribed in the food plan with other equivalents or changing the quantity of water consumed as a function of ambient temperature.

This case study investigates the problem of creating and monitoring diet plans in patients with dementia—such as those with Alzheimer's. It allows the creation of nutritional plans by the nutritionists and the follow-up of these plans by the ICs through a mobile app to significantly increase the patient's quality of life. The app will send the IC notifications regarding proper nutrition and hydration in the due moment. It also controls hydration using the smart water bottle. In addition, the application will suggest alternatives to plan foods if they are unavailable or rejected by the patient. Another feature important for this group is the dynamic adaptation of water administration to the patient as a function of environmental conditions observed by temperature and humidity sensors. This feature is vital when the patient is unable to express thirstiness.

7. Case Study: Sports

The recent growth in the pursuit of sporting activities, motivated by a widespread increase in the perception of the importance of maintaining physical fitness, campaigns explicitly aimed at combating physical inactivity, and opportunities created by the revelation of lesser-known modalities, has brought forward fundamental questions such as the correct nutrition of the practitioners. Several institutions and individuals involved in physical activity have integrated these concerns into their scope, including nutritionists.

Food plan elaboration and monitoring present two main challenges: (1) obtaining the person's biometric data, eating habits, and energy consumption, and (2) monitoring user food intake and providing dynamic adaptation of the food plan.

Sports nutrition is one of the most complex areas of nutrition. It requires observing a comprehensive set of metrics, encompassing the athlete's physical aspects, physical activity, and eating habits. Fortunately, devices for measuring specific physical parameters represent a common practice among athletes. The creation of data repositories to help nutritionists build the plan is only possible by automatically integrating data collected by these devices with other data not directly observable—such as dietary habits and subjective metrics. These repositories also contain data that can help adapt the food plan at its execution stage. For example, variations in temperature or physical intensity may demand quick changes in individual energy or hydration needs. In these scenarios, the support system uses data collected by devices to dynamically adjust the food plan and send alerts to athletes to eat food or water at the right time.

8. Interaction Results

This section presents the human–computer interaction cost associated with typical user tasks to visualize the food plan and log food intake.

Traditional methods used in the usability evaluation of an interface fall into two categories: (1) subjective opinion of users and experts—mainly applying questionnaires [42] and inspection methods [43,44]—and (2) objective techniques such as rules [45], analytics modeling [46], and automated testing [47,48]. Notwithstanding that these approaches provide important tools to determine the usability of the user interface, there is both cost and time needed to implement user interaction evaluation with acceptable coverage, coupled with the need to use experts to compensate for the user's faults.

8.1. Keystroke-Level Model

We applied the keystroke-level model (KLM) [49] to the user interface depicted in Figure 2, for testing the quality of the human–computer interaction and estimating the time spent in critical tasks. In this model, a unit task is defined with two parts: *task acquisition* and *task execution*. The total time to complete a unit task is given by $T_{task} = T_{acquire} + T_{execute}$.

At the execution level, KLM provides physical, mental, and response operators with predefined time values. These operators are defined by a letter and include K (keystroke ≈ 0.12 s), P (point ≈ 1.1 s), H (homing the hand(s) on the keyboard or other device ≈ 0.4 s), D (draw is measured in real time), B (button press ≈ 0.1 s), M (mental preparation for action ≈ 1.35 s), and R (system response, which is a parameter measured in real time). The execution time is the sum of the time for each of the operators from the final KLM string $T_{execute} = T_K + T_P + T_H + T_D + T_B + T_M + T_R$.

8.2. Interaction Results

Table 5 presents the time required to execute each application task. The KLM string generated is represented in the *sequence of operators* column and the respective time required to execute each task in the *estimated time* column. The task *"update food entries for train and competition"* allows the creation of periodic food requirements and is specific to the sports scenario. In contrast, the Alzheimer's and the sports scenarios share the other tasks. The results are presented for the user application since we aim to reduce user abandonment motivated by interaction costs resultant from food logging activities.

As expected, results show that tasks that change the original food plan for logging purposes manifest higher interaction costs. Food plan visualization requires 1.2 or 2.3 s, depending on the UI view. Logging one meal by confirming the original food plan only requires 1.2 s. On the other hand, logging tasks regarding food intake not present in the food plan are costly. Each extra food added to the food plan requires 8.66 s of the user's time.

Manual logging of water using the application requires 3.6 or 4.8 s, depending on the view. The adoption of smart bottles avoids that interaction, which may repeat dozens of times during the day.

The interaction time of tasks performed by the smartwatch (e.g., energy expenditure logging) is not presented in this section. Despite the automation of the data logging process, the user can not perform any equivalent task manually.

Table 5. Interaction results.

Actor	Tasks	Sub-Tasks	Sequence of Operators	Estimated Time (s)
User	Visualize food plan	Graphical representation of the meal	PB	1.20
		Composition of the meal (by food)	PBP	2.30
User	Log food intake (items of the sub-tasks column marked with * repeat in the meal)	Add new food to the meal *	PBPBMHKKKMHPBPB	8.66
		Add new extra food (snack between meals) *	PBPBMHKKKMHPBPB	8.66
		Remove food *	PBPBPB	3.60
		Specify percentage of food intake *	PBPBPBPB	4.80
		Change food plan food *	PBPBPB	3.60
		Confirm food intake from food plan with no changes *	PB	1.20
User	Log water intake	Through food plan	PBPBPB	3.60
		Through interaction menu	PBPBPBPB	4.80
Fitbit (bottle)	Update water intake		-	0.00
User	Visualize statistics		PBPB	1.20
System	Update food entries for train and competition		-	0.00
User	Change active food plan (train or competition)		PBPBPBPB	4.80
User	Connect watch API		PBPBPBMH42KMHPB	13.34
User	Provide consent to access Fitbit API		PBPBPBR	4.60

8.3. Analysis of Results

The observed results of human–computer interaction times pinpointed the tasks requiring improvement of interaction times. They provide a baseline for evaluating other interaction schemes and assessing the contribution of automation (e.g., using IoT devices) to the goals established in this article. The lower the interaction time, the lower the user discipline needed to maintain a food plan visualization and logging process, and the lower the user abandonment rate.

We designed the application to minimize human interaction with the support of UI experts. The most challenging tasks using a UI (those with more significant interaction times) require the search of new food manually. Although the interface implementation can still be questionable in terms of the specific design that may compromise the generalization of results, it is evident that there is little space for improvement when we need to perform a generic search for food using text.

Machine learning techniques are natural solutions to help reduce the time required for logging extra food in addition to—or in replacement of—those present in the food plan. As referred to in Section 2, there have been several attempts to recognize food objects in pictures taken with the mobile phone to reduce the burden of manually logging food. However, interaction is still required to take the picture, and an accuracy less than perfect could even increase the interaction time since the user would need to correct these data. The previous rationale leads to a different strategy for exploring machine learning for reducing interaction time. Creating a food preferences model customized to each user would likely lessen the food search interaction time considerably. By resorting to historical data and observable features (e.g., user location, day of the week, and weather), the system can anticipate the consumption of specific food. In that scenario, the interaction time would be equivalent to confirming a meal in the food plan.

9. Conclusions

This article unveils the concepts, requirements, and technologies needed to build a system that could support the nutritionist in creating food plans aligned with the individual profile. Further, it presents an architecture and software developed for smartphones (PWA) and smartwatches. The software furnishes food plan visualization logging of food and water intake, among other related features. It also integrates other devices, such as smart bottle technology and temperature sensors, to reduce human–computer interaction.

The availability of off-the-shelf devices has brought unprecedented ways of gathering data from physical phenomena without resorting to direct human–computer interaction. We propose an architecture that integrates the nutritionist back-office, the user application, and smart devices, focused on interaction cost reduction when users follow a food plan. We presented a baseline of the human interaction effort associated with several tasks pinpointing the most critical (expensive) operations. Such baseline sustains the evaluation of future machine learning and IoT approaches targeting the reduction of human interaction effort when completing critical operations.

As future work, we plan to explore machine learning techniques to reduce interaction times in two demanding user groups: Alzheimer's patients and athletes. The Alzheimer's group offers interaction challenges since several caretakers are elderly and have difficulties using apps or are not motivated to use apps as a data logging mechanism. On the other hand, athletes are very disciplined but need tight control of food intake before, during, and after physical activity.

Author Contributions: Conceptualization and project administration, C.A.S.C.; software interaction design, R.P.D. All authors have read and agreed to the published version of the manuscript.

Funding: This work is funded by National Funds through the FCT—Foundation for Science and Technology, I.P., within the scope of the project Ref. UIDB/05583/2020. Furthermore, we would like to thank the Research Centre in Digital Services (CISeD) and the Polytechnic of Viseu for their support. Moreover, the authors greatly thank IPV/CGD-Polytechnic Institute of Viseu/Caixa Geral de Depositos Bank, within the scope of the projects PROJ/IPV/ID&I/002 and PROJ/IPV/ID&I/007.

Institutional Review Board Statement: Not applicable.

Informed Consent Statement: Not applicable.

Acknowledgments: We want to thank the nutritionist Marco Pontes, Sao Mateus Hospital, Social Works of City Hall at Viseu, Valter Alves, Maria Joao Sebastiao, Maria Joao Lima, Carlos Vasconcelos, Lia Araujo, and Carlos Albuquerque for their support in this project.

Conflicts of Interest: The authors declare no conflict of interest.

Appendix A

This appendix presents the use cases described using the unified modeling language (UML) related to the application requirements.

Appendix A.1. Nutritionist Use Cases

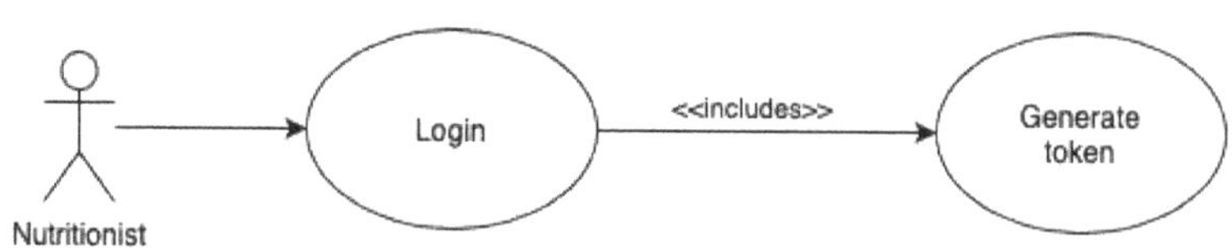

Figure A1. Nutritionist login.

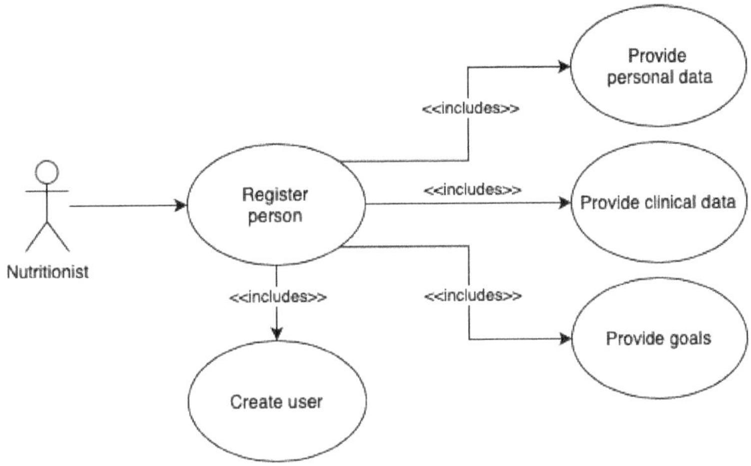

Figure A2. Person registration.

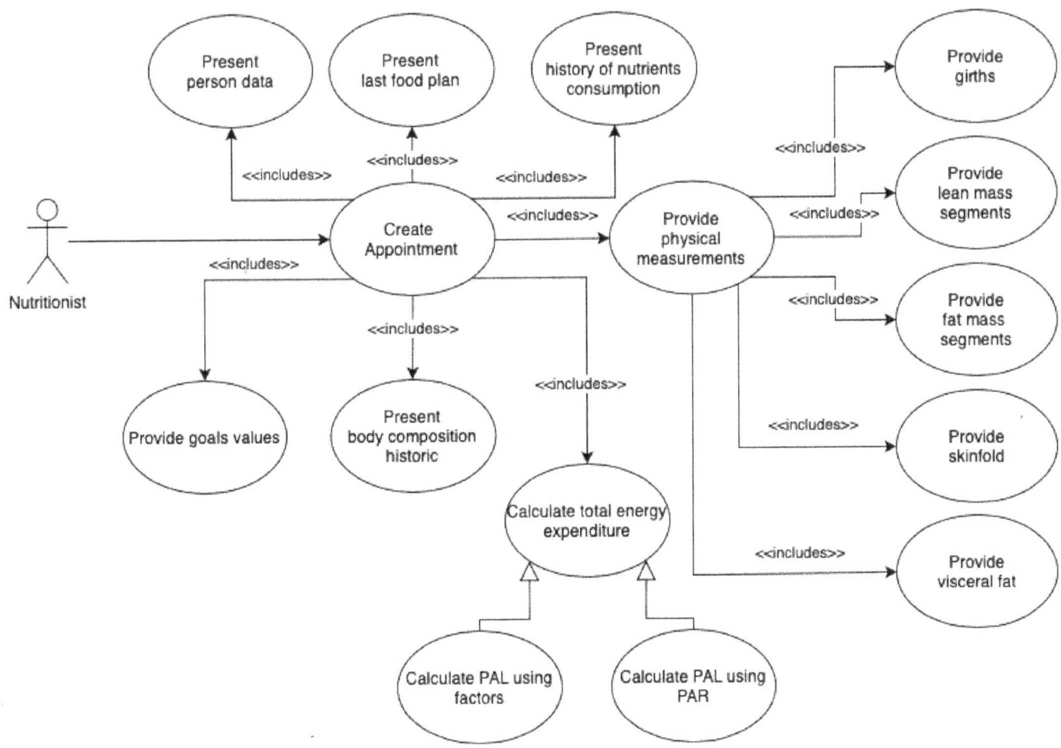

Figure A3. Appointment creation.

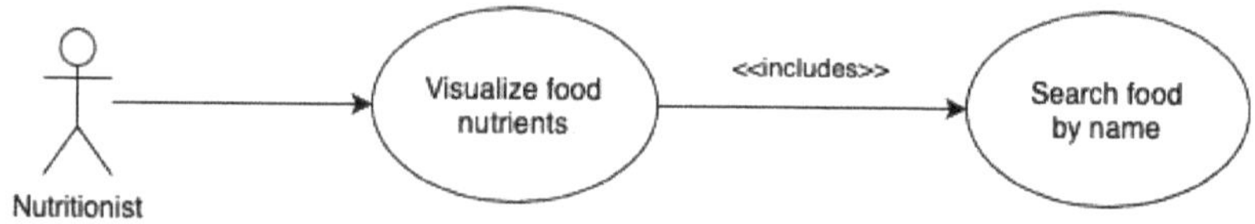

Figure A4. Visualize food nutrients.

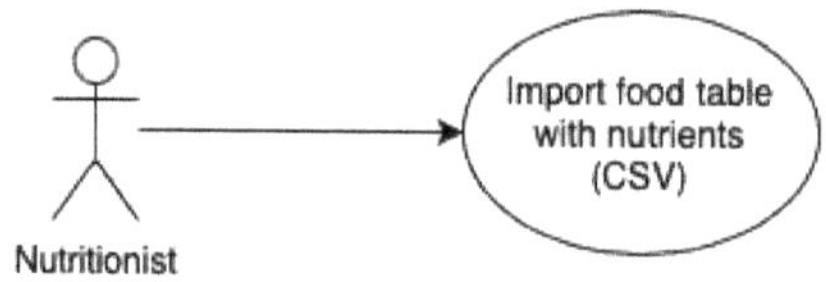

Figure A5. Import food table with nutrients.

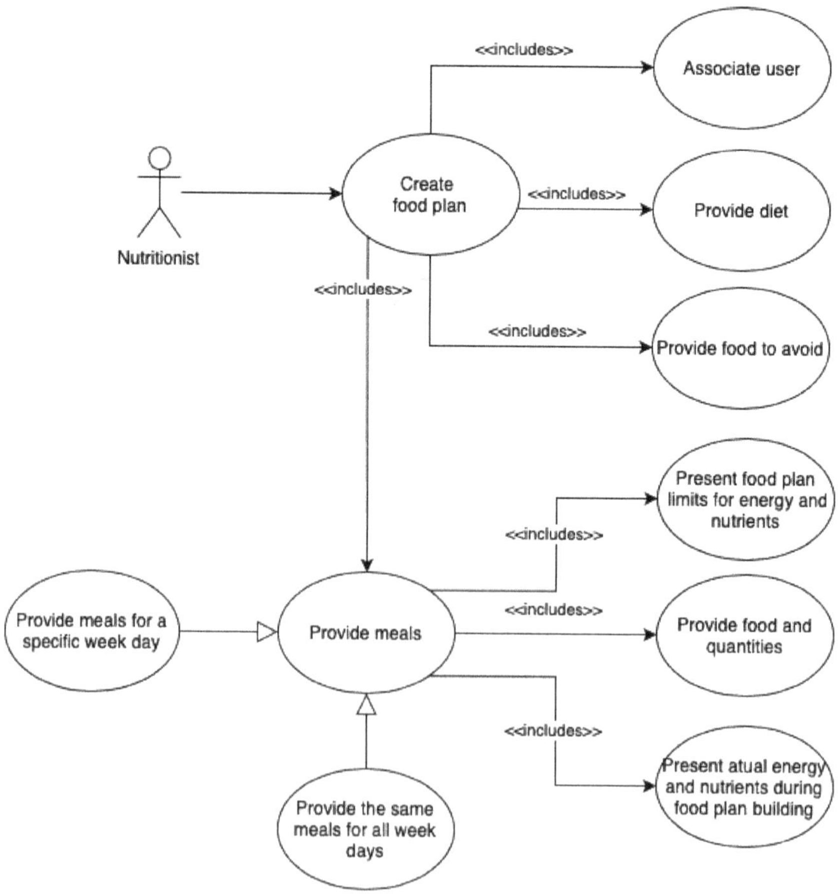

Figure A6. Food plan creation.

Appendix A.2. User Use Cases

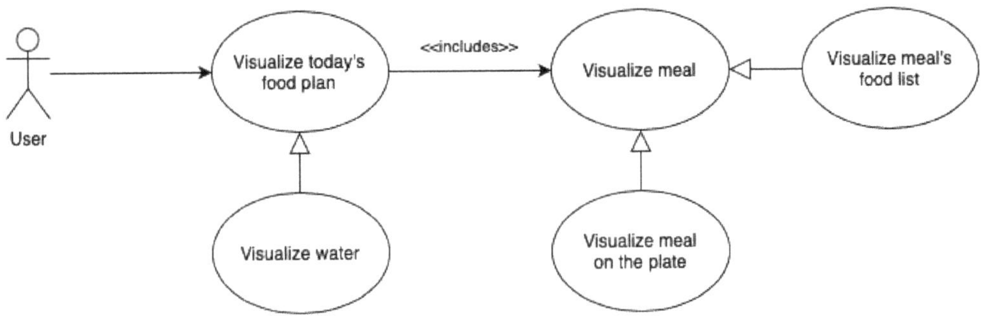

Figure A7. Visualize food plan.

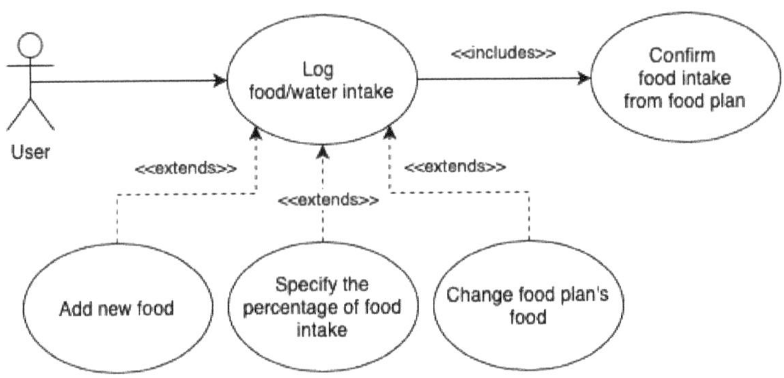

Figure A8. Log food and water intake.

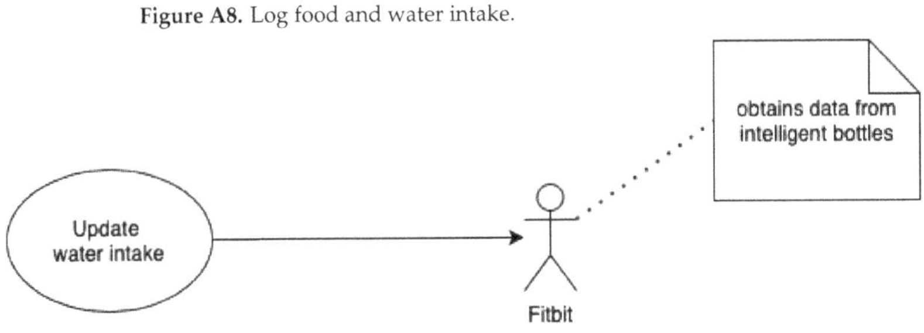

Figure A9. Update water intake.

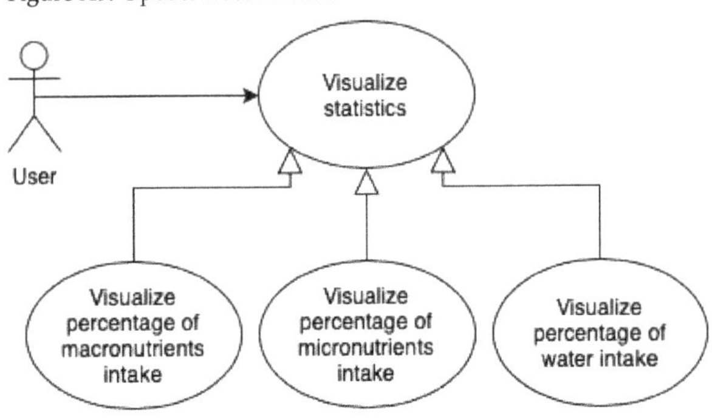

Figure A10. Visualize statistics.

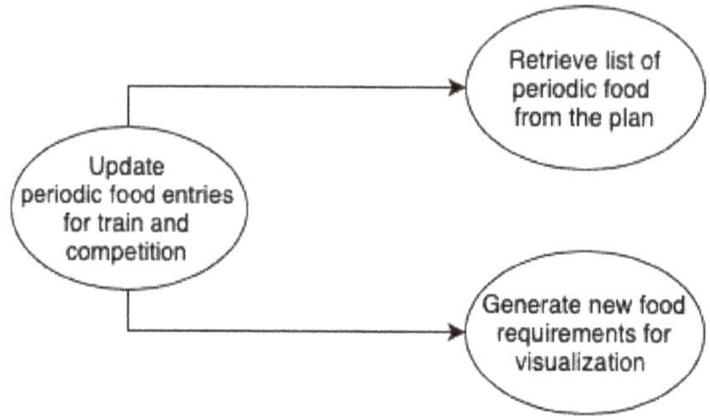

Figure A11. Update periodic food entries and train for competition.

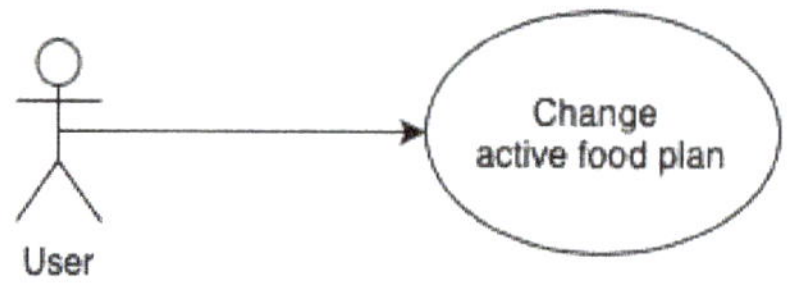

Figure A12. Change active food plan.

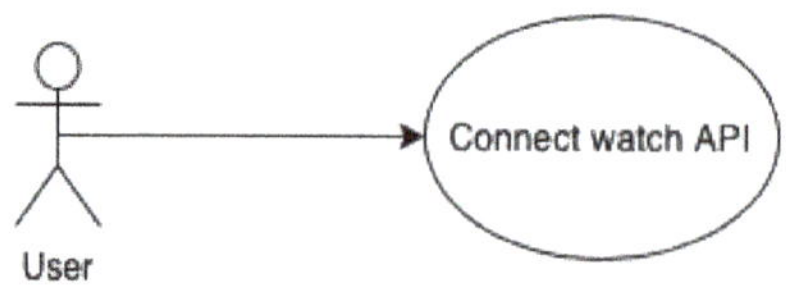

Figure A13. Connect watch API.

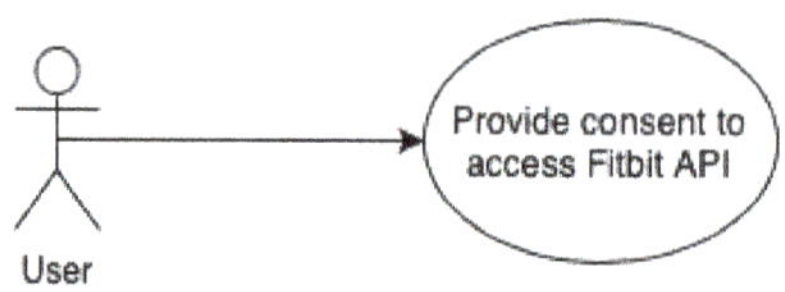

Figure A14. Provide Fitbit consent.

References

1. Afshin, A.; Sur, P.J.; Fay, K.A.; Cornaby, L.; Ferrara, G.; Salama, J.S.; Mullany, E.C.; Abate, K.H.; Abbafati, C.; Abebe, Z.; et al. Health effects of dietary risks in 195 countries, 1990–2017: A systematic analysis for the Global Burden of Disease Study 2017. *Lancet* **2019**, *393*, 1958–1972. [CrossRef]
2. Wohlers, E.M.; Sirard, J.R.; Barden, C.M.; Moon, J.K. Smart Phones are Useful for Food Intake and Physical Activity Surveys. In Proceedings of the 2009 Annual International Conference of the IEEE Engineering in Medicine and Biology Society, Minneapolis, MN, USA, 3–6 September 2009; Volume 2009, pp. 5183–5186. [CrossRef]
3. Tsai, C.C.; Lee, G.; Raab, F.; Norman, G.J.; Sohn, T.; Griswold, W.G.; Patrick, K. Usability and Feasibility of PmEB: A Mobile Phone Application for Monitoring Real Time Caloric Balance. *Mob. Networks Appl.* **2007**, *12*, 173–184. [CrossRef]
4. Zmora, N.; Elinav, E. Harnessing smartphones to personalize nutrition in a time of global pandemic. *Nutrients* **2021**, *13*, 422. [CrossRef] [PubMed]
5. Lo, F.P.W.; Sun, Y.; Qiu, J.; Lo, B. A novel vision-based approach for dietary assessment using deep learning view synthesis. In Proceedings of the 2019 IEEE 16th International Conference on Wearable and Implantable Body Sensor Networks (BSN), Chicago, IL, USA, 19–22 May 2019; pp. 1–4.
6. Fang, S.; Shao, Z.; Kerr, D.A.; Boushey, C.J.; Zhu, F. An End-to-End Image-Based Automatic Food Energy Estimation Technique Based on Learned Energy Distribution Images: Protocol and Methodology. *Nutrients* **2019**, *11*, 877. [CrossRef]
7. Kalantarian, H.; Mortazavi, B.; Alshurafa, N.; Sideris, C.; Le, T.; Sarrafzadeh, M. A comparison of piezoelectric-based inertial sensing and audio-based detection of swallows. *Obes. Med.* **2016**, *1*, 6–14. [CrossRef]
8. Kirk, D.; Catal, C.; Tekinerdogan, B. Precision nutrition: A systematic literature review. *Comput. Biol. Med.* **2021**, *133*, 104365. [CrossRef]
9. Yetisen, A.K.; Martinez-Hurtado, J.L.; Ünal, B.; Khademhosseini, A.; Butt, H. Wearables in Medicine. *Adv. Mater.* **2018**, *30*, 1706910. [CrossRef]

10. Hezarjaribi, N.; Mazrouee, S.; Ghasemzadeh, H. Speech2Health: A Mobile Framework for Monitoring Dietary Composition From Spoken Data. *IEEE J. Biomed. Health Inform.* **2018**, *22*, 252–264. [CrossRef]
11. D'Alessio, D.A.; Kavle, E.C.; Mozzoli, M.A.; Smalley, K.J.; Polansky, M.; Kendrick, Z.V.; Owen, L.R.; Bushman, M.C.; Boden, G.; Owen, O.E. Thermic effect of food in lean and obese men. *J. Clin. Investig.* **1988**, *81*, 1781–1789. [CrossRef]
12. Reed, G.W.; Hill, J.O. Measuring the thermic effect of food. *Am. J. Clin. Nutr.* **1996**, *63*, 164–169. [CrossRef]
13. Frankenfield, D.C.; Muth, E.R.; Rowe, W.A. The Harris-Benedict Studies of Human Basal Metabolism History and Limitations. *J. Am. Diet. Assoc.* **1998**, *98*, 439–445. [CrossRef]
14. Bendavid, I.; Lobo, D.N.; Barazzoni, R.; Cederholm, T.; Coëffier, M.; de van der Schueren, M.; Fontaine, E.; Hiesmayr, M.; Laviano, A.; Pichard, C.; et al. The centenary of the Harris–Benedict equations: How to assess energy requirements best? Recommendations from the ESPEN expert group. *Clin. Nutr.* **2021**, *40*, 690–701. [CrossRef] [PubMed]
15. United Nations University, Food and Agriculture Organization of the United Nations; World Health Organization. *Energy and Protein Requirements: Report of a Joint FAO/WHO/UNU Expert Consultation [Held in Rome from 5 to 17 October 1981]*; Scotti Bassani: Milan, Italy, 1985.
16. Conway, J.M. Human Energy Requirements: A Manual for Planners and Nutritionists. *Am. J. Clin. Nutr.* **1991**, *53*, 1506. [CrossRef]
17. National Agricultural Library. Available online: https://www.nal.usda.gov/ (accessed on 13 December 2021).
18. Lairon, D. Soluble fibers and dietary lipids. *Adv. Exp. Med. Biol.* **1997**, *427*, 99–108. [CrossRef] [PubMed]
19. Roza, A.M.; Shizgal, H.M. The Harris Benedict equation reevaluated: Resting energy requirements and the body cell mass. *Am. J. Clin. Nutr.* **1984**, *40*, 168–182. [CrossRef] [PubMed]
20. Mifflin, M.D.; Jeor, S.T.S.; Hill, L.A.; Scott, B.J.; Daugherty, S.A.; Koh, Y.O. A new predictive equation for resting energy expenditure in healthy individuals. *Am. J. Clin. Nutr.* **1990**, *51*, 241–247. [CrossRef] [PubMed]
21. McArdle, W.D.; Katch, F.I.; Katch, V.L. Exercise Physiology: Energy, Nutrition, and Human Performance, 3rd Edition. *Med. Sci. Sport. Exerc.* **1991**, *23*, 1403. [CrossRef]
22. Cunningham, J.J. A reanalysis of the factors influencing basal metabolic rate in normal adults. *Am. J. Clin. Nutr.* **1980**, *33*, 2372–2374. [CrossRef]
23. Deurenberg, P.; Weststrate, J.A.; Seidell, J.C. Body mass index as a measure of body fatness: Age- and sex-specific prediction formulas. *Br. J. Nutr.* **1991**, *65*, 105–114. [CrossRef]
24. vans, E.M.; Rowe, D.A.; Misic, M.M.; Prior, B.M.; Arngrímsson, S.A. Skinfold Prediction Equation for Athletes Developed Using a Four-Component Model. *Med. Sci. Sport. Exerc.* **2005**, *37*, 2006–2011. [CrossRef]
25. van der Ploeg, G.E.; Gunn, S.M.; Withers, R.T.; Modra, A.C. Use of anthropometric variables to predict relative body fat determined by a four-compartment body composition model. *Eur. J. Clin. Nutr.* **2003**, *57*, 1009–1016. [CrossRef] [PubMed]
26. Biørn-Hansen, A.; Majchrzak, T.A.; Grønli, T.M. Progressive Web Apps: The Possible Web-native Unifier for Mobile Development. In Proceedings of the 13th International Conference on Web Information Systems and Technologies, Porto, Portugal, 25–27 April 2017; pp. 344–351. [CrossRef]
27. Google. Twitter Lite PWA Significantly Increases Engagement and Reduces Data Usage. Available online: https://developers.google.com/web/showcase/2017/twitter (accessed on 21 December 2021).
28. Google. Ola Drives Mobility for a Billion Indians with Progressive Web App. Available online: https://developers.google.com/web/showcase/2017/ola (accessed on 31 January 2022).
29. LitElement. Available online: https://lit.dev (accessed on 31 January 2022).
30. Sheppard, D. *Beginning Progressive Web App Development: Creating a Native App Experience on the Web*; Springer: Berlin/Heidelberg, Germany, 2017. [CrossRef]
31. Hidratespark Smart Bottle. Available online: https://hidratespark.com (accessed on 31 January 2022).
32. Fitbit. Available online: https://www.fitbit.com (accessed on 31 January 2022).
33. El-Amrawy, F.; Nounou, M.I. Are Currently Available Wearable Devices for Activity Tracking and Heart Rate Monitoring Accurate, Precise, and Medically Beneficial? *Healthc. Informatics Res.* **2015**, *21*, 315–320. [CrossRef] [PubMed]
34. Lops, P.; Gemmis, M.d.; Semeraro, G. Content-based Recommender Systems: State of the Art and Trends. In *Recommender Systems Handbook*; Springer: Boston, MA, USA, 2011; pp. 73–105. [CrossRef]
35. Akrour, R.S. APRIL: Active Preference Learning-Based Reinforcement Learning. In *Joint European Conference on Machine Learning and Knowledge Discovery in Databases*; Springer: Bristol, UK, 2012; pp. 116–131.
36. Poleto, T.; Carvalho, V.D.H.d.; Silva, A.L.B.d.; Clemente, T.R.N.; Silva, M.M.; Gusmão, A.P.H.d.; Costa, A.P.C.S.; Nepomuceno, T.C.C. Fuzzy Cognitive Scenario Mapping for Causes of Cybersecurity in Telehealth Services. *Healthcare* **2021**, *9*, 1504. [CrossRef] [PubMed]
37. Sittig, D.F.; Belmont, E.; Singh, H. Improving the safety of health information technology requires shared responsibility: It is time we all step up. *Healthcare* **2018**, *6*, 7–12. [CrossRef] [PubMed]
38. Alami, H.; Gagnon, M.P.; Ahmed, M.A.A.; Fortin, J.P. Digital health: Cybersecurity is a value creation lever, not only a source of expenditure. *Health Policy Technol.* **2019**, *8*, 319–321. [CrossRef]
39. Ksibi, S.; Jaidi, F.; Bouhoula, A. Cyber-Risk Management within IoMT: A Context-Aware Agent-Based Framework for a Reliable e-Health System. In Proceedings of the 23rd International Conference on Information Integration and Web Intelligence, Linz, Austria, 29 November 2021–1 December 2021; pp. 547–552. [CrossRef]
40. Aghili, S.F.; Mala, H.; Shojafar, M.; Peris-Lopez, P. LACO: Lightweight Three-Factor Authentication, Access Control and Ownership Transfer Scheme for E-Health Systems in IoT. *Future Gener. Comput. Syst.* **2019**, *96*, 410–424. [CrossRef]

41. Knopman, D.S.; Amieva, H.; Petersen, R.C.; Chételat, G.; Holtzman, D.M.; Hyman, B.T.; Nixon, R.A.; Jones, D.T. Alzheimer disease. *Nat. Rev. Dis. Prim.* **2021**, *7*, 1–21. [CrossRef]
42. Paz, F.; Paz, F.A.; Pow-Sang, J.A. Evaluation of Usability Heuristics for Transactional Web Sites: A Comparative Study. In *Information Technology: New Generations*; Latifi, S., Ed.; Springer International Publishing: Cham, Switzerland, 2016; pp. 1063–1073. [CrossRef]
43. Paz, F.; Paz, F.A.; Villanueva, D.; Pow-Sang, J.A. *Heuristic Evaluation as a Complement to Usability Testing: A Case Study in Web Domain*; ITNG '15; IEEE Computer Society: Washington, DC, USA, 2015; pp. 546–551. [CrossRef]
44. Yushiana, M.; Rani, W.A. Heuristic evaluation of interface usability for a web-based OPAC. *Libr. Hi Tech* **2007**, *25*, 538–549. [CrossRef]
45. Lim, C.; Song, H.D.; Lee, Y. Improving the usability of the user interface for a digital textbook platform for elementary-school students. *Educ. Technol. Res. Dev.* **2012**, *60*, 159–173. [CrossRef]
46. Tonn-Eichstädt, H. Measuring Website Usability for Visually Impaired People-a Modified GOMS Analysis. In Proceedings of the 8th International ACM SIGACCESS Conference on Computers and Accessibility, Portland, OR, USA, 22 October–27 October 2011; Assets '06; Association for Computing Machinery: New York, NY, USA, 2006; pp. 55–62. [CrossRef]
47. Theofilou, S.; Vardas, N.; Katsanos, C. FLM-2A: Towards Automated HCI Modeling of Android Applications Based on a Modified Version of the Keystroke Level Model. In *Human-Computer Interaction. Theory, Methods and Tools*; Kurosu, M., Ed.; Springer International Publishing: Cham, Switzerland, 2021; pp. 329–342. [CrossRef]
48. Cunha, D.; Duarte, R.P.; Cunha, C.A. KLM-GOMS Detection of Interaction Patterns Through the Execution of Unplanned Tasks. In Computational Science and Its Applications—ICCSA 2021, Cagliari, Italy, 13 September–16 September 2021; Gervasi, O., Murgante, B., Misra, S., Garau, C., Blečić, I., Taniar, D., Apduhan, B.O., Rocha, A.M.A., Tarantino, E., Torre, C.M., Eds.; Springer International Publishing: Cham, Switzerland, 2021; pp. 203–219. [CrossRef]
49. Card, S.K.; Moran, T.P.; Newell, A. The Keystroke-Level Model for User Performance Time with Interactive Systems. *Commun. ACM* **1980**, *23*, 396–410. [CrossRef]

sensors

MDPI

Article

Empowering People with a User-Friendly Wearable Platform for Unobtrusive Monitoring of Vital Physiological Parameters

Maria Krizea [1,2], John Gialelis [1,2,*], Grigoris Protopsaltis [1], Christos Mountzouris [1] and Gerasimos Theodorou [1]

[1] Applied Electronics Laboratory, Department of Electrical and Computer Engineering, University of Patras, 26504 Patras, Greece; mkrizea@ece.upatras.gr (M.K.); g.protopsaltis@upnet.gr (G.P.); mountzou@ece.upatras.gr (C.M.); gtheodorou@upatras.gr (G.T.)

[2] Industrial Systems Institute, ATHENA RC, 26504 Patras, Greece

* Correspondence: gialelis@ece.upatras.gr

Abstract: Elderly people feel vulnerable especially after they are dismissed from health care facilities and return home. The purpose of this work was to alleviate this sense of vulnerability and empower these people by giving them the opportunity to unobtrusively record their vital physiological parameters. Bearing in mind all the parameters involved, we developed a user-friendly wrist-wearable device combined with a web-based application, to adequately address this need. The proposed compilation obtains the photoplethysmogram (PPG) from the subject's wrist and simultaneously extracts, in real time, the physiological parameters of heart rate (HR), blood oxygen saturation (SpO_2) and respiratory rate (RR), based on algorithms embedded on the wearable device. The described process is conducted solely within the device, favoring the optimal use of the available resources. The aggregated data are transmitted via Wi-Fi to a cloud environment and stored in a database. A corresponding web-based application serves as a visualization and analytics tool, allowing the individuals to catch a glimpse of their physiological parameters on a screen and share their digital information with health professionals who can perform further processing and obtain valuable health information.

Keywords: wrist-wearable device; PPG processing; physiological parameters; web-based applications; data analysis

Citation: Krizea, M.; Gialelis, J.; Protopsaltis, G.; Mountzouris, C.; Theodorou, G. Empowering People with a User-Friendly Wearable Platform for Unobtrusive Monitoring of Vital Physiological Parameters. *Sensors* **2022**, *22*, 5226. https://doi.org/10.3390/s22145226

Academic Editors: Ivan Miguel Serrano Pires and Georg Fischer

Received: 26 May 2022
Accepted: 8 July 2022
Published: 13 July 2022

1. Introduction

Average life expectancy has increased over the years, resulting in a rise in senior populations [1]. The attitude of society towards senior citizens and their well-being is an indicator of its organization and civilization. Elderly people are a demographic that needs expert care and dedicated assistance, sometimes even on an everyday basis. They tend to feel even more vulnerable especially after experiencing health issues and having been released from a health care facility. This is a crucial point in their recovery and well-being, and they need all the help they can get, in either physical or virtual form. Assistive technology based on the Internet of Things (IoT) can support unobtrusive health monitoring at home with the use of electrical devices, such as sensors and other gadgets (wearable or not) that provide feedback and remote access to the end user, aiming at improving inhabitants' quality of life by providing more independence and better care [2]. According to [3], existing smart home health monitoring technologies include physiological monitoring, functional monitoring/emergency detection and response, safety monitoring and assistance, security monitoring and assistance, social interaction monitoring and assistance, and cognitive and sensory assistance.

Treading on the groundwork of assistive technology, the aim of minimizing the hospitalization days of the elderly and sending them home without compromising their safety seems to be doable. The achievement of this goal has at least two advantages. First, the elderly benefit from returning home to a safe environment as soon as possible and this can work in favor of nurturing a positive psychology for them. Second, the health care system also profits from early yet safe discharge of the elderly to their home. The financial resources dedicated to the health sector are not unlimited [4] and the struggle to mitigate the consequences of the pandemic is ongoing. Unburdening the health care system is a tangible positive ramification of assistive technology development, and the work we envisaged could be a major abetment to this line of action.

Wrist-wearable devices are a common accessory of everyday life, worn by many people just to tell time in the beginning. With the integration of proper sensors, they evolved to non-invasive monitoring units that aggregate vital signals. With the application of befitting embedded algorithms, several physiological parameters critical for health status assessment can be extracted. These devices can also log activities, steps, calories and sleep patterns. The autonomy and portability of such an apparatus grants the user the possibility to wear it everywhere and at any time, which is a notable advantage. Identifying outliers or anomalies in heart rates and other features can help establish patterns that play a significant role in understanding the underlying cause of the demise of physical well-being. Additionally, the accumulation of this valuable information in a backend system, in a secure manner, can be leveraged to the advantage of the end users for optimizing their living standard. This can be accomplished by providing medical experts with the tools to interact with the data and draw valuable conclusions regarding health status and everyday lifestyle.

Light, the essence of photoplethysmography, known as PPG, is used to measure the volumetric variations of blood circulation. This measurement provides invaluable information regarding humans' physiological parameters entailing their health status. PPG technology has been proven less complex operationally, more comfortable for the user and more cost efficient compared to other monitoring techniques [5]. The continuous detection and monitoring of human physiological parameters such as heart rate (HR), blood oxygen saturation (SpO_2), and respiratory rate (RR) are of paramount importance. A wearable device that unobtrusively grants the elderly an opportunity to continuously and without any intervention extract those parameters constitutes a major achievement. The state-of-the-art techniques in modern practices approach the task of obtaining such measurements using validated pulse oximeters, which are worn on individuals' fingers. These pulse oximeters are based on the transmission-mode PPG. The continuous monitoring of the human vital signals has shifted from simply an appealing idea to fitness enthusiasts into an everyday habit for a plethora of people using a smartwatch. Sensors integrated within smart watches use reflectance-mode PPG to gather vital signals. Although they are widespread amongst sportspeople, they have not been widely used in clinical practice [6,7]. This can be attributed to the fact that PPG signals are vulnerable to Motion Artifacts (MAs) caused by hand movements, which affect considerably the accuracy of the entailed physiological parameters [8,9].

PPG signal acquisition becomes increasingly challenging when additional factors such as environmental noise or sensor misplacement are considered, thus further affecting the accurate assessment of its features. Measuring physiological parameters utilizing wrist-wearable devices can be a more perplexing procedure compared with the finger or another part of the body due to the low blood perfusion in the wrist area. The design of a wrist-wearable device must consider factors such as the spacing of the light-emitting diodes (LEDs), photodiode surface, as well as biological factors such as skin tone. It is clear that efficient PPG processing must be performed to enable reliable reading of its features and consequently accurate extraction of the physiological values. All the aforementioned challenges highlight the need for the implementation of a device capable of monitoring physiological parameters continuously and in an unobtrusive manner, that can easily be operated without need for excess previous training.

This work is about presenting the architecture and the constituent components of a comprehensive user-friendly system which provides well-being monitoring services promoting peace of mind to senior citizens. The overall solution consists of two subsystems that are integrated via well-defined interfaces, but each one performs autonomous functions in an opaque manner: The subsystem of bio signal recording and physiological parameters extraction and the back-end subsystem of storage, data processing and services. The novelty of the proposed system predominantly lies on the minimal design of the wearable sensing device while entailing accurately all the parameters, allowing a least disturbing interaction with the end user. This system has the capacity to continuously extract in real time the physiological parameters of HR, SpO_2 and RR, to perform well-being status assessment, and to provide personalized feedback to improve health status. Furthermore, in comparison with similar solutions, the proposed system has the advantage of functioning without the aid of a paired smartphone for collecting and delivering the bio parameters to the back-end system, thus enhancing its user-friendly attribute for users with low digital literacy.

The remainder of this paper is organized as follows: Section 2 introduces the main challenges regarding the monitoring of physiological parameters using wearable devices and gives details on related works and their limitations. Section 3 presents design and implementation aspects of the proposed integrated system at both physical and operational level as well as performance evaluation features of the proposed wrist-wearable device. Section 4 describes the implementation and the core components of the web-based application, which comprises a front-end to display physiological parameters to end-users, and a back-end to allow data processing and service provisioning. Finally, Section 5 concludes this paper with future research directions.

2. Related Work

As shown further, research has been conducted towards producing simultaneously more than one physiological parameter using the PPG method, through wrist-wearable devices as enablers of continuous health monitoring in everyday life. With the evolution of medical science and technology, the wearables are incorporated with multiple different sensors so that they can keep track of a wide range of measurements such as heart rate, blood oxygen level, body temperature and activity monitoring including many others. Nowadays, wearables play an important role in making health care practices more efficient and cost-effective. These devices can be connected to smart phones or web apps allowing people to store their data for future reference. An overview of such wearables follows.

Li et al. [10] predicted the outbreak of Lyme disease and inflammation by combining sensor data along with medical measurements. This work gave evidence of how wearables can monitor activity along with physiology.

Mishra et al. [11] detected COVID-19 by utilizing physiological (HR) and activity (steps) data acquired by wearable devices. 5200 subjects participated in the analysis, including individuals with COVID-19. The study indicated elevated resting heart rates relative to the subject's baseline. Two algorithms—resting heart rate difference (RHR-diff) and heart rate over steps anomaly detection (HROS-AD)—were developed. The first algorithm was based on standardizing the resting heart rate over a fixed time frame to observe baseline residuals.

Seshadri et al. [12] performed a data-driven COVID-19 prediction employing an early detection algorithm (EDA) based on HR, HRV and RR collected from wearables devices. The EDA can detect physiological changes and alert users of possible infection with SARS-CoV-2 before they develop clinical symptoms.

Downey et al. [13] showed that only 16% of the subjects remained connected to obtrusive monitoring systems after 72 h. Furthermore, the cost for a complete vital sign monitoring system can be quite significant.

Zenko et al. [14] proposed a battery powered wearable device along with a simple algorithm for the extraction of the physiological parameters of HR, Pulse Rate Variability and SpO_2. This work evaluates the acquired HR parameter while calibration and verification of the SpO_2 parameter still needs to be performed.

Son et al. [15] introduced a wearable device which measures oxygen levels in the blood using a light reflection method while it integrates hardware for wireless data transmission. Experimental results were compared to the Texas Instruments development board (SpO_2

AFE44x0 EVM), and the maximum deviation was 6.7% in HR measurement and 4.4% in SpO_2 measurement.

Jarchi et al. [16] integrated the AC components of red and infrared PPG signals in a complex waveform and then by applying the bivariate empirical mode decomposition algorithm, the SpO_2 value is estimated with an approximate error of 3%.

Preejith et al. [17] developed a wrist-based optical heart rate device which, in order to eliminate the noise, ignores the measurements when motion is detected. The accuracy of the HR measurements equals to 0.9, expressed as Pearson's r. The coefficient r indicates the strength of a correlation between estimated and real values, and the magnitude of 0.9 nominates that the variables can be considered highly correlated.

Eugene et al. [18] designed a wearable device equipped with PPG sensors for extracting bio-information and the Centralized State Sensing (CSS) algorithm was developed for estimating HR. After comparisons on readings taken across sensors, it was proved that this specific algorithm achieved more accurate HR measurements.

Wojcikowski [19] proposed an algorithm for real-time HR estimation by a wrist-wearable device. The device incorporates PPG and accelerometer sensors. The acceleration signal is used to detect body movements which distort the PPG signal. The evaluation results evidenced that the developed algorithm for HR measurements outperformed the other algorithms from the literature.

Münzner et al. [20] presented methods for development of robust deep learning (DL) methods for human activity recognition (HAR) addressing the problems of normalization and fusion of multimodal sensor HAR data. The results show that sensor-specific normalization increases the prediction accuracy of the convolutional neural networks (CNN). In the context of multimodal HAR, further normalization techniques should be investigated which focus on other modalities such as physiological sensors.

Tang et al. [21] proposed a new CNN that uses hierarchical split (HS) for a large variety of HAR tasks, which can enhance multi-scale feature representation ability via capturing a wider range of receptive fields of human activities within one feature layer. Benchmarks demonstrated that the proposed HS module is an impressive alternative to baseline models with similar model complexity and can achieve higher recognition performance.

Zhang et al. [22] presented a Deep Neural Network (DNN) to detect lumbar-pelvic movements (LPMs), including flexion, lateral flexion, rotation, and extension, locally on-device, where the data were collected from a clinically validated sensor system. Continuous monitoring of these movements can provide real-time feedback to both patients and medical experts with the potential of identifying activities that may precipitate symptoms of low back pain (LBP) as well as improving therapy by providing a personalised approach.

Aside from the prototypes that emerged from literature research, as depicted in Table 1, there are wrist-worn commercial products available which utilize PPG sensors for obtaining physiological measurements, as shown in Table 2. Empatica E4 [23] is a wearable wireless multisensory device for real-time data acquisition and computerized biofeedback. E4 comprises four embedded sensing modules: a photoplethysmography (PPG) module, an electrodermal activity (EDA) module, a 3DOF accelerometer module, and a temperature sensing module. E4 offers the readings of HR, activity status and temperature while being capable of characterizing the function of the autonomic nervous system, EDA for assessing the sympathetic activation and HRV for assessing the parasympathetic activation. The device is compliant with international safety and emissions standards. MaxRefDes103# [24], is a physiological signal sensing band reference design available to the research community for further development. It is a wrist-worn wearable exhibiting high sensitivity and algorithmic processing capabilities comprising an enclosure and a biometric sensor hub with an embedded algorithm that processes PPG signals in real time for extracting HR and SpO_2 only. Eventually, its corresponding output and raw data are streamed via Bluetooth to an Android application or PC GUI for demonstration, evaluation, and further elaboration. In addition to displaying the extracted HR and SpO_2, the Android application furnishes additional algorithms for calculating RR, HRV, and sleep quality. Other wrist-worn wearables used for health monitoring are the Fitbit Versa 3 [25], Samsung Galaxy Watch 4 [26] and Apple Watch Series 7 [27]. The Fitbit smartwatch records the physiological features: HR, SpO_2, skin temperature variation, sleep stages and RR measurement during sleep. Fitbit utilizes the BLE communication protocol and has mobile applications compatible with Android and iOS. Both the Samsung Galaxy Watch 4 and the Apple Watch Series 7 hold the capability for ECG and sleep monitoring. Moreover, both wearables calculate the HR and SpO_2 physiological parameters and provide their data wirelessly through BLE/Wi-Fi and BLE correspondingly. The Samsung wearable is compatible with Android while the Apple wearable is compatible with iOS. As can be seen, among the approaches and the wearable solutions, there are limitations concerning the set of physiological values provided.

Table 1. Indicative functionalities and features of prototypes emerged from literature research.

Device Functionalities and Features	Proposed Device	Zenko et al. [14]	Son et al. [15]	Jarchi et al. [16]	Preejith et al. [17]	Eugene et al. [18]	Wojcikowski [19]
PPG	X	X	X	X	X	X	X
SPO_2	X	X			X	X	X
HR	X	X	X	X			
RR	X						
Acceleration	X						
Continuous Measurement	X		X		X	X	
Communication Protocol	Wi-Fi	N/A [1]	BLE	BLE	BLE	BLE	BLE
Battery life in streaming mode	48 h	NS [2]	NS	NS	40 h	NS	NS
RT raw data	X	N/A	N/A	N/A	N/A	N/A	N/A

[1] N/A: non-available. [2] NS: non-stated.

Table 2. Indicative functionalities and features of prototypes emerged from literature research.

Device Functionalities and Features	Proposed Device	Empatica E4 [23]	MaxRef Des103# [24]	Fitbit Versa 3 [25]	Samsung Galaxy Watch 4 [26]	Apple Watch Series 7 [27]
PPG	X	X	X	X	X	X
SPO_2	X	X	X	X	X	X
HR	X	X	X	X	X	X
RR	X			X		
Acceleration	X	X		X	X	X
Continuous measurement	X	X	X	X	X	X
Communication protocol	Wi-Fi	BLE	BLE	BLE	BLE/Wi-Fi	BLE
Autonomy in streaming mode	48 h	24 + h	NS	12 + h	40 h	18 + h
RT raw data	X	N/A	X	N/A	N/A	N/A

Our proposed comprehensive system introduces a non-invasive wrist-wearable device capable of capturing the PPG signal and extracting in real time the HR, RR, SpO_2 physiological parameters simultaneously as well as rendering the raw PPG signal available which allows further processing for the sake of new bio parameters and features assessment. The extraction process takes place in situ ensuring the optimal utilization of its resources as well the network ones. It is a lightweight and embedded device with minimum add-ons exhibiting optimized memory capacity and processing power. It supports direct connection to 802.11.xx communication infrastructures which makes it an ideal candidate for instantaneous unhindered use in existing communication infrastructures offering high speed information sharing. Additionally, a cloud based back-end infrastructure offers all the required means to securely store the aggregated via https data in a time-series manner where end users and health professionals can perform visualization and algorithmic processing, respectively.

3. Proposed Device Components

This section focuses on the design and development aspects of the proposed wrist-wearable device for the continuous monitoring of the PPG physiological signal and the extraction in real time of the HR, SpO_2 and RR physiological parameters. Its scope is to provide an unobtrusive means to accurately assess the physiological data of the users' enabling them to monitor their well-being status. The proposed design follows a modular approach both at physical (hardware modules) and at operational level (software modules) as described below.

3.1. Hardware Modules

The wrist-wearable device is a microcontroller-based device designed for continuous monitoring of PPG. The device extracts the HR, SpO_2 and RR physiological parameters by implementing dedicated algorithms and transmitting the information over Wi-Fi to a developed web-based platform. The device is powered by a Lithium Polymer (LiPo) battery which can be charged using a USB cable. It is also equipped with an on-off switch for turning on or shutting down the device accordingly. Figure 1 depicts the block diagram of the proposed embedded device along with its external peripherals.

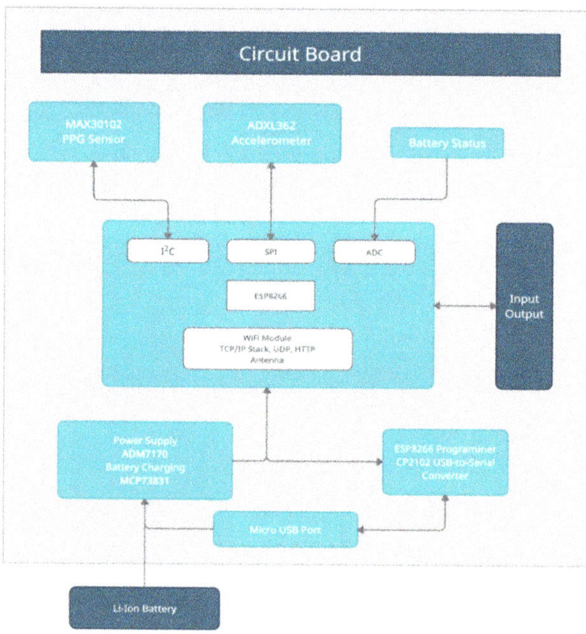

Figure 1. Block diagram of the developed wearable device.

The hardware components of the device are surface mounted on a custom printed circuit board (PCB) which was designed considering effortless repair, analysis, and field modification of circuits with dimensions of 66.0 × 42.0 mm.

- Microcontroller and Radio The device's microcontroller board incorporates the Espressif ESP-WROOM-02D, which is based on the ESP8266 chip implementing the Wi-Fi communication protocol [28]. Specifically, the core of the platform is the ESP8266 processor of Espressif systems, which is a Wi-Fi SoC integrating the full TCP/IP stack. The developed firmware code for the acquisition and the processing of the PPG signal, the extraction of physiological measurements and the code for the wireless transmission is being executed on the ESP8266 microprocessor. The ESP8266 integrates a Tensilica L106 32-bit RISC processor, achieving ultra-low power consumption and reaches a maximum clock speed of 160 MHz Moreover, ESP-WROOM-02D integrates an RF switch, matching balun, and a PCB antenna.

- Sensing Components Two sensing components are integrated into the embedded device, an optical sensor, and an accelerometer. MAX30102 is a high-sensitivity optical sensor able to continuously obtain the PPG signal and it is mounted on the standalone MAXREFDES117# reference board [29]. This low-power sensor board with a size of 12.7 mm $\times$ 12.7 mm is placed on the downside of the device's PCB. In addition, the main board of PCB incorporates the Analog Device's ADXL362 sensor, which is an ultra-low-power, 3-DOF MEMS accelerometer with measurement ranges of $\pm$2 g, $\pm$4 g, and $\pm$8 g [30].
- Power Supply The board is powered by either a Li-Ion 1600 mAh battery or the USB port. The power supply source is automatically switched on by the hardware. The battery is being recharged when a common USB power supply is connected to the USB port with a charging current set at 350 mA, controlled by the Microchip's MCP73831 chip. The voltage of the power supply is stabilized by the Analog Device's chip ADM7170 at 3.0 Volts and the board can harness all the available battery charge. The ADM7170 monitors constantly the voltage across the battery and cuts off the power supply of the circuit when the battery is not able to provide the correct voltage and current to the circuit.
- USB to Serial programmer The USB port can also be used to program the ESP8266 without the need of an external programmer. Furthermore, it is possible to manually download the code using a switch and two buttons mounted on the PCB. The necessary translations of the USB protocol to Serial are performed by Silicon Labs, CP2102 chip.

A custom casing was designed to enclose the device using a 3D printer with PLA material. Figure 2 presents the printed circuit board of the device as well as the complete wearable device mounted on the wrist. The device is designed to be worn on the left hand and the placement is approximately 2 cm from the beginning of the wrist. Constant pressure between the PPG sensor and the skin is applied with the aid of the attached wrist strap. Inappropriate device placement results in insufficient light detected by the photodetector, a condition which activates a notification for proper alignment on the interface of the web-based platform.

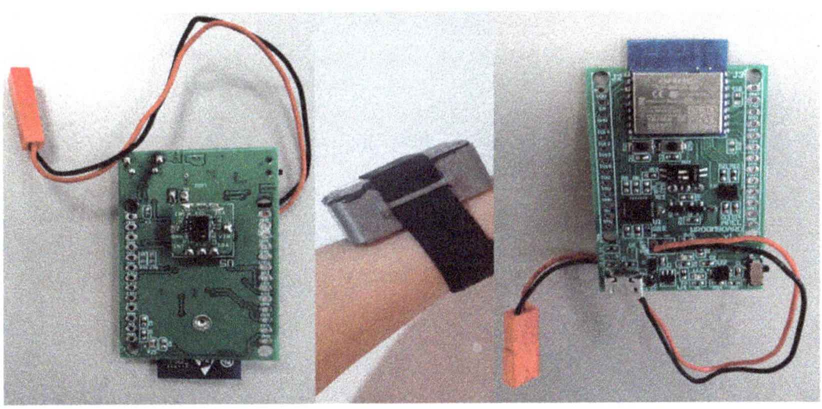

Figure 2. The proposed wearable device.

3.2. *Software Modules*

The main operation of the proposed SW modules include

- The adaptation of the optical sensor configurations to each individual user's wrist and skin physiology;
- The processing of the accelerometer's signal to detect the respective subject's motion
- The appropriate filtering of PPG signals;
- Their further processing for extracting HR, SpO_2 and RR physiological parameters.

Figure 3 overviews the device's algorithmic operation.

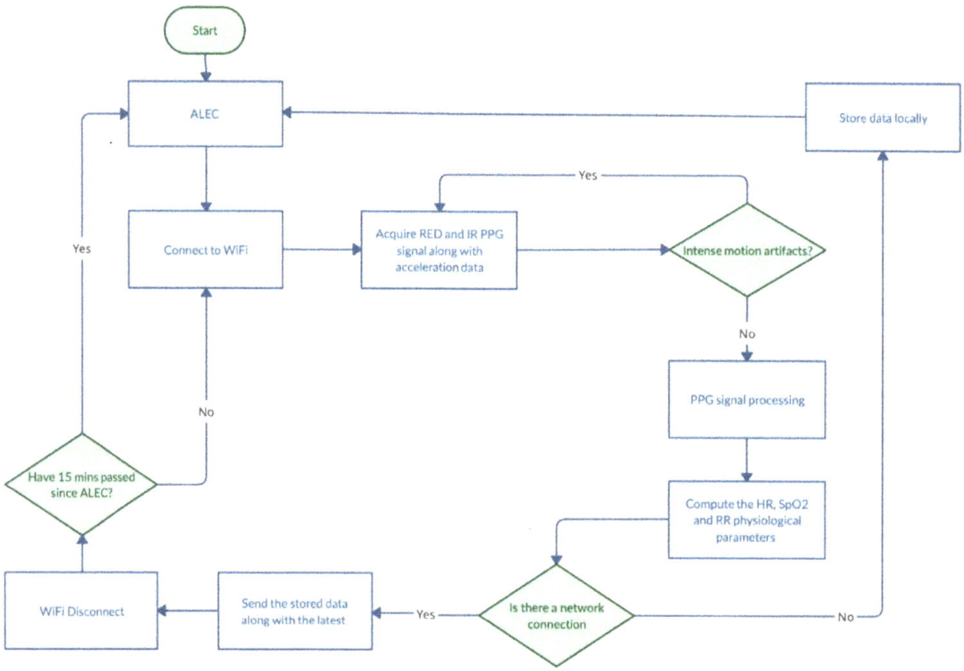

Figure 3. Flow chart of the algorithm.

3.2.1. PPG Acquisition

The principle of the device's operation is PPG signal recording. During the initial stage, the Automatic Led Emission Control (ALEC) technique is performed, which algorithmically mimics the Automatic Gain Control closed-loop feedback circuit. ALEC automatically adjusts the system to the specific characteristics of each user, as it regulates the LEDs luminosity depending on skin tone and the diameter of the wrist of each user. During experimentation it was observed that the ADC output tends to alter in time. The former observation was attributed to the fact that even with perfect placement, the device is subject to slight movement, whereas the latter, to the unique physiology of each subject's wrist and skin tone. To eliminate this problem, the LED level is adjusted until the ADC output reaches a satisfactory level that is optimal for the imminent signal processing and for not reaching saturation. This technique was developed to lead the system to a consistent response.

Subsequent to ALEC the device starts recording the IR and RED PPG signals along with acceleration data. The acceleration is monitored with the aim to detect conditions of intense movement. The segments of the PPG signal captured in circumstances of extensive movement impose catastrophic Motion Artifacts (MAs) onto the raw PPG signal, deeming the extraction of any physiological parameter unfeasible [31]. To overcome this, segments of the PPG where motion is detected are excluded and the sampling restarts. To detect motion, the built-in logic of the ADXL362 is utilized, whose activity and inactivity events are used as triggers for manipulating the PPG sampling. An activity event is triggered when acceleration of the device remains above a predetermined threshold for a specified time period. The accelerometer features two modes of operation, the absolute and referenced configuration. During the absolute type of operation, each incoming acceleration sample is compared with a user defined threshold, which when surpassed for a certain amount of time, signals that activity is detected. On the contrary, during the referenced mode of operation, each acceleration sample undergoes a regularization, to remove the effects of gravity, able to reach 1 g, and account for the status of the device prior to sampling. This is achieved by subtracting an internally determined value, captured regularly during inactive periods, from the acceleration sample. The corrected acceleration value is then compared with the user defined threshold, and in cases it is surpassed, an activity event is issued. Consequently, activity is detected only when the acceleration has deviated sufficiently from the initial orientation. The threshold selected for the activity event was set at 350 mg for at least 1 s. This helps to eliminate only movements whose intensity can obscure the physiological parameters estimation and allows lower intensity motion to be handled by the signal processing algorithm.

3.2.2. Signal Processing

The PPG signal consists of AC and DC components. The DC component corresponds to non-pulsatile tissue, while the AC component alternates according to the heart cycle. Only the variable part of the signal is relevant for HR and RR determination; thus, the mean value is usually subtracted from the signal used in the HR and RR measurement. The recorded raw PPG signal is shown in Figure 4.

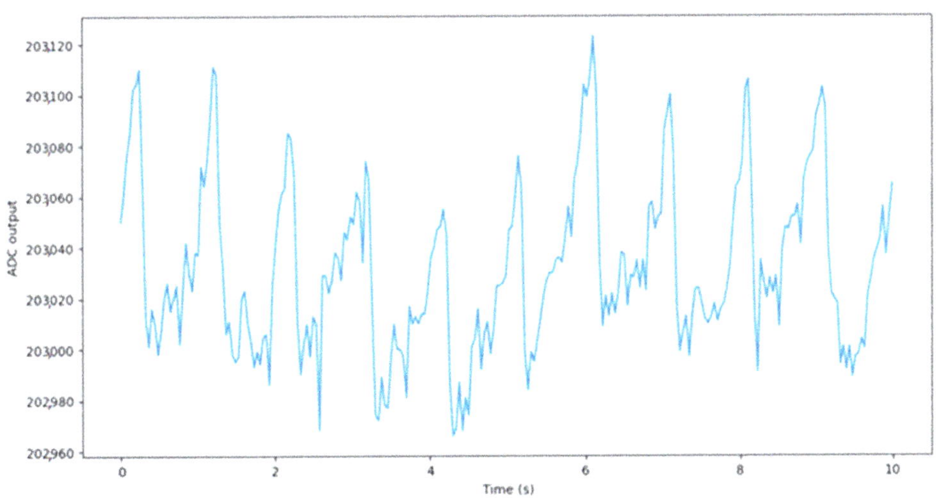

Figure 4. Raw PPG signal.

The implemented device, which is thoroughly described in [32], deploys an algorithm for the digital processing of the PPG signal in the time domain to remove the effect of MAs and the DC component. Given the fact that the appropriate to our analysis frequencies of the PPG signal are ranging from 0.1 to 5.0 Hz [33], an IIR Butterworth bandpass filter with a passband of [0.1, 5] Hz is applied prior to signal manipulation. The variable component of the signal after filtering is shown in Figure 5.

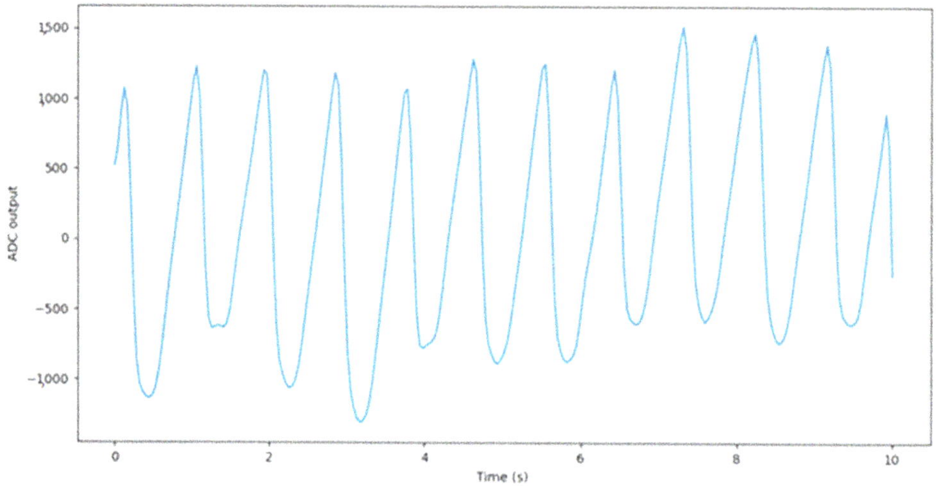

Figure 5. Filtered PPG signal.

Aiming at producing reliable estimations of the physiological parameters, extensive experiments were performed which resulted in the requirement for a PPG signal collection of at least 30 s. Within this time, useful raw PPG signal is certainly included, enabling the procedures of processing and physiological data extraction. Consequently, the measurements are produced every 30 s in cases of absence of considerable movement. As long as there are new measurements, the data are sent to the backend through Wi-Fi. Implementing the mentioned operating specifications, the device sustains a battery life of 48 h in continuous operation.

3.2.3. Heart Rate Estimation

For the estimation of the HR, the acquired signal of the infrared (IR) led source is utilized. After filtering the IR PPG signal, the Slope Sum Function (SSF) is applied [34], as shown in Equation (1). The procedure amplifies the peaks of each pulse and suppresses the noise represented by lower amplitudes.

$$\mathrm{SSF} = \sum_{k=0}^{n} \Delta x_k{}^2 \text{ where } \Delta x_k = \{\Delta s_k : \Delta s_k > 0,\ 0 : \Delta s_k \leq 0\}, \quad (1)$$

At this stage, the emphasized peaks of each window of the SSF output signal are identified as local maxima, as shown in Figure 6. After locating the peaks and estimating the time difference -d- between them, the instantaneous HR is computed for every pair of successive peaks using the Formula (2) [35]:

$$\mathrm{HR_{inst}} = \frac{6 \times 10^4}{d}, \quad (2)$$

Finally, the HR measurement is reckoned as the average of the instantaneous HR values in a 30 s time window.

3.2.4. Blood Oxygen Saturation Estimation

To enable the assessment of the blood oxygen saturation in the blood, two LEDs, operating at the RED and IR wavelengths are utilized [36]. The principle of pulse oximetry is based on the comparison of the two waveforms, whose deviation is a direct indicator for the oxygen saturation. Their deviation occurs due to the different amount of light absorbed and emitted by the two types of hemoglobin, namely oxyhemoglobin and deoxyhemoglobin. Regarding the red wavelengths, deoxyhemoglobin absorbs a higher amount of light than oxyhemoglobin, while the opposite happens in the infrared region. Hence, the responses from the RED and IR LEDs captured from the photodetector are different.

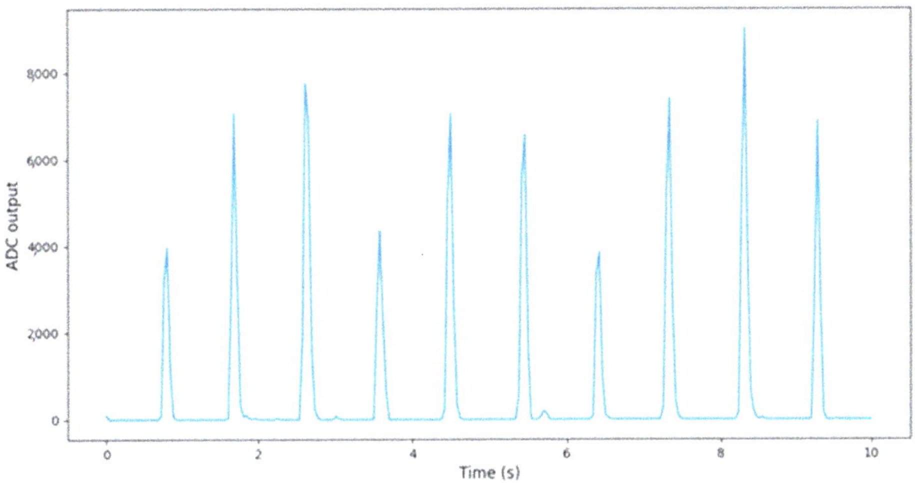

Figure 6. SSF output.

The PPG waveform consists of two different components: the DC component corresponding to the light diffusion through tissues and non-pulsatile blood layers, and the AC (pulsatile) component due to the diffusion through the arterial blood. The developed algorithm locates the existing peaks and valleys and subsequently calculates the AC and DC components of both RED and IR PPG waveforms [37]. The DC component fluctuates slightly with respiration, while the AC component oscillates in concurrence with the changes appearing in the volume of arterial blood during the cardiac cycle [38]. Given the AC and DC components, a ratio R is calculated by the Equation (3):

$$R = \frac{AC_{Red}/DC_{Red}}{AC_{IR}/DC_{IR}}, \tag{3}$$

Eventually, the SpO_2 value is estimated using the Equation (4) provided by Maxim Integrated:

$$SpO_2 = -45.006 \times R^2 + 30.354 \times R + 94.84, \tag{4}$$

3.2.5. Respiratory Rate Estimation

To perform RR estimation, two modulated signals need to be extracted from the original PPG signal obtained from the IR signal [39]. The two components, namely the Frequency Modulation (FM) and the Amplitude Modulation (AM), illustrate the effects of respiration as a physiological process on the PPG signal. Respiration is a complex process consisting of various mechanisms which cause many subtle changes to the original PPG signal. The most prominent of those effects regard FM which is the manifestation of the spontaneous increase in the heart rate during the inspiration and the corresponding decrease during expiration [40] and AM which is the result of reduced stroke volume during inhalation reducing the pulse's amplitude [41].

Following the raw PPG acquisition, a bandpass filter is applied to eliminate frequencies not related to respiratory information. The process of peak characterization includes separating the waveform in individual pulses and detecting their maximum value. The FM can be defined as the time difference between two consecutive peaks as described in Equation (5), whereas the AM is formed by each individual amplitude peak of the signal as shown in Equation (6). Each time value assigned to the FM and AM signal samples is calculated as the mean of the time of occurrence of two peaks as shown in Equation (7).

$$x_{FM} = \left| t_{peak_{i-1}} - t_{peak_i} \right|, \ i = 2, .., N, \tag{5}$$

$$x_{AM} = \left| y_{peak_i} \right|, \ i = 1,\ 2, .., N, \tag{6}$$

$$t = \frac{\left| t_{peak_{i+1}} + t_{peak_i} \right|}{2}, \ i = 2, .., N, \tag{7}$$

The values of the modulated signals are not homogenous, thus inhibiting the signal processing. To evenly sample the two waveforms, Shape-Preserving Piecewise Cubic Interpolation is performed on the acquired data points and the sampling rate is set at 4 Hz. Prior to the Fast Fourier Transform (FFT), a Hamming window is applied to minimize the side lobes of the frequency response.

At this stage, the two power spectra are combined to amplify the potential peaks, the dominant frequency (F_d) in the plausible range is identified and the final RR value is then computed by Equation (8):

$$RR = F_d \times 60 \ (\text{breaths/min}), \tag{8}$$

3.2.6. Data Transmission

The wearable device transmits data to an API endpoint via Wi-Fi, implementing the HTTPS communication protocol which follows the HTTP protocol over a secure and encrypted connection. A time window of 30 s interposes between two data transmissions. A payload in URL encoded format is generated, which includes the values of physiological parameters collected by the wearable device; HR, SpO_2, and RR. In addition, the payload includes the MAC address of the Wi-Fi's Access Point (AP) and the battery level of the wearable device. The parameters of the payload are described in Table 3.

Table 3. Parameters of payload.

Variable	Description
mac	MAC address of the Wi-Fi AP
bat	Battery level of wearable device
timestamp	The time when the measurements are extracted (in the form of Epoch Unix Timestamp)
hr	Heart rate
spo2	Blood oxygen saturation
rr	Respiratory rate

The API service, located on the server-side of the proposed web-based application, provides an endpoint to wearable device in which it can POST requests. Therefore, a parametric URL, presented in Figure 7, is utilized by the API service. By the time a POST request is applied to the API service, it parses the URL, deploys the GET variable to extract the value of each parameter, validates that each parameter has the appropriate format and stores the parameters into the database.

Figure 7. API service parametric URL.

The MAC information is included as the proposed device has the capability of connecting to the Wi-Fi AP with the best signal strength in the case there is more than one

Wi-Fi AP in the surrounding space. The AP with the highest signal strength will be the closest to the user. Thus, the MAC address of a particular AP can be exploited for assuming the approximate location of the user, given the location of each AP.

Typically, the SSID and the password of the Wi-Fi network are assigned into variables within the device's code. This would require the end-users to enter their Wi-Fi credentials and upload new code on the device. To overcome this, the Wi-Fi Manager library is implemented, which allows end-users to connect to different APs without having to interact with the firmware. More specifically, when the device is activated by a user, a connection to a previously saved AP is attempted. If this process fails, the AP mode is enabled, allowing the user to configure a new set of SSID and password. The user has to navigate to a web page with default IP address 192.168.4.1 and enter his Wi-Fi credentials into a form. Once a new valid set of SSID and password is set, the device automatically reboots and establishes a connection.

The device also has the capacity to locally store data in cases of dropped or lost Wi-Fi connection. ESP8266 module provides the user with a flash memory of 1 MB, from which the 0.4 MB are occupied by the device's firmware. The remaining available memory is utilized for storing the measurements produced during the absence of network connection. The saved data are transmitted when the Wi-Fi connection is restored and then the device continues its regular operation. The device extracts the physiological parameters every 30 s and transmits them along with other data. As mentioned, the 30 s time period is a specification, which emerged during the trials, since within this time period, useful raw PPG signal is definitely included, facilitating the production of reliable physiological data values.

3.2.7. Measurements Evaluation

Aiming to evaluate the performance of the suggested wrist-wearable device and the accuracy of the corresponding extracted physiological parameters, commercial off-the-shelf certified devices were used. The values obtained by these devices are considered as reference and are compared with the values of the proposed wrist-wearable device.

Regarding the evaluation process of HR and SpO_2 physiological parameters a medical finger pulse oximeter was utilized. The commercial finger pulse oximeter chosen is a certified medical device manufactured by Berry: BM2000D Bluetooth Pulse Oximeter [42]. The accuracy of RR determination methodology was evaluated utilizing the chest worn Zephyr BioHarness device, which is a physiological monitoring system with proven reliability in determining RR [43].

Ten healthy subjects with varying wrist circumferences and skin tones were provided with the wrist-wearable device and the reference devices. In particular, the subjects were equipped with the proposed wrist-wearable device and the Berry Pulse Oximeter along with the Zephyr BioHarness as ground truth devices. The experiments were performed at a sedentary state and the total duration of the experiment for each subject was 1 h, yielding an aggregation of 10 h of data.

To analyze the alignment between the data acquired from our proposed device and those from the reference instruments, the Bland–Altman plot was deemed ideal. The Bland–Altman graph consists of a plot of the difference between paired readings of two variables, in our case the derived and the reference values, over the average of these readings. Incorporated into the plot are the ± 1.96 SD lines (the Confidence Interval) parallel to the mean difference line.

Figures 8–10 present the comparative analysis of our data and render the proposed device a reliable system for the extraction of the desired physiological parameters.

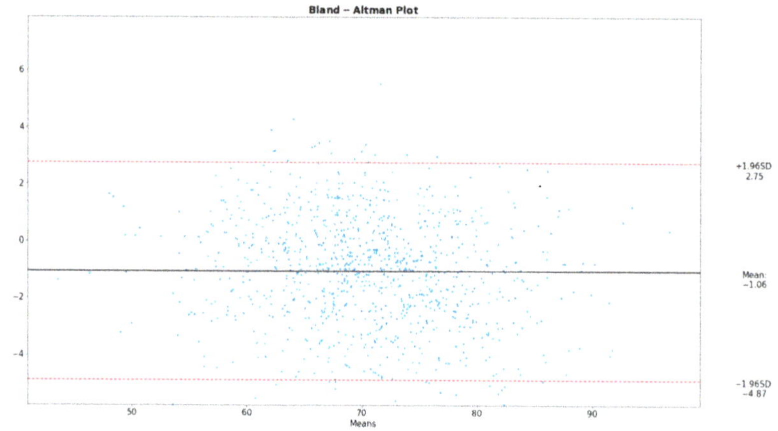

Figure 8. Bland–Altman plot for HR.

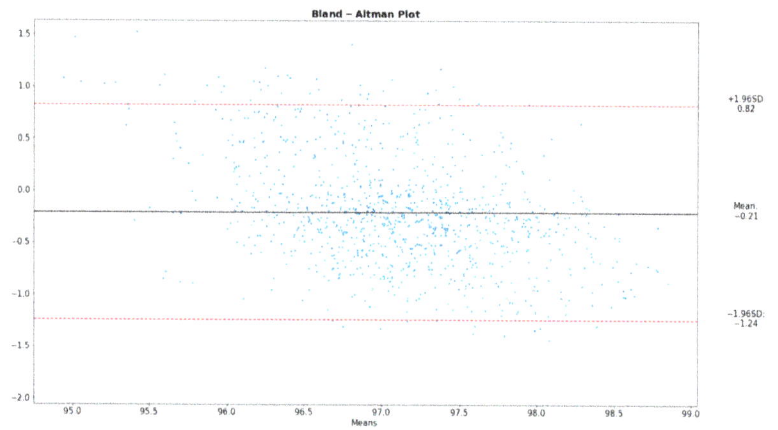

Figure 9. Bland–Altman plot for SpO_2.

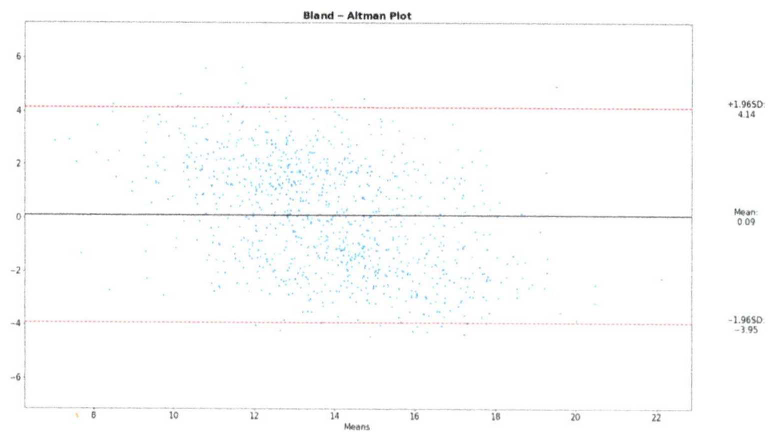

Figure 10. Bland–Altman plot for RR.

The Bland–Altman plot displays four types of data misbehavior: systematic error (mean offset), proportional error (trend), inconsistent variability, and excessive or erratic variability. It should be noted that the Bland–Altman comparison method innately assumes that the two methods compared are portraying close results, a condition satisfied in our analysis.

4. Web-Based Application

Nowadays, web-based applications dominate markets as they offer considerable advantages over traditional desktop applications. Web-based applications run directly on web browsers, demand minimal computer resources, do not require any installation process, are highly scalable, and maintenance tasks are performed centrally on the web server. In addition, they are portable and cross-platform available, which allows users to access them from any place and any type of device. Therefore, web-based applications are deemed to be the optimal solution for flexible projects.

Web-based applications can be structured with various architectural patterns. The proposed one adopts the client-server model. The client-side refers to the visible and interactive part of the application, whereas the server-side is responsible for processing client requests and responses. In most cases, the server-side includes a database to support transactions with data and API services to support interoperability with end-devices and systems.

4.1. Front-End Implementation

Front-end stands for the client-side of a web-based application. It provides a graphical interface, which allows end-users to interact with the web-based application and visualizes the information acquired by the wearable device. The front-end of the proposed web-based application employs the core web technologies; HTML5 structures the UI components, CSS styles the UI components, and JavaScript enables interactivity between the UI components and the end-users. In addition, a series of external frameworks and libraries were utilized to optimize the development process, i.e., to meet the business objective of the web-based application with minimal effort cost, achieve high performance standards, and ensure the appropriate infrastructure for further scalability. More specifically, Bootstrap framework, an open-source framework for front-end development, employed to accelerate the development process, optimize the overall performance in terms of computational resources, and provide a set of user-friendly and highly customizable UI components. This framework supports responsive design for web-based applications and incorporates web-accessibility standards. Chart.js, an open-source JavaScript library, employed for advanced chart implementation within the front-end. It offers pre-built and highly customizable UI chart components, allows efficient handling of data objects, optimizes the performance of charts' drawing process, and improves the style of visualized data. jQuery, a lightweight and open-source JavaScript framework, employed to simplify event handling, improve the manipulation of UI components, and empower the interactivity capabilities.

The dashboard, shown in Figure 11, is the core interface of the front-end, and serves as personalized analytics overview and real-time monitoring tool. The primary objective of the proposed dashboard relates to the visualization of the user's latest measurements, captured by the wearable device, in an intuitive and user-friendly way. Therefore, it employs UI components, mainly embedded line charts within cards, to depict the trend of each physiological parameter over the last one hour. Charts establish an asynchronous connection with the database; thus, they are automatically updated in a real-time manner with the latest measurements recorded by the wearable device. The horizontal axis of each chart represents the time, whereas the vertical axis represents the captured values of the specific physiological parameter displayed. Below each chart stands a card footer which informs end-users for the time of the chart's data latest update.

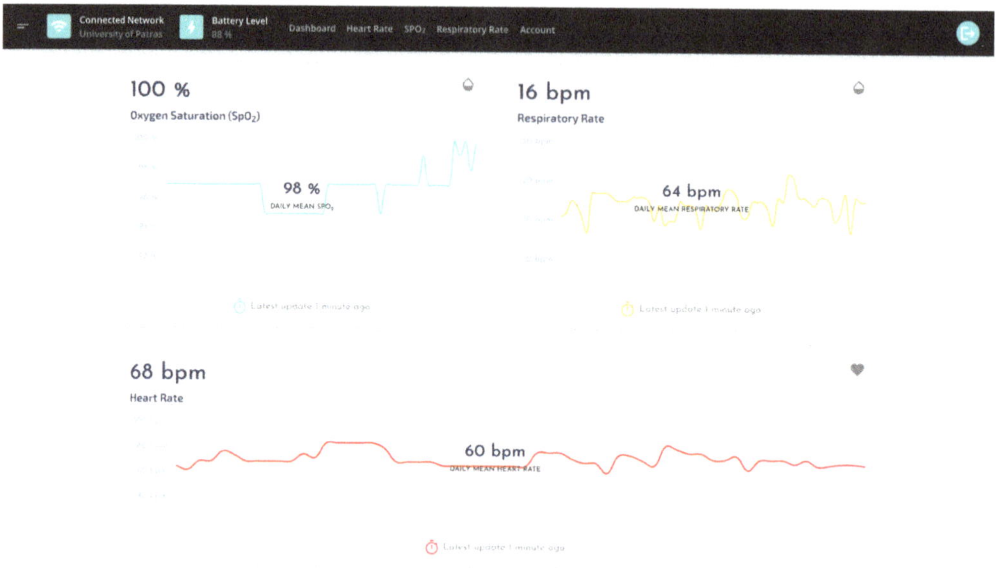

Figure 11. The UI of dashboard.

Another core component of the front-end is the horizontal navigation menu bar, placed at the UI's header area. It comprises a minimal information area, which displays the connection status of the wearable device and the battery level, as well as a list that maps the analytics and account areas of the web-based application. Analytics areas (Figure 12) visualize historical data for the corresponding physiological parameter in a user-friendly way. In addition, they implement a data picker filter to allow end-users apply the desired time span of the displayed data.

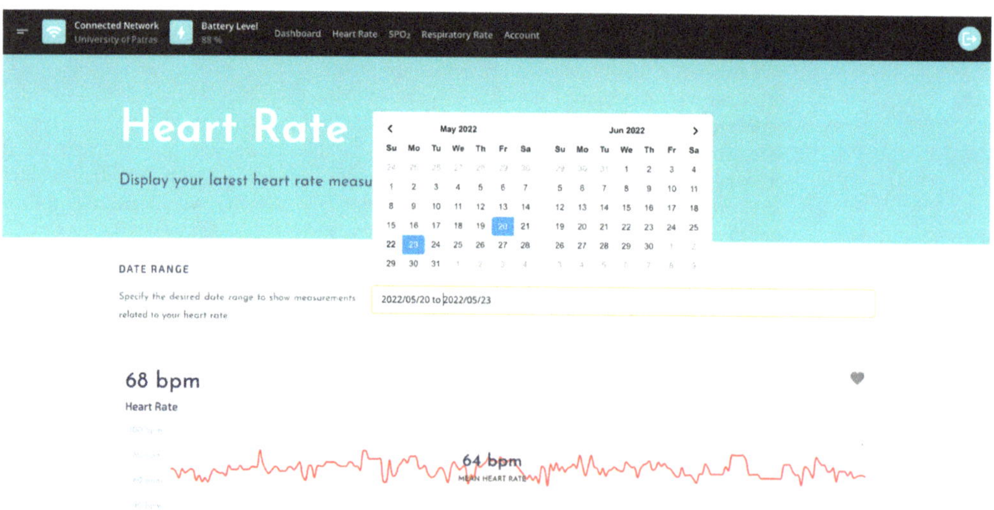

Figure 12. The UI of heart rate analytics.

Account area (Figure 13) displays a list with the available Wi-Fi networks and permit users to perform two (2) actions; edit network and delete network. The 'edit network' action allow users to change credentials of an existing Wi-Fi connection and 'delete network' action allows users to remove a specific network from the saved networks list. Moreover, from the 'Account' page, users can configure a connection with a new network.

Figure 13. The UI of user account page.

4.2. Back-End Implementation

Back-end stands for the server side of a web-based application. On the back-end side, the web-based application implements embedded PHP scripts within the front-end code to provide dynamic functionalities. The dynamic aspect of the web-based application enables the capability to retrieve and visualize data from the database in a near real-time manner. The back-end comprises a RESTful API that allows submission of collected data from the wearable device into the database. More specifically, it parses a JSON object with the payload's data and decodes it to extract each value and insert it into the database. A data validation process takes place to ensure the quality of data, such as checks for duplicate entries, appropriate data format and type, null values. In addition, SQL queries are executed to confirm that the extracted identifiers exist on the database to avoid conflicts on the data selection process.

A user authentication mechanism stands at the back-end, too. It has been implemented with Keycloak, an efficient, reliable, and extendable authentication and resource access management framework for web-based applications. The proposed application supports rights for two (2) roles, the administrator and the user. Administrators can access only an administrative dashboard page from which they can perform administrative tasks such as manual password resets, user account management, and service health monitoring. Users can access only the application UI resources—dashboard, analytics pages and account configuration page—from which they can keep up with visualized insights, apply changes related to their account information, and configure connections with networks.

The web-based application employs a MySQL database to perform transactions with data automatically collected by the wearable device and data manually provided by the end-users. Automatically collected data are included within the transmitted payload, i.e., physiological parameters, timestamp and MAC address. The database adopts a relational schema which structures the data into tables, as shown in Figure 14. The core functionality of the web-based application, i.e., data visualization, user authentication, and connection configuration, relies on four (4) tables; "Users", "Devices", "Connections" and "HealthRecords".

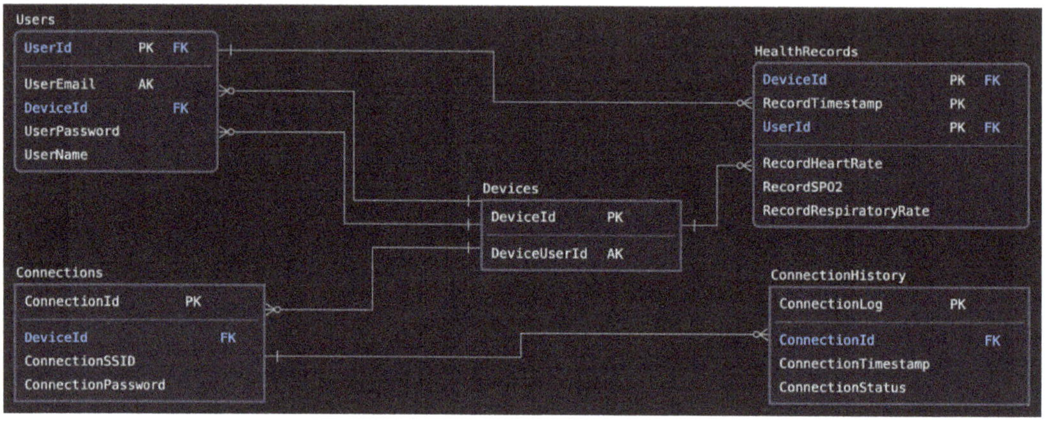

Figure 14. Database schema.

1. The table of "Users" represents the entity of end-user. It stores information related to each end-user's authentication credentials and the identifier of wearable device which is registered by the specific end-user. The primary key (PK) of this table is 'UserId' (int) which is automatically produced when a new end-user registers the web-based application. Therefore, auto-increment indexing has been declared to this attribute.
2. The table of 'Devices' represents the entity of wearable device. It stores minimal information related to the matching between end-users and wearable devices. The primary key (PK) of this table is the attribute 'DeviceId' (varchar) which corresponds to the MAC address of each wearable device, as parsed from the transmitted payload.
3. The table of 'Connections' stores information related to the Wi-Fi connection established between the wearable device and the local network. It stores information about the credentials of each connection, i.e., SSID and password. For each set of credentials, stands a 'ConnectionId' (int) which is automatically produced at the server-side with auto-increment indexing.
4. The table of 'HealthRecords' stores information related to the physiological parameters. It has composite primary key (PK) which consists of the user identifier, wearable device identifier, and timestamp of the collected measurements. The attributes of user and wearable identifiers are foreign keys (FK) 'Users' and 'Devices', respectively. In addition, it includes attributes that represent all the physiological parameters collected by the wearable device, i.e., heart rate, SpO_2, and respiratory rate.

5. Discussion

Physiological parameters provide critical information about individuals' well-being status and signal early signs of a body dysfunction. For example, detecting an abnormally high HR could be an indicator that actionable measures should be taken to achieve healthy levels in order to reduce the risk of cardiovascular disease, while monitoring RR can detect early signs of a respiratory illness or allergy. SpO_2 is useful in any setting where an individual's oxygenation may be unstable or low for determining the sufficiency of oxygen or the need for supplemental oxygen.

Digital health technologies facilitate an individual-focused preventative approach through continuous monitoring of physiological parameters. This approach paves the way for personalized treatment, better care access and quality of service.

Providing that the subject be in a sedentary state, the proposed wearable apparatus introduced here, can unfailingly detect PPG signals and then reliably extract the physiological parameters of HR, SpO_2 and RR. Specifically, the device achieves a mean percentage error equal to 2.47% and 0.8% for HR and SpO_2, respectively, while estimating the RR parameter with a deviation of ± 1.4 breaths per minute. The evaluation procedures showed that the wrist-wearable device can accurately detect fluctuations of the physiological parameters in a sedentary state. Consequently, it can be used to effectively monitor well-being status and provide valuable information. Moreover, a portable and cross-platform available web-based application has been developed, which serves as an informative and Wi-Fi connection establishment tool, from which individuals and health professionals can access the entailed parameters in a user-friendly and efficient way regardless the location, the type of device, and the operating system, and handle connections with networks in a simple way.

Impetus for future work is to enhance the accuracy of the extracted parameters in all respects. Our endeavor aspires to adopt an accelerometer-based detection and removal of faulty PPG segments or a signal preprocessing approach for MAs elimination. Alongside this, more trials with different subjects should be further performed. Collecting a larger amount of sample data from users of a broader spectrum in terms of age or skin color would also provide an opportunity for better algorithm calibration. Our future research interest is also focused on a multi wavelength photoplethysmography approach, which has proven superior performance than the single wavelength. Recently, advances on the sensing modality for detecting light from multiple sources enabled the development of a single chip sensor, removing the need for spectrometers, which have a prohibitive size for a wearable device. Moreover, in an effort to expand our knowledge on the effect of the light wavelength on the quality of the PPG signal, experiments are being conducted at wavelengths other than the red and infrared, which are currently used. Last but not least, such a device paves the way for injury prevention, early detection of illnesses or disorders, as well as early interventions with the aim to avoid the deterioration of health conditions.

6. Conclusions

This work presents in both physical and operational level all the discrete components of a comprehensive, user-embracive system able to unobtrusively record and process vital physiological parameters. It comprises a non-invasive wrist-wearable device able to incessantly detect PPG signals and solidly extract the physiological parameters of HR, SpO_2 and RR and a multimodal web-based application via which the end users visualize real-time or historical data while allows health professionals to interact with that data for further algorithmic processing. The configuration of the system and its Wi-Fi connection was designed to be effortless even for older individuals that are not so much accustomed to high-end technology. The wrist-wearable device is a lightweight modular embedded device with a microcontroller based main board exhibiting optimized memory capacity and processing power, as well as long autonomy, and portable mounted off-the-shelf sensors for capturing the PPG signal. Moreover, by supporting direct connection to 802.11.xx communication protocols, it is an ideal device for the utilization of existing communication infrastructures that offer high speed information sharing. The cloud based back-end infrastructure offers all the required means to securely store the transmitted data from the wearable device over HTTPS protocol in a time-series manner. Both health professionals and end users themselves have easy access to historical and real-time data. The professionals can utilize the collected historical data to perform statistical analysis or execute AI/ML methods, aim to obtain valuable health information either for an individual or a group of them, thus unlocking a vast field of possibilities. The end users can glance at their data according to their preferences, simply by applying filters to adjust the friendly UI chart components of the front-end.

The accuracy assessment of the extracted physiological parameters, along with the evaluation of the system performance, were carried out against commercial off-the-shelf certified equipment, which was worn by healthy subjects with different anatomical characteristics.

Future actions include the increase in accuracy of the extracted parameters in all respects, the enhancement of the algorithmic processing capabilities and the execution of more trials on diverse subjects.

Author Contributions: Conceptualization, J.G.; Data curation, G.P., C.M. and G.T.; Methodology, J.G. and G.T.; Resources, J.G.; Software, M.K., J.G., G.P., C.M. and G.T.; Writing—review & editing, M.K., J.G. and C.M. All authors have read and agreed to the published version of the manuscript.

Funding: This research was co-funded by the Greek General Secretariat of Research and Technology, through ESPA 2014–2020, under the project T1EDK-02489 entitled "Intelligent System in the Hospitals ED and Clinics for the TRIAGE and monitoring of medical incidents—IntelTriage" and the European Union, under the Horizon 2020 project Nª 957736 entitled "Intelligent Interconnection of Prosumers in PEC with Twins of Things for Digital Energy Markets—TwinERGY", H2020-LC-SC3-2020-EC-ES-SCCRIA.

Institutional Review Board Statement: Not applicable.

Informed Consent Statement: Not applicable.

Data Availability Statement: Not applicable.

Acknowledgments: J.G. and G.T. made substantial contributions to the conception, the design and implementation of the hardware components and embedded software modules of the proposed integrated system, provided approval for publication of the content, and agreed to be accountable for all aspects of the work. G.P. and M.K. made substantial contributions to the design and implementation of the embedded software modules, provided approval for publication of the content, and agreed to be accountable for all aspects of the work. C.M. made substantial contributions to the design and implementation of the web-based application, provided approval for publication of the content, and agreed to be accountable for all aspects of the work.

Conflicts of Interest: The authors declare no conflict of interest.

References

1. United Nations. Department of Economics and Social Affairs. Population Division (2017). In *World Population Ageing, 2017 Highlights*; United Nations: New York, NY, USA, 2017.
2. Thapliyal, H.; Nath, R.K.; Mohanty, S.P. Smart Home Environment for Mild Cognitive Impairment Population: Solutions to Improve Care and Quality of Life. *IEEE Consum. Electron. Mag.* **2018**, *7*, 68–76. [CrossRef]
3. Maswadi, K.; Ghani, N.B.A.; Hamid, S.B. Systematic Literature Review of Smart Home Monitoring Technologies Based on IoT for the Elderly. *IEEE Access* **2020**, *8*, 92244–92261. [CrossRef]
4. Eurostat. Healthcare Expenditure Statistics—Statistics Explained. [online] ec.europa.eu. 2021. Available online: https://ec.europa.eu/eurostat/statistics-explained/index.php?title=Healthcare_expenditure_statistics (accessed on 22 May 2022).
5. Zhang, Y.; Song, S.; Vullings, R.; Biswas, D.; Simões-Capela, N.; van Helleputte, N.; van Hoof, C.; Groenendaal, W. Motion Artifact Reduction for Wrist-Worn Photoplethysmograph Sensors Based on Different Wavelengths. *Sensors* **2019**, *19*, 673. [CrossRef]
6. Tamura, T.; Maeda, Y.; Sekine, M.; Yoshida, M. Wearable Photoplethysmographic Sensors—Past and Present. *Electronics* **2014**, *3*, 282–302. [CrossRef]
7. Castaneda, D.; Esparza, A.; Ghamari, M.; Soltanpur, C.; Nazeran, H. A review on wearable photoplethysmography sensors and their potential future applications in health care. *Int. J. Biosens. Bioelectron.* **2018**, *4*, 195–202.
8. Clarke, G.W.J.; Chan, A.D.C.; Adler, A. Effects of motion artifact on the blood oxygen saturation estimate in pulse oximetry. In Proceedings of the IEEE International Symposium on Medical Measurements and Applications (MeMeA), Lisbon, Portugal, 11–12 June 2014.
9. Cornacchia, M.; Ozcan, K.; Zheng, Y.; Velipasalar, S. A Survey on Activity Detection and Classification Using Wearable Sensors. *IEEE Sens. J.* **2017**, *17*, 386–403. [CrossRef]
10. Li, X.; Dunn, J.; Salins, D.; Zhou, G.; Zhou, W. Digital Health: Tracking Physiomes and Activity Using Wearable Biosensors Reveals Useful Health-Related Information. *PLoS Biol.* **2017**, *15*, e2001402. [CrossRef]
11. Mishra, T.; Wang, M.; Metwally, A.A. Wearable Sensors for COVID-19: A Call to Action to Harness Our Digital Infrastructure for Remote Patient Monitoring and Virtual Assessments. *Nat. Biomed. Eng.* **2020**, *4*, 1204–1220.
12. Seshadri, D.R.; Davies, E.V.; Harlow, E.R.; Hsu, J.J.; Knighton, S.C.; Walker, T.A.; Voos, J.E.; Drummond, C.K. Pre-symptomatic detection of COVID-19 from smartwatch data. *Front. Digit. Health* **2020**, *4*, 1208–1220.

13. Downey, C.L.; Chapman, S.; Randell, R.; Brown, J.M.; Jayne, D.G. The impact of continuous versus intermittent vital signs monitoring in hospitals: A systematic review and narrative synthesis. *Int. J. Nurs. Stud.* **2018**, *84*, 19–27. [CrossRef]
14. Zenko, J.; Kos, M.; Kramberger, I. Pulse rate variability and blood oxidation content identification using miniature wearable wrist device. In Proceedings of the International Conference on Systems, Signals and Image Processing (IWSSIP), Bratislava, Slovakia, 23–25 May 2016.
15. Son, L.P.; Thu, N.T.A.; Kien, N.T. Design an IoT wrist-device for SpO2 measurement. In Proceedings of the International Conference on Advanced Technologies for Communications (ATC), Quynhon City, Vietnam, 18–20 October 2017.
16. Jarchi, D.; Salvi, D.; Velardo, C.; Mahdi, A.; Tarassenko, L.; Clifton, D.A. Estimation of HRV and SpO2 from wrist-worn commercial sensors for clinical settings. In Proceedings of the IEEE 15th International Conference on Wearable and Implantable Body Sensor Networks (BSN), Las Vegas, NV, USA, 4–7 March 2018.
17. Preejith, S.P.; Alex, A.; Joseph, J.; Sivaprakasam, M. Design, development and clinical validation of a wrist-based optical heart rate monitor. In Proceedings of the IEEE International Symposium on Medical Measurements and Applications (MeMeA), Benevento, Italy, 15–18 May 2016.
18. Eugene, L.; Chen, Y.L. PPG-Based Smart Wearable Device with Energy-Efficient Computing for Mobile Health-Care Applications. *IEEE Sens. J.* **2021**, *21*, 13564–13573.
19. Wojcikowski, M. Real-Time PPG Signal Conditioning with Long Short-Term Memory (LSTM) Network for Wearable Devices. *Sensors* **2022**, *22*, 164. [CrossRef]
20. Münzner, S.; Schmidt, P.; Reiss, A.; Hanselmann, M.; Stiefelhagen, R.; Dürichen, R. CNN-based sensor fusion techniques for multimodal human activity recognition. In Proceedings of the 2017 ACM International Symposium on Wearable Computers, New York, NY, USA, 11–15 September 2017; pp. 158–165.
21. Tang, Y.; Zhang, L.; Min, F.; He, J. Multi-scale Deep Feature Learning for Human Activity Recognition Using Wearable Sensors. *IEEE Trans. Ind. Electron.* **2022**. [CrossRef]
22. Zhang, Y.; Haghighi, P.D.; Burstein, F.; Yao, L.; Cicuttini, F. On-Device Lumbar-Pelvic Movement Detection Using Dual-IMU: A DNN-Based Approach. *IEEE Access* **2021**, *9*, 62241–62254. [CrossRef]
23. Empatica. Available online: https://www.empatica.com/en-eu/research/e4/ (accessed on 22 May 2022).
24. Maxim Integrated. Available online: https://www.maximintegrated.com/en/products/sensors/MAXREFDES103.html (accessed on 22 May 2022).
25. Fitbit. Available online: https://www.fitbit.com/global/us/products/smartwatches/versa3 (accessed on 30 June 2022).
26. Samsung. Available online: https://www.samsung.com/us/watches/galaxy-watch4/buy/?modelCode=SM-R840NTKAXAR (accessed on 30 June 2022).
27. Apple. Available online: https://www.apple.com/watch/?afid=p239%7C196318&cid=aos-us-aff-ir (accessed on 30 June 2022).
28. Espressif. Available online: https://www.espressif.com/en/products/socs/esp8266 (accessed on 22 May 2022).
29. Maxim Integrated. Available online: https://www.maximintegrated.com/en/design/reference-design-center/system-board/6300.html/tb_tab2 (accessed on 22 May 2022).
30. Analog Devices. Available online: https://www.analog.com/en/products/adxl362.html (accessed on 22 May 2022).
31. Ismail, S.; Akram, U.; Siddiqi, I. Heart rate tracking in photoplethysmography signals affected by motion artifacts: A review. *EURASIP J. Adv. Signal Process.* **2021**, *2021*, 5. [CrossRef]
32. Krizea, M.; Gialelis, J.; Kladas, A.; Protopsaltis, G.; Theodorou, G.; Koubias, S. Accurate Detection of Heart Rate and Blood Oxygen Saturation in Reflective Photoplethysmography. In Proceedings of the IEEE International Symposium on Signal Processing and Information Technology (ISSPIT), Louisville, KY, USA, 9–11 December 2020.
33. Biswas, D.; Capela, N.S.; Hoof, C.; Helleputte, N. Heart Rate Estimation from Wrist-Worn Photoplethysmography: A Review. *IEEE Sens. J.* **2019**, *19*, 6560–6570. [CrossRef]
34. Jang, D.; Park, S.; Hahn, M.; Park, S. A Real-Time Pulse Peak Detection Algorithm for the Photoplethysmogram. *Int. J. Electron. Electr. Eng.* **2014**, *2*, 45–49. [CrossRef]
35. Ram, M.R.; Madhav, K.V.; Krishna, E.H.; Reddy, K.N.; Reddy, K.A. Adaptive reduction of motion artifacts from PPG signals using a synthetic noise reference signal. In Proceedings of the IEEE EMBS Conference on Biomedical Engineering and Sciences (IECBES), Kuala Lumpur, Malaysia, 30 November–2 December 2010.
36. Jubran, A. Pulse oximetry. *Crit. Care* **2015**, *19*, 272. [CrossRef]
37. GitHub—MaximIntegratedRefDesTeam. Available online: https://github.com/MaximIntegratedRefDesTeam/ (accessed on 22 May 2022).
38. Tamura, T. Current progress of photoplethysmography and SPO2 for health monitoring. *Biomed. Eng. Lett.* **2019**, *9*, 21–36. [CrossRef]
39. Charlton, P.H.; Birrenkott, D.A.; Bonnici, T.; Pimentel, M.A.F.; Johnson, A.E.W.; Alastruey, J.; Tarassenko, L.; Watkinson, P.J.; Beale, R.; Clifton, D.A. Breathing Rate Estimation from the Electrocardiogram and Photoplethysmogram: A Review. *IEEE Rev. Biomed. Eng.* **2018**, *11*, 2–20. [CrossRef]
40. Berntson, G.G.; Cacioppo, J.T.; Quigley, K.S. Respiratory sinus arrhythmia: Autonomic origins, physiological mechanisms, and psychophysiological implications. *Soc. Psychophysiol. Res. Psychophysiol.* **1993**, *30*, 183–196. [CrossRef] [PubMed]

41. Meredith, D.J.; Clifton, D.; Charlton, P.; Brooks, J.; Pugh, C.W.; Tarassenko, L. Photoplethysmographic derivation of respiratory rate: A review of relevant physiology. *J. Med. Eng. Technol.* **2012**, *36*, 1–7. [CrossRef] [PubMed]
42. Berry. Available online: https://www.shberrymed.com/wrist-pulse-oximeter-p00040p1.html (accessed on 22 May 2022).
43. Jono, H.; Andrew, E.K. Reliability and Validity of the Zephyr™ BioHarness™ to Measure Respiratory Responses to Exercise. *Meas. Phys. Educ. Exerc. Sci.* **2020**, *15*, 293–300.

sensors

MDPI

Article

PATROL: Participatory Activity Tracking and Risk Assessment for Anonymous Elderly Monitoring

Research Dawadi [1], Teruhiro Mizumoto [2], Yuki Matsuda [1] and Keiichi Yasumoto [1,*]

[1] Graduate School of Science and Technology, Nara Institute of Science and Technology, Ikoma 630-0192, Nara, Japan
[2] Graduate School of Information Science and Technology, Osaka University, Suita 565-0871, Osaka, Japan
* Correspondence: yasumoto@is.naist.jp; Tel.: +81-90-2460-3965

Abstract: There has been a subsequent increase in the number of elderly people living alone, with contribution from advancement in medicine and technology. However, hospitals and nursing homes are crowded, expensive, and uncomfortable, while personal caretakers are expensive and few in number. Home monitoring technologies are therefore on the rise. In this study, we propose an anonymous elderly monitoring system to track potential risks in everyday activities such as sleep, medication, shower, and food intake using a smartphone application. We design and implement an activity visualization and notification strategy method to identify risks easily and quickly. For evaluation, we added risky situations in an activity dataset from a real-life experiment with the elderly and conducted a user study using the proposed method and two other methods varying in visualization and notification techniques. With our proposed method, 75.2% of the risks were successfully identified, while 68.5% and 65.8% were identified with other methods. The average time taken to respond to notification was 176.46 min with the proposed method, compared to 201.42 and 176.9 min with other methods. Moreover, the interface analyzing and reporting time was also lower (28 s) in the proposed method compared to 38 and 54 s in other methods.

Keywords: elderly monitoring; successful aging; mobile application; gerontechnology

Citation: Dawadi, R.; Mizumoto, T.; Matsuda, Y.; Yasumoto, K. PATROL: Participatory Activity Tracking and Risk Assessment for Anonymous Elderly Monitoring. *Sensors* **2022**, *22*, 6965. https://doi.org/10.3390/s22186965

Academic Editors: Ivan Miguel Serrano Pires and Antoni Martínez Ballesté

Received: 4 July 2022
Accepted: 12 September 2022
Published: 14 September 2022

1. Introduction

Advancements in medicine and health care technologies have led to an increase in life expectancy over the years. It is expected that, by 2050, there will be at least 2 billion people over the age of 60 years [1]. The statistical handbook of Japan released in 2021 by the Statistics Bureau, Ministry of Internal Affairs and Communications, Japan has revealed that, in 2015, there were about 22 million households with residents aged 65 and above, including 6 million who lived alone [2]. Living independently, especially for the elderly, is risky because, in addition to mental problems such as memory loss, depression, and loneliness, there can be physical problems such as falling down, issues with eyesight, hearing loss, back pain, etc. [3]. Though different remedies have been developed for different types of physical and mental ailments, with an increasing number of elderly people, it is apparent that there is a need for monitoring and anomaly detection mechanisms. A lot of research has thus contributed to recognizing, predicting, and monitoring activities inside smart homes [4,5].

As people get older, their involvement in different physical and mental activities decline [6]. They go out less, engage in activities related to physical fitness less, have difficulty with reading for a long time due to weakened eyesight, and so on. Similarly, they deal with issues they had not dealt when they were younger, such as the need to take medication every day and the adverse effects of missing a meal. Similarly, falls or any similar incidents tend to make the elderly cautious in their activities, impacting their confidence, activity completion, and social interactions. Therefore, it becomes imperative to track whether the elderly has completed basic day-to-day activities every day in order to detect any abnormal conditions that might have occurred or might occur [5,7]. There have been many advancements in human monitoring, collecting vital health statistics and tracking human behavior over the recent years [8]. Off-the-shelf sensors can be now used in houses that can provide information about light intensity, temperature, and usage of doors and appliances of houses [9], making it possible to determine activities inside the house.

Research has also been carried out in health care centers, but implementing such technology in the home environment is more suitable for the elderly. The elderlies have made memories over the years in their home and have possessions they cherish [10]. Hence, they feel more comfortable to live in their own home as well as conduct their basic everyday activities. Moreover, hospitals and health care centers are either expensive or overbooked. The cost can be reduced by up to 52% when patients receive treatment and help in their home compared to hospitals [11]. It is therefore necessary to develop systems that can help to enhance elderly care in their own home rather than hospitals or support homes. Professional caretakers are expensive as well, and with the increasing number of elderly people they tend to be overbooked and busy [5]. Home monitoring technologies can help family members and relatives who are far away be assured about the safety and contentment of the elderly [1]. However, their busy schedule may not allow them to monitor the activities regularly, which is why personnel dedicated to remote monitoring such as remote caretakers or volunteers should be assigned the monitoring responsibilities.

With these issues in consideration, in this paper, we propose a monitoring system, PATROL (Participatory Activity Tracking and Risk assessment for anOnymous eLderly monitoring) that can track basic activities of the elderly anonymously inside their home and detect or prevent any potential risks in their day to day activities using a smartphone application. For the successful implementation of the PATROL system, the following requirements need to be fulfilled: *(Req. 1) anonymous monitoring*, *(Req. 2) timely monitoring and report of activities*, and *(Req. 3) easy and intuitive risk detection* because of the following reasons.

Home monitoring can be considered intrusive as in some cases, the elderly may prefer to hide things in their house if there is a video based monitoring or surveillance system [12]. Similarly, they are also usually concerned about privacy and security, and the types of information about them that are disclosed [1]. This is why we propose anonymous monitoring (Req. 1), where any personal details of the elderly being monitored is not disclosed to the monitoring person. Smartphones are a suitable device for regular tracking and monitoring since many people carry them the whole day or they are always in the vicinity of the users. Furthermore, notifications have become an essential feature of most of the smartphone applications [13]. This is why we propose a smartphone application that can be used by volunteers for tracking and monitoring activities of elderly people. Similarly, we send frequent notifications in the smartphone application, which ensures that the monitors can quickly access information about the activities of the elderly, compared to using web pages (Req. 2). Continuous usage of smartphone applications in general has been attributed to factors such as ease of navigation, ease of carrying out actions within the application, and appropriate visual clues [14], which is why we focus on the visualization of activities and propose a method for visualizing activities and detecting risks in the daily activities that not only helps to identify risks in the activity visualization easily, but also incurs a lesser burden to the monitoring person (Req. 3).

Therefore, in this paper, we propose an elderly monitoring system that can be used by anonymous volunteers to check everyday activities of the elderly and determine if there are any risky situations in their day to day activities. The anonymity is maintained by not disclosing any personal or private information of the elderly to the volunteers, and similarly by not disclosing any personal or private information of the volunteers to the elderly person. Using volunteers for elderly care is a very common practice in Japan [15] where part-time civil servants committed by the Minister of Health, Labor, and Welfare as volunteers, locally known as minsei-iin, are assigned to regularly check the elderly people personally, have a conversation with them, etc. These part-time civil servants are people who volunteer themselves in the area of helping children, elderly people, people with disabilities, etc. and have no mandatory obligation to serve in such areas. We believe that our system is an extension of such practice in the field of elderly care. Instead of visiting the elderly, our volunteers can check the elderly by using the smartphone application even if they are not in the vicinity of the elderly. This is helpful in cases when the elderly might not prefer an unknown person to visit them personally, and also in cases where the number of people serving as minsei-iin might not be enough. Since in our system, we aim to use multiple monitors, we ensure that the activities of the elderly are regularly checked. To maintain anonymity, even if the volunteers discover a risky situation in the daily activities of the elderly, the handling of such a situation, in person, is carried out by emergency contacts of the elderly, and not the volunteers themselves. For our system, we define risk as a deviation in start/end time and duration of activities from the usual routine of the elderly people.

We developed an Android based smartphone application that provides information about the completion of seven basic activities: sleep, shower, medication, breakfast, lunch, dinner, and entertainment (use of television (TV)). We created a dataset by including some risky situations in the elderly activity dataset [16] to determine if those situations can be detected using our application design. To make the monitoring process less burdensome and intuitive, we also included visualization features such as a candlestick chart representation of activities, single interface design, and textual and color codes for their current state, through which it is easy to infer any deviation in the completion time and duration of activities. Similarly, we focused on quick tracking and monitoring of activities by including two types of notifications to trigger frequent use of the smartphone application: one sent every two hours, and another sent immediately after the elderly completed an activity.

The main contributions of this paper are the following:

1. First, we proposed a novel system that can be used by volunteers to anonymously monitor completion of daily activities of elderly people, and report if they detect any deviation in the activities compared to the usual routine of the elderly. We developed an Android based smartphone application that is designed with numerous visualization features and two types of notification strategies to make activity monitoring and detection of anomalies easy, intuitive, quick, and less burdensome.
2. Second, we evaluated our smartphone application with visualization features and a two notification strategy by comparing it with baseline methods (the method without the notification strategy or the visualization features) and confirmed that our proposed method not only provided better risk identification, but also incurred lesser burden on the monitoring person. We also show that our proposed method resulted in quick tracking and monitoring of activities.

The rest of the paper is organized as follows: Section 2 introduces some available research and how they relate to our study. In the next section, Section 3, we introduce our system followed by the explanation of our smartphone application. We explain the evaluation study and findings of the study in Section 4 and in Section 5, and we discuss the significance of the results for our system along with the limitations of this study. Finally, we conclude with our contributions in Section 6.

2. Related Studies and Challenges

Increasing demands in safe, secure, and smart homes for the elderly have led to many research and advances in the field of home monitoring and home automation [4,5,17]. Similarly, with increasing use of smartphone notifications to provide various information to users, we look into studies that explored reliable triggers to inspire people to respond early to mobile notifications. With these factors in mind, we studied existing research, which are divided into two subsections that deal with activity detection and remote monitoring, and importance of smartphone notifications.

2.1. Activity Recognition and Remote Monitoring

In recent years, research dedicated to monitoring people and their activities inside their house has been increasing rapidly since activities of people can be identified with the help of various sensors that can be attached to different household objects [18]. Most home monitoring methods utilize camera or video captures to learn about the activities of the elderly [7]. Video and microphone based monitoring can be time consuming for monitoring, burdensome, and also intrusive [12], and also restrict the area of the house the elderly can occupy to regular monitoring [5]. Numerous research studies have been carried out to tackle not only such problems, but also improve recognition accuracy and reduce the burden of using wearable sensors. The daily activity pattern of elderly people was identified using only motion and domotic sensors by identifying the duration of occupancy of a certain room by the elderly [1]. Similarly, using energy harvesting PIR (passive infrared sensor) and door sensors, an activity recognition system was developed that was efficient as well as cost effective [19].

Many other activity recognition systems utilise non-wearable sensors such as motion sensors [20], Bluetooth Low Energy (BLE) beacon [4,17], wireless accelerometers [21], a combination of temperature, humidity, and illumination sensors [22], and a combination of ECHONET Lite appliances and motion sensors [8]. Similarly, deploying a system that used motion sensors, environmental sensors, and a button to be pressed at the start and end of an activity, daily activities of the elderly were collected for a period of about two months in houses consisting of elderly people [16]. All these studies help to highlight that it is possible to collect activities in the house using sensors such as motion sensors, environmental sensors, etc. accurately without the use of any wearable sensors in a cost-effective way and handling concerns for privacy and security.

Activity recognition systems also allow the elderly to live an independent life in their own house whilst their activities are monitored remotely [5]. There have been measures to monitor vital signs and biomedical signals of adults with medical conditions [23] or people working in extreme conditions such as firefighters [24]. The Allocation and Group Awareness Pervasive Environment (AGAPE) system used on-body sensors to monitor the elderly and contacted nearby caregiver groups in case it detected an anomaly in sensor data [25]. Systems can also contact the emergency contact, or caregivers for the elderly if any anomaly in the collected data are observed, for example, when the data exceed a predefined threshold [26,27]. When it comes to elderly remote monitoring, fall recognition systems are also very important, with some systems recording the average response time of fall detection between 7 min and 21 min [28]. The systems can detect falls using various types of sensing strategies such as acoustic sensors [29], wearable sensors [30], or accelerometers in smartphones [31].

Many commercially available products are also available that are used to monitor the elderly remotely. Systems such as Mimamori [32] and Canary [33] are specially designed to monitor activities of elderlies by their children and close family members who live in a distant location. Another system, GreatCall Responder, uses a physical button, called a responder, that the elderly can press in case they feel they have an emergency, and the system contacts their caregiver [34]. Similarly, there are systems that track numerous activities using motion sensors that remote caregivers can monitor using a private and secure webpage [35,36]. There are also systems that include secure video communication between doctor and patients for regular or emergency situations, remote health monitoring, and emergency care services [37,38].

Many elderly people, however, regard new technologies as an invasion of their privacy and security [10], and tend to accept technologies only if it is beneficial to them or it adheres to their day to day activities without providing any hindrance [39]. A study revealed that being monitored in their house, conducting their day to day activities did not affect regular daily behavior of the elderly [40]. Their extensive study requested the elderly to answer online questionnaires weekly and included daily activities of sending, reading and deleting emails, along with tracking their total everyday activities, walking speed, and time spent outside their home. Hence, if issues of privacy and security are tackled, and the elderly feel that the activity recognition system will be valuable to them, then there is higher chance of acceptance of such a system.

These systems also provide some areas of concern. The alerts are sent to caretakers of health professionals via text or email [28] or direct phone calls [26]. However, the number of false alarms, which can be as high as 5 in one hour [29], can cause annoyance to the caretakers. Similarly, even though the accuracy of fall detection systems is high such as 97.5% [28] or 94% [30], the information regarding the time it takes such systems to inform the caretaker or the time it takes caretakers to respond are not explicitly evaluated. In another system, the activities of elderly were divided into critical, stable, scheduled and overlooked, and alerts for them were generated in a smartphone application as per the type such as after 5 min of usual time for critical activity such as medication and after 30 min for other activities [41]. These alerts were first sent to the elderly, and if they failed to respond, the caretakers were alerted. However, it is difficult to determine the exact time the elderly might prefer to do their daily activities. Similarly, in the case of emergency, the elderly may not be physically able to respond to alerts [41] or press the emergency button [34].

2.2. Smartphone Notifications

Smartphones have become a daily necessity as it helps to tackle isolation, as well as helping to stay in contact with family and friends easily [42]. Smartphones have become an essential tool to be updated about personal health, work, and news updates [43]. Smartphone owners interact with their phones an average of 85 times a day [44] which makes them a befitting tool for remote monitoring. Notifications are essential to keep the users updated about news, emails from work, and information from social media [45]. Although initially they were intended for short message services (SMS) or emails, these days, notification features are used by almost all of the applications to attract attention of the users. A study determined that notifications can be divided into two categories: personal notifications like emails, SMS, or those from social networking sites; and mass notifications like news and advertisements [46]. They concluded that people tend to attend to personal notification faster and more frequently than mass notifications.

The response to notifications depends on different factors such as sender, type of alert, and the visual representation of the alert [47]. In a recent study, it was shown that users receive approximately 64 notifications each day [48], hence the context of a notification plays an important role in the response of the notification. Time of notification reception, activeness of the user, and amount of time the user will take to respond to the received notification are influential for opening the notification promptly [46]. From a study of about 200 million notifications from more than 40,000 users [13], it was discovered that users view each notification differently and prefer to respond to notifications from social networking sites quickly over those from the smartphone system or emails.

Notifications can however lower task performance and affect attention of the user negatively [45]. Response time and response rate of notifications were determined by analyzing the current context of the user through audio from their smartphones [49]. They concluded that the present context of the user plays a very vital role in the response time as well as response rate of the notifications. Similarly, a systematic review on the effects of context aware notification management systems found that context aware notifications increase the response rate [50]. However, it is difficult to predict what time and context can be considered as appropriate for interruption. Since remote monitoring technologies can send multiple notifications in a day, it is essential to determine if such notifications will be viewed as disruptive. Similarly, to our knowledge, the effectiveness of smartphone notifications in remote monitoring systems, especially using multiple types of notification strategies, has not been investigated.

2.3. Challenges

We found out that there are many methods with which activities can be detected accurately. However, in the case of elderly people, it is also necessary to monitor such activities on a regular basis [5]. A smartphone application, equipped with adequate notification strategies, can provide a quicker remote monitoring compared to most of the remote monitoring platforms that are currently web based [35,37,38]. The smartphone application that we have designed can be used to instantly monitor completed activities and receive quick feedback from the monitoring person. It is essential not only to track activities, but also check if any risks that have occurred, and predict or prevent any potential risks in the daily life of elderly. Hence, at first, it is necessary to determine what activities to monitor and if those activities can be properly visualised in the application, and, furthermore, if any deviation in the routine of the elderly can be distinguished so that any potential risky situation of the elderly can be detected. Similarly, it is essential to identify if using the application, and monitoring activities regularly will put a burden on the monitoring person. With all this in mind, we propose the following research questions (RQs), which we try to verify with an experimental study:

- RQ1: Is it possible to identify daily routine of individuals using a smartphone application?
- RQ2: Can a monitoring person detect potential risks in day to day activities based on visualization of activities in our application?
- RQ3: Is constant notification and using the application a burden for the monitoring person?

3. System Design

In this section, we first explain the overview of the proposed PATROL (Participatory Activity Tracking and Risk assessment for anOnymous eLderly monitoring) system. Then, we describe the design and interface of our smartphone application in detail.

3.1. System Overview

The architecture of PATROL system is shown in Figure 1, where we denominate the elderly being monitored as *Target* and the person conducting monitoring as *Monitor*.

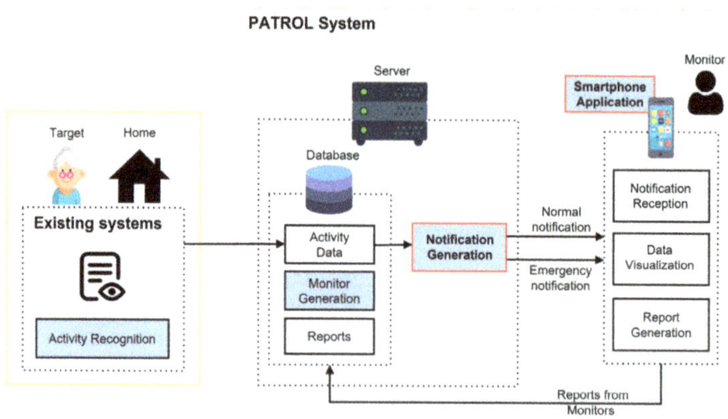

Figure 1. System architecture of PATROL.

The monitoring can be conducted in different ways. One Target can be monitored by a single or multiple Monitors and one Monitor can conduct monitoring of a single or multiple Targets. Consequently, multiple Monitors can be used to monitor multiple Targets.

The overall system can be further divided into four sections: activity recognition, monitor generation, notification generation, and smartphone application, as highlighted in

Figure 1. In this research, we focus mainly on the two sections: notification generation and smartphone application. We will now discuss each of the sections and their application in our overall system.

3.1.1. Activity Recognition

Most elderly people have a definite time and duration for their activities, and follow a routine set of activities throughout the day [51]. It is important to check for everyday basic activities because, with old age, these important basic daily activities can sometimes be missed or incomplete or not properly carried out [7]. For the purpose of our research, we assume that the Target is residing in a smart home equipped with an activity recognition system, where it is possible to collect information related to daily activities like eating, sleeping, watching TV, taking medicine, etc. through the use of different kinds of sensors and power consumption meters available in the house [1,8,16]. We have designed our system in a way that it can incorporate any available activity recognition systems. Therefore, it is easy to integrate in houses which already have an activity recognition system. Activities that we showcase in the smartphone application are shown in Table 1. We believe that the state of everyday basic activities can be used as criteria to determine the wellness of the elderly person. There can be instances when anomalies can occur whilst conducting activities that are not listed in Table 1. However, such incidences will subsequently impact the occurrence of basic activities that we aim to monitor. Therefore, our system can detect anomalies that can occur doing activities that are not directly monitored in our application. Since our aim is to disclose as less information about the Target as possible, whilst making it possible to determine their current status, we only use time of completion and duration of the activities to provide information about them. We assume that the activity recognition system outputs events (i.e., start and end times of activities performed by the resident) which are utilized for data visualization and notification generation, as shown in Figure 1. This feature will be further discussed in Section 3.2.

Table 1. Areas and activities to monitor.

Area	Home Objects with Sensors	Activities
Kitchen	• Stove • Microwave	• Breakfast, Lunch or Dinner
Bedroom	• Bed • Medicine Bottle	• Sleep time • Medication
Living room	• TV	• Entertainment
Bathroom	• Water consumption	• Shower

3.1.2. Monitor Generation

The PATROL system is designed to be used especially for monitoring the elderly, and to be deployed in nursing homes, elderly residential areas, care homes, municipalities, etc. The overall system needs to be handled by a system administrator who can be the head of the residence association or personnel who work in such institutions. In case of changes in the system administrator, then the outgoing system administrator under the authority of the local welfare committee (and/or residents' association) will have to train the new system administrator immediately. In our context, Monitors are usually volunteers who work in the field of helping elderly in care homes, elderly residential areas, etc. The Monitors participate in tracking the activities and determining risky situations in the activities of the elderly. The system administrators have the responsibility of training the Monitors to use the smartphone application, assigning Monitors for each Target, assessing the performance of Monitors and determining if any change needs to be done. In case of changes in Monitors as well, the training of new Monitors is handled by the system administrators. Similarly, the initial testing and assessment of our application is handled by the system administrators as well who check if the system is working properly, and the application is generating activity reports and notifications regularly. Since our application shows activities not just of the current day, but of a period of days (e.g., week), including previous days, a new user can still be familiar with start/end and duration of activities of a range of days and deduce a pattern or routine of the target easily.

The number of Targets assigned for each Monitor may vary based on the preference of each volunteer. The volunteers are free to choose a minimum or maximum number of Targets to monitor, after which the system administrator will assign them Targets. Therefore, the number may vary from a single Target to multiple ones based on each volunteer.

3.1.3. Notification Generation

To encourage regular usage of the application, frequent notifications are sent to the Monitors. This functionality helps to timely track the recent activities of the Target and detect any change in the usual routine. We think there should be two types of notifications generated: emergency and general. General notifications are sent to remind monitors about using the application and check current activities of the target. Emergency notifications are sent when the system itself detects abnormalities in the recent activities of the target. We do not generate or analyze emergency notifications in this research because we aim to determine how often general notifications are responded by the Monitors, if they motivate the monitors to frequently use the application or not, and if constant notifications will be burdensome or disturbing.

The notification scheduling techniques that are commonly used can be divided into three types: randomized time points in a day, timed at specific intervals, and event dependent times [52]. In our system, general notifications are generated by using two types of notification strategies: timed at specific intervals and event dependent notifications. This ensures that the monitors are notified regularly to use the application, and can instantly check information about the activity completed.

3.1.4. Smartphone Application

The information collected from the house of the Target is utilised to create graphical representation of activity completed in a time series form which helps to identify a pattern in the time of completion of activity and its duration, so that any deviation from the usual pattern can be identified with ease. Hence, we develop an Android based smartphone application, PATROL, which can be used to view the activities completed and send reports. Since smartphones have become a common gadget among the elderly as well [53], our application can be used by the young volunteers as well as the elderly. For our research, we conduct an experiment using smartphones, but the application can also be used in any other Android based devices like tablets.

The interaction between the Monitor and the application is shown in Figure 2. We have tried to minimize the number of actions required to be carried out by Monitors. In the application, the Monitors receive notifications as a trigger so that they can check the time of completion and duration of the activity for the current day and previous days, after which they can judge whether the Target is in a risky situation or not, and submit a report. If the Monitor reports that the Target is in a high risk situation, then the application can notify the system administrator and emergency contacts of the elderly via text, email, or automated phone calls, who can take necessary actions immediately. The Monitors did not disclose any details of the Target even in such situations to maintain the anonymity of our system. The system administrators, who are in the vicinity of the Target, will take the responsibility for checking the Target as soon as such reports are received. The report sent by the Monitors are saved and analyzed to evaluate their monitoring capability.

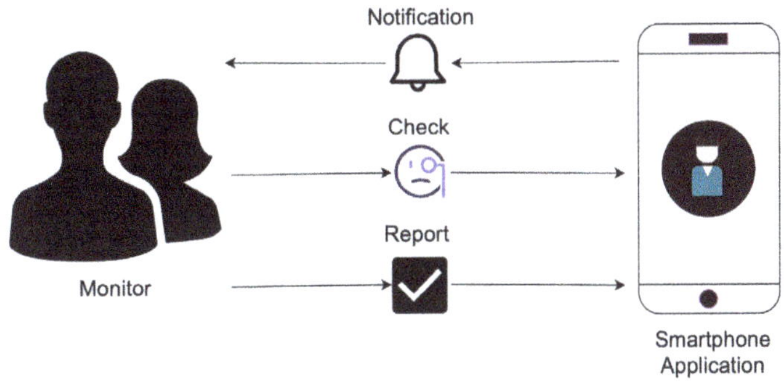

Figure 2. Interaction between Monitor and smartphone application.

For the accurate analysis of our application, it is necessary that risky situations of the Targets are identified correctly. We, however, at first need to define what these risks are, and how they can be related to real life situations. We created a total of four risk stages, as shown in Table 2. These risks are based on the changes in the routine of the Target. If there is no change in their routine i.e., no noticeable deviation in their activity, then we regard the risk as None. Low and Medium risks are defined based on the amount of deviation from the usual start/end time or duration of the activities. High risks refer to situations when the activity has not started, or completed indicating that the Target needs urgent attention.

Table 2. Risks used in the application and their description.

Risk	Description
None	Everything seems to be okay with the elderly.
Low	There is some problem, but can be handled by elderly themselves.
Medium	There is a problem and the elderly should be assisted/checked by caretaker, family members or doctor.
High	Elderly is in emergency and requires immediate medical care.

We have used standard deviation to define *low* and *medium* level risks. We calculated standard deviation of duration and time of completion of each activity, for each targets. Then, we defined *low* and *medium* level of risks as follows:

- Low risk
  - duration $\pm 1.5 \times$ standard deviation of duration
  - time $\pm 1.5 \times$ standard deviation of activity completion time
- Medium risk
  - duration $\pm$ 3 $\times$ standard deviation of duration
  - time $\pm$ 3 $\times$ standard deviation of activity completion time

The purpose of using this technique is that it gives us a wide range of duration and activity start/end times that we can relate with risks in real life scenarios. The low risk indicates that the deviation in time or duration was not so concerning, which meant that the elderly had some problems but were able to deal with them themselves. Medium risk indicates a higher deviation in time or duration of activity, which indicates that the elderly might not be doing so well and need to be attended to personally. Since we have ourselves defined these ranges of duration and start/end times for low and medium risks, they are flexible, and hence can be modified based on the activity data of the elderly.

3.2. Application Design

Even though recent technologies have been designed and developed targeting with an average young user in mind, who is efficient at handling new systems or devices [54,55], we have tried to make the interface simple and intuitive so that it can be used by people of all ages conveniently. As shown in Figure 2, the number of tasks to be carried out by the Monitor in the application are very minimal. Therefore, we believe that the application will be easy to use, and the burden of using the application will be low for the Monitors. Our final goal is to achieve remote elderly care and prompt identification of risky situations; however, we believe that to achieve them, the design and interface of the application should be favorable to the monitors. We aim for our concern of providing a continuous and detailed elderly care system, and an easy and intuitive interface for monitors does not remain mutually exclusive. The actions in the application to be carried out are: respond to notifications, check activity, and submit a report. Below, we will explain different features in the interface of the smartphone application, and the notification strategy that we developed.

3.2.1. Features of the Application Interface

We have designed the application with various features in the interface that is aimed at helping the monitoring process. All the activities are shown in a single interface to reduce the burden of going back and forth between interfaces to monitor the activities. We will now discuss the features of the application interface.

Activity Report

The application shows the option to choose whom to monitor among a list of Targets, as shown in Figure 3a. Since our application is anonymous, the real names of the Targets are not shown. We used three commonly used names in Japan (Taro, Watanabe, and Yamazaki) to denominate the Targets in our application. Once the Target is chosen, then the activity report interface is shown, as shown in Figure 3b.

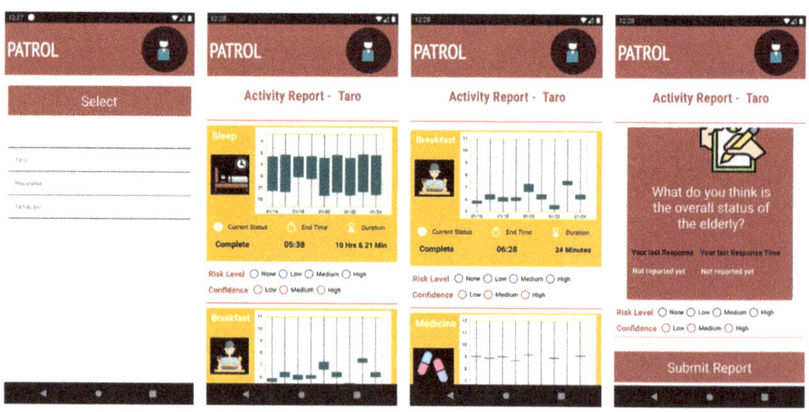

(a) (b) (c) (d)

Figure 3. Snippet of the smartphone application for: (**a**) choosing Targets, (**b**) sleep card, (**c**) breakfast card, and (**d**) submitting report.

The activity report interface breaks down each activity into different cards, with each card showcasing the current status of the activity (incomplete, ongoing, or completed), activity completion time (in graph as well as text), and duration of the activity, as shown in Figure 3b,c. In case of activities like TV and medication that can occur multiple times in a day, each separate activity is represented by separate cards. The Candlestick chart style helps to identify a pattern in the time of completion of activity and its duration, so that any deviation from the usual routine can be recognized with ease. We use a candlestick chart to show activities because it can showcase the time as well as duration with clarity, and the difference between consecutive days is also understandable.

The Monitor, ideally, should be able to submit only one report per activity per day as well as provide the report for an activity only after the activity has been completed. Hence, in order to prevent multiple and erroneous reporting, we use two techniques: color codes in activity cards; and radio button for reporting. In the cases of activities that occur multiple times in a day (such as TV, medication), multiple activity cards of the same activity are shown. To avoid confusion for the users, only one activity card is shown at the start of the day, when no multiple activities have occurred. The activity cards are then subsequently added soon after their occurrence.

Colors Codes in Activity Cards

Traffic light colors have been used in various research studies, from labelling traffic colors on food to indicate their edibility or freshness [56,57], to using traffic colors as a means of self-monitoring by recording the weight and shortness of breath in a diary [58]. We use traffic color codes for the activity cards in order to make the current status of activities of the Target clear, as shown in Figure 4.

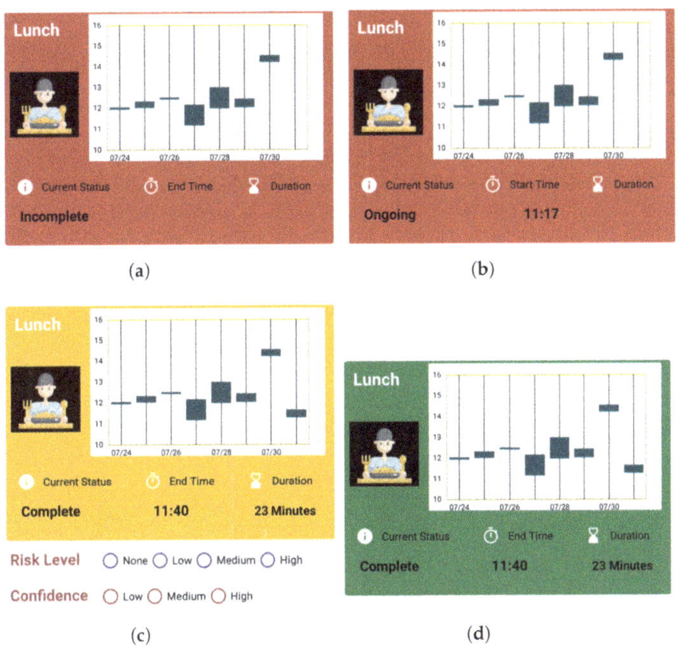

Figure 4. Use of color for representing activity state for: (**a**) activity not complete, (**b**) activity ongoing, (**c**) activity complete, and (**d**) activity reported.

The background color of the activity card is represented by *red* when the Target has not completed the activity, as shown in Figure 4a. The current status information, shown as *Incomplete* , also gives an update that the activity has not been finished for the current day. The information about end time and the duration of the activity is also empty at this stage.

The background color of the activity card is represented by *red* when the target starts the activity, as shown in Figure 4b. The current status information is changed to *Ongoing* in this case, and the information about the start time of that activity is updated. The information about the duration of the activity is also empty at this stage.

The background color changes to *yellow* when the activity is finished by the Target. The current status is also updated, to *Complete*, along with information about end time and duration of the activity. Along with the change in color, the radio buttons for reporting the status are also shown below the card, as shown in Figure 4c.

When the Monitor reports about the activity, then the background color of the card is changed to *green*. Along with that, the radio buttons for reporting are hidden, as shown in Figure 4d. Thus, when the Monitor opens the application again after submitting a report, the option to report again is not available, and the color codes help them identify the activities they have already reported.

We believe that, since people are familiar with traffic colors and their functions, this feature in the application is intuitive, and helpful in clearly distinguishing the states of activity. The colors are also directly related to the state of the elderly as well as the necessity of Monitor's attention. When the background color is red, activities are either ongoing or not started at all, which means that the elderly has not completed any activity. This state requires a higher amount of attention from the monitor because if the background color does not change from red for a prolonged time, then it should be deduced by the Monitors that the elderly might be in a risky situation and thus the Monitor should report, via an overall report card. When the background color of the card changes to yellow, it indicates that the elderly has completed an activity, and the monitor should now check the activity and submit a report. This state requires lower attention from the the Monitor compared to the red background color state. Similarly, a green color gives Monitors a confirmation that they have completed the reporting task already and should not pay any attention to that particular activity anymore.

Overall Report

Along with the activity cards for each activity, there is a separate card called Overall card. This, in general, is to report about overall impression about the status of the elderly. This can be reported multiple times by the Monitor throughout the day, and has the same reporting option of risks and confidence as in other activity cards, as shown in Figure 4d. Thus, when submitting reports for activities, the Monitor has the option to choose what they feel is the overall status of the Target based on their judgement of activities completed or not completed. In cases of High risk situations such as no activity or long deviation, the target will not register completion of activities regularly, which means that no notifications are sent and the activity cards are not updated. If no activity has been updated for a significant time, then the monitors can deduce that there is something wrong with the target. In such situations, they can report the emergency situation using the Overall card. The card also shows the type and time of previous response for the Overall card, to make it easier for the Monitor to recall their previous impression, as shown in Figure 5b.

Submit Report

The task for the Monitor is to check the activity report of the Target and analyze the information shown and then submit their report. The report can be submitted for one activity at a time, as well as for multiple activities at the same time. To submit the report for each activity, the Monitor needs to scroll down in the activity report interface and click the submit button at the end of the activity report interface as shown in Figure 3d.

If the monitor responds with *high* risk and *high* confidence to any activity, then the application can infer that the elderly might be in an emergency situation, and can promptly notify the emergency contact of the Target (friends, family or health professionals) via text message, email, or automated phone calls, and they can take necessary actions. Similarly, if more than two subsequent *medium* risks are reported with *high* confidence, then their emergency contact can be notified immediately. Thus, to provide a base to analyze the confidence of the report, we divided the confidence level for each report as *Low*, *Medium* and *High*, as seen in Figure 4c. The confidence levels hence act as reference points of risks for each activity, especially when there are multiple Monitors. The confidence level provides a perception of each of the Monitors and their report, and also helps to analyze their monitoring capabilities.

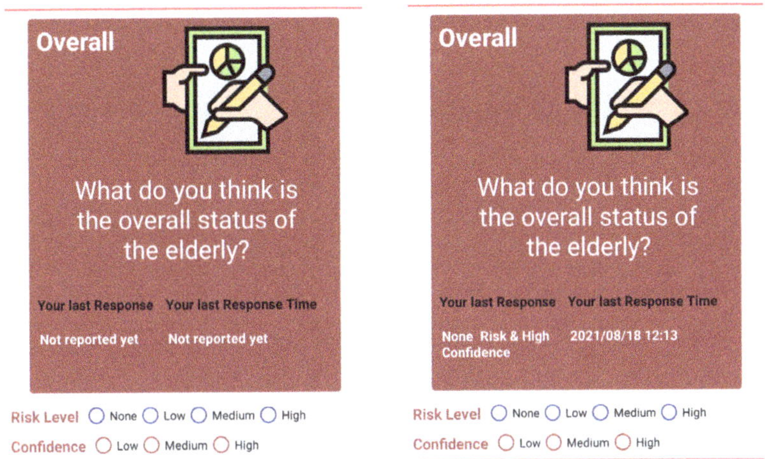

(a) (b)

Figure 5. Overall report card: (**a**) before report submission and (**b**) after report submission.

3.2.2. Notification Strategy

We deploy two kinds of notification patterns in our application: recurring notifications (rN) and activity based notification (abN). We send notifications every two hours (rN) to provide a trigger to the targets to use the application. The period for recurring notification is two hours because we feel that two hours is an appropriate time gap for reminding users, as sending a notification every 30 min or an hour will be too disruptive. Analyzing the activity completion times and usual gap between activities, we feel that two hours is an appropriate gap to send a recurring notification. We have also analyzed the perception of users towards recurring notifications of two-hour intervals, and have empirically proved that they are not perceived as disturbing and were responded to about 87% of the time [59].

Apart from this, we also send a notification, abN, which is sent as soon as a target completes an activity. We mentioned in Section 2.2 that it is necessary to provide contextual information in notifications for quick responses. We provide the name of the target and the activity completed in the notification, to provide context of the notification to the monitors, as shown in Figure 6. To make distinction between the two types of notifications, we indicate abN with a red icon of notification (see Figure 6a) and rN with a blue icon (see Figure 6b).

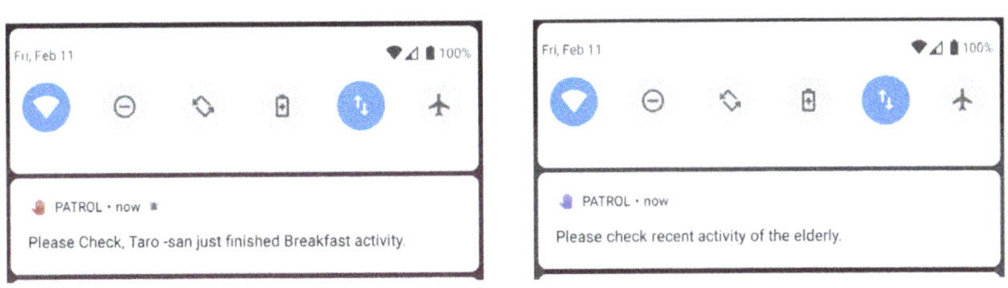

(a) (b)

Figure 6. Example of notifications generated: (**a**) activity based notification (abN) and (**b**) recurring notification (rN).

4. Implementation and Evaluation

In this section, we will explain the details of the experiment conducted to analyze the application, including the dataset used for the application, multiple versions of PATROL application that we created, and finally explain the result of our study.

4.1. Multiple Versions of PATROL Application

In order to concretely determine that our proposed method of a graphical interface (GI), as shown in Figure 3b, is intuitive and has a higher degree of user acceptance, we needed to compare that interface with commonly used activity representation techniques. To make that distinction, we created a separate version of our application where activities were shown in a textual interface, rather than graphs. Figure 7 shows the activity report interface of this kind of version of the application. All the features of the application mentioned in Section 3.2 are included in this version as well, so the working principle is the same regardless of the interface. This helps create less confusion for the participants and ensures that the performance and perception of users is solely based on the type of interface, and not on other features of the application.

(a) (b) (c)

Figure 7. Example of tabular interface (TI): (**a**) activity incomplete, (**b**) activity complete and (**c**) activity reported.

Similarly, we created a third version of our application (GR), in which we did not send notifications to the monitors when the activity was completed by a target. We only send them recurring notifications every two hours. With this version of the application, we aim to determine if the monitors are able to report about activities of the elderly even if they do not receive activity based notifications (abN) and thus our strategy of providing both abN (activity based notification) and rN (recurring notification) can be effective to encourage and motivate monitors to use the application frequently and receive continuous reports of activities of the target.

Table 3 summarizes the three versions of the application created, and we will use the same label for versions (GAR, TAR, and GR) in future discussions. GAR refers to the proposed version of PATROL, which consists of a Graphical interface, Activity based notification, and Recurring notification. We investigate the accuracy of risk identification, and the burden of use of our application by comparing the versions GAR and TAR (Tabular interface, Activity based notification, and Recurring notification). Similarly, we compare the effectiveness of using activity based notifications by comparing GAR with GR (Graphical interface and Recurring notification).

Table 3. Types of versions of PATROL application.

Version	Interface		Notification	
	Graphical	Tabular	Activity-Based	Recurring
GAR	✓	–	✓	✓
TAR	–	✓	✓	✓
GR	✓	–	–	✓

4.2. Dataset

The dataset used in our experiment is taken from a real life experiment conducted in the houses of elderly residents over the age of 60 [16]. The activity dataset was obtained by Matsui et al. through an extensive research conducted over a period of two months, where motion and environmental sensors were installed in each of the houses. Along with that, a physical button was installed in each of the houses, and the residents were requested to press the button whenever they started and ended an activity [16]. The original dataset consists of activity recognition data from single as well as two-person households. For the purpose of this research, we selected only single resident households that were three in total. We use cleaned and collected data from the above-mentioned study, and consider that the activity recognition system is 100% accurate (we used ground truth labels of activities in the dataset as the output of the activity recognition method).

The daily activities of the elderly that we want to track and monitor are mentioned in Table 1. The original dataset, however, does not contain data related to the Medication activity. Similarly, we also wanted to include multiple activities related to frequent use of TV. To fulfill our desired dataset, we added aforementioned activities into the original dataset. The total period of experiment of the two-month study was longer than our intended experiment period of 10 days. Hence, we only selected data for a 10 day period from the available two months of data. We included data from the same time period section for all the three single-resident households.

We included some risky situations into the dataset based on the definition shown in Table 2. For the purpose of our research, we included only *low* and *medium* level risks. As defined, *none* risk indicates that there is no problem with the elderly. Hence, we do not need to alter the dataset for such risk, since they concur with the regular routine of the elderly. If the level of risk is *high*, it indicates that the elderly person is in a serious condition and in need of immediate medical care. In such cases, no activity will be completed by the elderly, and the activity report in the application will not be updated.

However, our aim is to determine if any deviation from regular routine of the activities could be determined using our application. Though *high* risks can occur suddenly, we also think that, if we regularly monitor and determine *low* and *medium* level risks, then *high* level risks can be prevented or predicted. Because of this, we did not include *high* level risks in our dataset.

4.3. Experiment Details

We recruited a total of nine participants (gender: 6 Male, 3 Female; age range: 25–34 years old, average age: 28.6 years) to take part in our evaluation study. The participants were playing the role of 'Monitors' throughout the experiment. The modified dataset of the three single-person households were used for the three 'Targets' in the application. The participants were divided into three groups each. Thus, we had three participants each in three study groups. This was carried out to implement random distribution of our application in a way that each group, with an equal number of participants, will use a different application at a given time compared to other study groups. To implement that, we divided the experiment period into three phases in total. Table 4 simplifies the study group and application interface division.

Table 4. Study groups and division of version of PATROL application.

Phase	Date (MM/dd)	Number of Days	StudyGroup A	StudyGroup B	StudyGroup C
1	08/25–08/27	Three	GAR	GR	TAR
2	08/29–09/01	Four	TAR	GAR	GR
3	09/03–09/05	Three	GR	TAR	GAR

The three versions of the application were uploaded to Google Play Store. Before the start of the experiment, we conducted a research and experiment introduction session that all the participants were requested to attend compulsorily. We explained the theme of the study and experiment in detail, their role as monitors, and the tasks they have to complete while using the application. They were also provided a document containing all the information about the working principles of the different versions of the application, along with QR codes for each version. The documents also indicated the version of the application they were supposed to use in each phase of the experiment. As a reward for participation in the experiment, the participants were provided with a gift card worth 2000 JPY.

To make the transition between interfaces easier for the participants, we included a one day gap between each phase. The participants were asked to take a break for a day in between the phases. The phases were designed to be of three days each. However, at the start of phase 2, we encountered some complications with the server connected to our application, and the application did not work properly until mid-day. Hence, we asked the participants to continue phase 2 for one day more. Thus, in total, the experiment period consisted of 12 days, with breaks of two days in total. After the end of each phase, we asked the participants to fill in a questionnaire developed using Google Forms. Most of the questions had to be rated on a five-point Likert scale (1 = strongly disagree, 3 = neutral, 5 = strongly agree), while some of them were open-ended. The participants were asked to respond to questions or statements related to their perception of the version of the application, as well as the effect of change in the version of the application, such as "*The activity related notifications were helpful in monitoring the elderly as it reminded me to check the application regularly.*", "*I found the change in the interface confusing.*", and "*I feel the new interface needed more mental effort.*" At the end of the experiment, the participants were asked to fill out a final questionnaire. The purpose of these questionnaires is to gain insight into the impression of the participants for different versions and different notification types.

4.4. Results

The results of our study are analyzed based on the following three conditions:

1. Accurate detection of risky situations;
2. Low burden of monitoring on Monitors;
3. Timely Detection of risky situations.

4.4.1. Accuracy of Risk Detection

In order to verify the effectiveness of our visualization technique, it is necessary to check if the risks included in the application, as mentioned in Section 4.2, will be identified correctly. In this section, we report the rate with which the risks included in the dataset were correctly identified in each phase, using different versions of the application. Table 5 and Table 6 show rate of correct identification of risks based on study groups and interfaces, respectively.

From Table 5, we can observe that StudyGroup C was the most consistent group, with the highest risk identification rate during all of the three phases of the experiment. The rate of correct identification also increased along with the experiment, which proves that familiarity with the application helped to analyze the activity reports and submit reports.

Table 5. Risk identification based on study groups.

Study Group	Phase	Interface	Risk Identification	Average
StudyGroup A	1	GAR	68.4%	72.6%
	2	TAR	67.7%	
	3	GR	81.9%	
StudyGroup B	1	GR	32.4%	46.1%
	2	GAR	64.7%	
	3	TAR	41.3%	
StudyGroup C	1	TAR	88.4%	90.7%
	2	GR	91.3%	
	3	GAR	92.6%	

There was a slight decrease in risk identification for StudyGroup A when the interface changed from graphical (GAR) to tabular (TAR) in phase 2 of the experiment. All of the participants in StudyGroup A agreed that the new interface needed more time to analyze in their questionnaires after phase 2, with 66.7% agreeing that the tabular interface (TI) needed more mental effort than graphical interface (GI). When the interface changed to graphical layout (GR) in phase 3 of experiment, there was an increase in the correct rate identification. When asked about the change, participants claimed that it was easier to understand the routine with the graph compared to tabular layout (66.7% agree, 33.3% strongly agree).

StudyGroup B showed a considerable increase in correct risk identification, in phase 2, as shown in Table 5, even though they had graphical layout for both phase 1 (GR) and 2 (GAR). We can predict that familiarity with the application was the reason for such change. In their questionnaire after phase 2, 66.7% strongly agreed that they were familiar with the application and found it easier to use the application during this phase. However, in phase 3, their interface changed to tabular layout (TAR). This led to reduction in risk identification, with 33.3% strongly agreeing that the change in interface was confusing.

As shown in Table 6, we found out that, in total, using GAR, on average about 75.2% of the time the risks were identified correctly. In comparison, the risks were identified correctly about 65.8% of the time using TAR. GR, which in this context, is the same in visualization as GAR had a risk identification accuracy of about 68.5%. The average rate of risk identification is lower for tabular interface (TI), compared to both of the graphical interfaces (GI). This can help to identify that graphical interfaces (GI) provide better understanding or identification of risks.

Table 6. Risk identification based on interface types.

Phase	GAR	TAR	GR
1	68.4%	88.4%	32.4%
2	64.7%	67.7%	91.3%
3	92.6%	41.3%	81.9%
Average	75.2%	65.8%	68.5%

We also found statistically-significant differences between the average risk identification rates of the three interfaces using the one-way ANOVA method ($p = 0.037$). A Tukey-HSD post-hoc test revealed a significant pairwise difference between interfaces GAR and TAR ($p = 0.032$) whilst no difference was observed between GAR and GR ($p = 0.2$).

To investigate this further, we combined the results of GAR and GR into a single group and compared it with TAR, to clearly determine differences between graphical and tabular interfaces for risk identification. Through the paired t-test analysis, we found that there is a significant difference between the two ($p = 0.047$).

4.4.2. Low Burden Evaluation

We define burden as the time taken by the participants between opening the application to check the activity report of targets and submitting the report. We logged the time of opening of the application as well as the time of reporting using "Shared preference" functionality available for Android developers. These time periods were saved together in the Firebase database. We analyzed the burden time for each participant using this data and calculated an average burden time for each participant over the whole experiment period, which is shown in Figure 8. The average burden time for each of the versions is also shown.

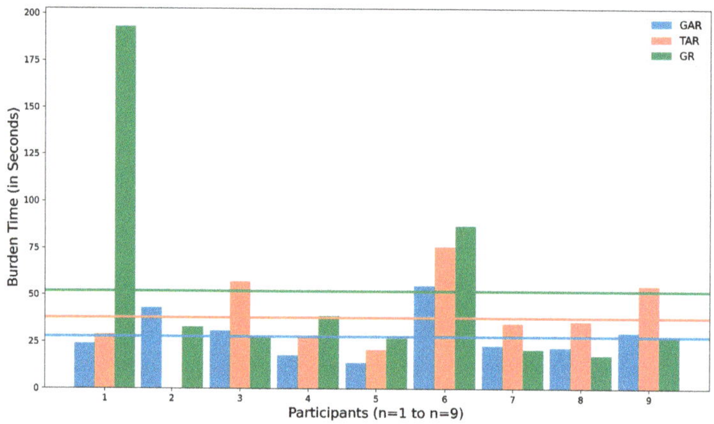

Figure 8. Average burden time of participants.

We can see that the burden time for GAR, on average, is always less than TAR. The mean burden time for GAR, TAR, and GR were observed to be 28 s, 38 s, and 52 s, respectively. As seen in Figure 8, the burden for participant 1 while using GR is very high compared to other participants, and other interfaces used by the same participant. Upon inspection, it was discovered that, while using GR, for one particular report, the participant recorded an unusually high burden time, which was uncharacteristic for the participant based on his other responses. Discarding the unusually high burden time, the average burden time of the participant 1 was reduced from 193 s to almost 20 s. However, for the final analysis, the skewed data are kept as it is. Similarly, the burden for participant 2 while using TAR is zero because the participant did not record any response during phase 2 of the experiment.

To analyze the link between burden of using the application, and engagement with the application over time, we calculated the average time it took to report based on the phases of the experiment. The results are shown in Figure 9. When the interface changed from graphical (GAR) to tabular (TAR), in phase 2 for StudyGroup A, we can see that the burden time was higher. In phase 3, when their interface changed back to graphical (GR), the burden time was observed to be extremely high (94 s) due to the unusual reporting by participant 1 as explained above. Discarding that particular incident, the burden time was observed to be lower than in phase 2 (28 s).

For StudyGroup B, the burden time was highest in phase 1, with 47 s, when using GR. However, the burden time decreased in phase 2 (25 s) when using GAR. This can be attributed to the participants getting familiar with the interface. In phase 3, however, when the interface changed to tabular (TAR), we can see that the average burden time increased to 37 s.

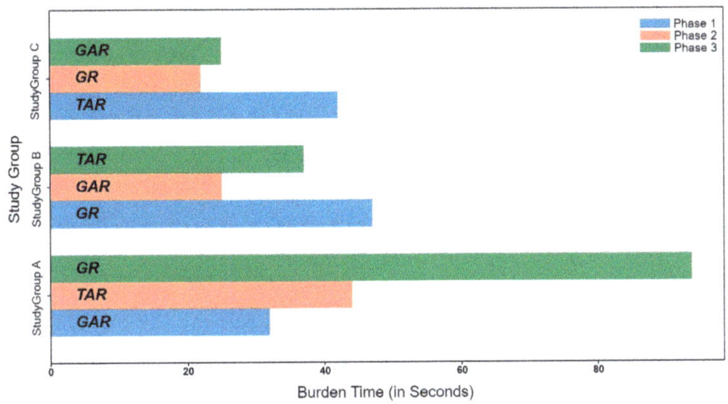

Figure 9. Average burden time of study groups per phase.

Similarly, when the interface was changed from tabular (TAR) to graphical (GR), for StudyGroup C in phase 2 of experiment, we can see that the average burden time was lower (22 s). Even though the burden time increased in phase 3 (25 s), using GAR, it was still lower than the burden time in phase 1 (42 s). Therefore, over the course of the experiment period, we can observe that change in interface had some effect on the engagement with the application and burden time. Familiarity with the application lowered the burden time, especially using a graphical interface (GI).

We found a statistically-significant difference in the burden time for the three interfaces using a one-way ANOVA method (p = 0.012). A Tukey-HSD post-hoc test revealed a significant pairwise difference between interfaces GAR and TAR (p = 0.039) whilst no difference was observed between GAR and GR (p = 0.13).

For further investigation, we combined the results of GAR and GR into a single group and compared it with TAR and through a paired t-test analysis; we found that there is a significant difference between the two (p = 0.049). This analysis, along with the results from Figures 8 and 9, help to show that there is a significant difference between tabular and graphical interfaces for the burden faced while using the application, with a graphical interface resulting in a lower burden for the participants.

Lesser burden also resulted in higher engagement with the application. Figure 10 shows that the total number of reports received using GAR across different phases were almost consistent across the three phases, and on average higher than when using TAR. There was a significant decrease in reports using TAR in phase 2 for StudyGroup A. This can be attributed to change in their interface because, in an earlier phase, they used graphical interface (GI). They also mentioned in the questionnaire after phase 2 that tabular interface (TI) was difficult to understand, which resulted in a lower number of reports.

We can thus conclude that GAR provides lesser burden to participants, in comparison with TAR, and on average has higher engagement and reporting. This further strengthens our proposal that graphical interface (GI), with adequate textual information, can be helpful for monitors to identify the routine of targets and distinguish risky situations whilst spending less time and effort analyzing the interface.

4.4.3. Timely Detection

Figure 11 shows the time taken to report about a completed activity during each phase, based on types of interface. Over the three phases of experiment, we can observe that using a graphical interface (GI), the reports for activities were received quicker compared to tabular interface (TI): GAR (average = 176.46 min, median = 115.01 min), TAR (average = 201.42 min, median = 118.85 min), and GR (average = 166.9 min, median = 121.12 min). Even though such high response times for the report are not favorable, we think that there were many factors that affected the reporting time for activities.

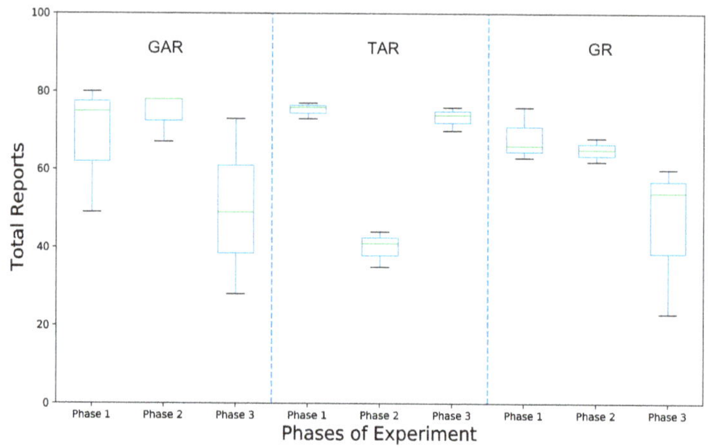

Figure 10. Total number of reports received.

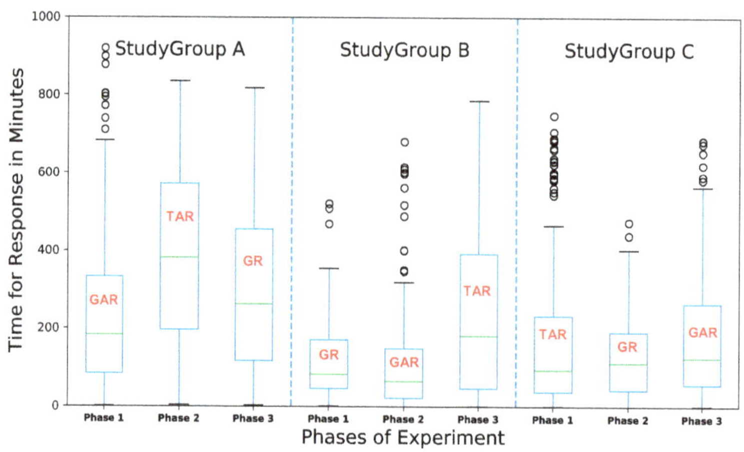

Figure 11. Response time for activities per phase based on study groups.

The time of notification generated, which is also the time when the activities were completed, was saved using "Shared preference" functionality, as mentioned in Section 4.4.2. Similarly, we also saved the time when the activity report was submitted. We determine the time taken to report an activity by calculating the time difference between report submission and notification generation. For StudyGroup A, when the interface changed from graphical (GAR) to tabular (TAR) in phase 2, the reporting time was higher compared to phase 1, even if they had received both rN (recurring notifications) and abN (activity based notifications) in both of the phases. This can be attributed to the change in interface because, when their interface changed back to graphical (GR) in phase 3, the time of response also was observed to be lower than on phase 2, even though they did not receive abN. This shows that type of visualization can have an effect on the response time for notifications received.

StudyGroup B were almost consistent in their performance throughout the first two phases of the experiment period. In phase 2, when their interface changed from GR to GAR, there was no significant change in their response time even if they did not receive abN. However, when their visualization changed to tabular (TAR) in phase 3, the time of responses was higher than in the previous two phases.

In contrast, StudyGroup C did not show any significant differences in response time for activities based on changes in interface as well as reception of abN. When their interface changed from TAR to GR in phase 2 and from GR to GAR in phase 3, their response time for notifications did not show any high amount of significant differences. StudyGroup C thus did not show any conclusive effect for the change in visualization or notification strategies for the reception of reports to activities.

Table 7 shows the average response time of each participant while using each of the interfaces, where the lowest response time taken among the three interfaces is highlighted. Even though TAR consisted of both abN and rN notifications, we found that none of the participants responded quickly while using it. Moreover, the mean response time using TAR is highest across all the participants (except participant 2, who did not register any response during phase 2). We found that, even though they did not receive abN, some of the participants (4) recorded lowest mean response time using GR. GAR and GR recorded mean response times of about 176.46 min and 166.9 min respectively, while TAR had a mean response time of 201.42 min. Even though GR had lower average response time, we observed that the median response time for notification was lower for GAR (115.01 min) compared to GR (121.12 min) and TAR (118.85 min). This shows that reports were received quicker using GAR than GR or TAR.

Table 7. Mean response time (in minutes) of each participant.

Participant	GAR	TAR	GR	Total Average
1	**225**	349	243	272
2	**335**	No response	341	225
3	**182**	418	260	286
4	146	287	**135**	189
5	**33**	146	59	79
6	**154**	301	167	207
7	187	188	**146**	173
8	114	171	**109**	131
9	205	146	**110**	156

The quickest mean response time for each participant is highlighted in bold text.

Upon further analysis, we found statistically-significant differences between activity response time for the three interfaces using a one-way ANOVA method ($p = 0.005$). A Tukey-HSD post-hoc test revealed a significant pairwise difference between interfaces GR and TAR ($p = 0.05$) whilst no difference was observed between GAR and GR ($p = 0.64$) or between GAR and TAR ($p = 0.055$).

We then combined the results of interfaces that received abN, i.e., GAR and TAR, into a single group and compared it with GR, and found that a paired t-test shows a significant difference between the two ($p = 0.022$).

This shows that reception of abN does indeed have an effect on the time for response to the activities. To investigate this further, we determined the time range within which the responses to the activity notifications were received. Table 8 shows the cumulative percentage of reports received within the given time ranges for the three versions of the application. We divide the time into 30 min intervals; however, the table only shows until 210 min, since the highest average time of response is within the 180–210 min range. We can see that the amount of responses received does not vary by a large amount if graphical interfaces are compared. However, for tabular interfaces, the response rate is lower even if abN was received. This shows that abN, when used with a graphical interface, provides a better result than compared with tabular interface. We then tried to investigate which interface provided the quickest response for activities.

Table 8. Cumulative percentages of responses received per time range (in minutes).

Time Range	GAR	TAR	GR
0–30	18.4%	15.45%	13.24%
30–60	30.55%	23.21%	29.66%
60–90	42.36%	31.22%	41.79%
90–120	51.56%	37.09%	49.44%
120–150	57.63%	41.42%	58.39%
150–180	63.88%	44.04%	65.29%
180–210	68.92%	47.29%	70.52%

We divided the notifications into those that were for regular activities and those that were for the risky situations. By using the time taken to report to activities, we determined the minimum time taken to submit a report for an activity among all the participants, and the version of the application used to submit that report. Thus, we found that, using which particular version of the application, we received the quickest response for each of the activities. The results are shown in Figures 12 and 13. We can see that the risky situations responded quicker when using interfaces that consisted of abN, even though there is not much difference between interfaces for the quickest time of response to non-risky notifications.

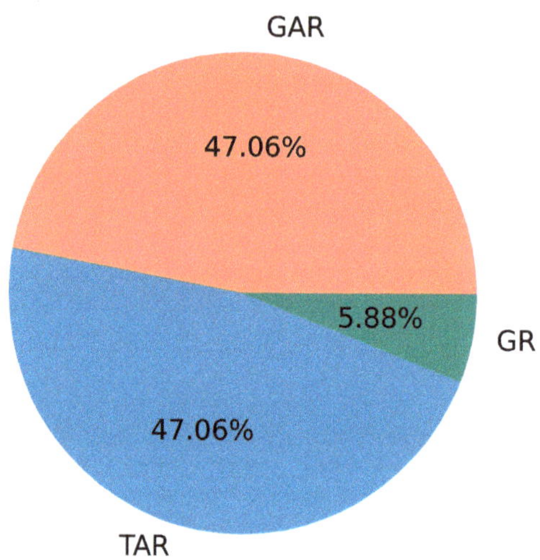

Figure 12. Quickest response for risky situations.

In the final questionnaire, the participants responded with the reasons that could also provide the reason for such higher response time. Almost 45% participants (n = 4) mentioned that they were busy with their research/private work and could not respond to the notifications on time. We received responses such as: *"I was so busy with my work"; "Busy with my research work or play a game"; "mentally busy with my own work"; "sometimes i was busy"*. Similarly, two of the participants mentioned that they often forgot to check the application. This can be attributed to the different interface types used and notifications received.

Two of the participants responded in the questionnaire that they did not use the application if they did not receive any notifications, while six (66%) of them said they did not wait for the notifications to use the application but were busy with their work and could not respond immediately. We also wanted to know if the notifications received were perceived as distracting or disturbing, to analyze if their perception played any role in the response time. When asked if the notifications received from the application were distracting, 2 (22%) of them strongly claimed they were not disturbed, 5 (55%) said they were not disturbed, while 1 of them was neutral, and 1 agreed that he was distracted. Similarly, 8 (88%) (strongly agree: 4; agree: 4) agreed that they prefer to receive abN so that they can be regularly notified monitor frequently, while 1 of them was neutral.

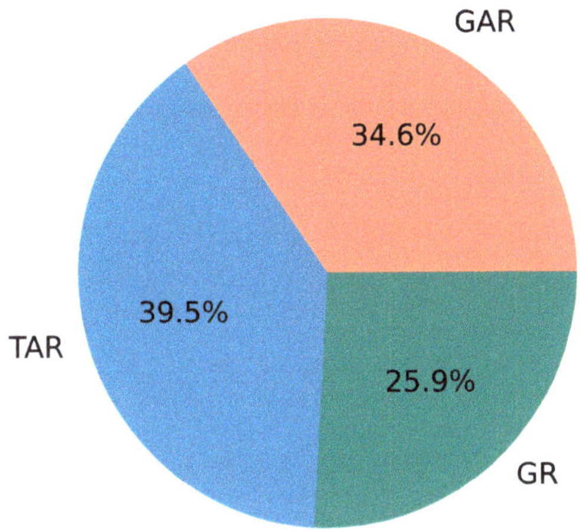

Figure 13. Quickest response for non-risky situations.

5. Discussion and Limitations

In this section, first we discuss the results and verify research questions RQ1–3 mentioned in Section 2.3, then we show some remaining issues as limitations.

5.1. Discussion

When considering user engagement and their ability to identify routine of individuals with the interface, we can conclude that the results are fairly positive towards GAR, as compared to TAR. Using GAR, we found that 75.2% of risky situations were correctly identified as risks, compared to 65.8% and 68.5% for TAR and GR, respectively. Though identification of risk varied between study groups using GAR (68.4% for StudyGroup A; 64.7% for StudyGroup B, and 92.6% for StudyGroup C), the overall identification rate is higher for GAR. This shows that risks can be identified using graphical interface and the style of graph that we used. A response from a participant , *"I can see the difference of the duration directly from the graph. The table one need to scroll up and down to see all the information, which sometimes kind of annoying"* also suggests that our visualization is effective. These findings justify our research questions, RQ1 and RQ2, that it is possible to identify the daily routine of individuals using a smartphone application, and it is possible to detect potential risks in such routine based on the visualization provided.

Using GAR, participants faced the lowest burden of 28 s, compared to 38 s in tabular (TAR). Similarly, none of the participants claimed that the application demanded a lot of time and effort from them. Regarding notifications, only one of the participants found them distracting, and 88.8% mentioned that they will prefer to receive activity based notifications for monitoring purposes. Similarly, all of the participants (77.8% strongly agree, 22.2% agree) responded that the use of traffic colors was useful to identify the state of the activities quickly. Therefore, we can verify RQ3, that constant notifications and using the application was not troublesome for the users.

We received a total of 1680 responses from participants over the experiment period. We can claim that such interaction is a result of their willingness to use the application. When interface of participants changed from graph to table, there was a reduction in the number of reports obtained (45.6% for StudyGroup A in phase 2, and 9.8% for StudyGroup

B in phase 3). Similarly, when the interface changed from tabular to graph, we obtained an increase in the number of reports by 96.7% for StudyGroup A in phase 3 and reduction by 11.5% for StudyGroup C in phase 2. In total, the engagement with the application is high, which along with the lower interface analyzing time, verifies RQ3, that using the application is not a burden for the monitoring person.

At the end of the experiment, we asked the participants which representation of activities they preferred: table or graph. All of them agreed that graphical representation was better. Some of the responses we received, such as, *"Got on a quick glance the exact duration of past activities and could check exact time of the day"; "With graph, it's easy for me to compare the length of activity at the glance."*, further strengthens our proposal that the graphical interface we proposed can help to identify a daily routine in a clear and intuitive manner and further justifies RQ1, that a smartphone application can be a good tool for identifying daily activities.

5.2. Limitations

Our system evaluation requires that there are certain risky situations in the activity of the elderly. We did not conduct a real-time activity recognition of elderly, but instead, we used a pre-existing activity dataset because, in real-time scenarios, there is no surety of receiving such risky situations, and we would need to request someone to deliberately change their activity pattern so that others could detect it. Such a situation can invoke unfavorable reactions. Similarly, since activity recognition systems are not perfectly accurate, sometimes the activities may not be correctly identified, or falsely identified, which would hamper our evaluation. Moreover, we recruited students for the experiment, but they are always busy because of their academic work, and/or personal lives which might have affected the number and time of reception of reports.

6. Conclusions

In this study, we proposed a system, PATROL, that can be used to anonymously track everyday activities of the elderly and identify any potential risks in their daily routine using a smartphone application. Our system is aimed to be deployed in elderly residential areas or communities and does not disclose any private information such as age, location, etc. to the monitoring person to maintain the privacy and security of elderly residents. The monitoring person receives recurring notifications every two hours and activity-based notifications whenever an elderly person completes an activity from the service server and assesses elderly condition by a smartphone application visualizing elderly activity history. We designed our application with features such as single interface design, intuitive graphical user interface for activity and anomaly detection, and color and textual information for state of activities. These features altogether help not only to conduct quicker monitoring of activities of elderly, but also to induce a low amount of burden to the monitoring person, who at once may be responsible for monitoring single or multiple elderly people.

We added risky situations in an activity dataset obtained from a real-life experiment with elderly residents and conducted a user study using the proposed method and two other baseline methods varying in visualization and notification techniques for three groups consisting of nine participants. We found that with our proposed method, 75.2% of the risks were successfully identified, while 68.5% and 65.8% were identified with other methods. The proposed method also provided a better result for the timely reception of activities: GAR (median = 115.1 min), TAR (median = 118.85 min), and GR (median = 121.12 min). Moreover, the interface analyzing and reporting time was also lower (28 s) in the proposed method compared to 38 and 54 s in other methods. As future work, we will conduct real-time activity recognition and monitoring using our application. To achieve that, we will also research/work on activity recognition systems using other kinds of sensors that can not only potentially provide better activity recognition in real time but also remove dependency on the elderly person for data collection. Moreover, we will explore the possibility to assess the elderly's activity state and detect anomalies by using measurements from ambient sensors (temperature, humidity, illumination, etc.). We will also include high risk situations such as Fall (and no activities after the incident) and try to determine if participants will be able to deduce such emergency situations quickly. We will also aim to increase the number of participants to receive more reports and analyze the results based on age, gender, etc.

Author Contributions: Conceptualization, R.D., T.M., Y.M. and K.Y.; Methodology, R.D., T.M., Y.M. and K.Y.; Software, R.D. and Y.M.; Validation, T.M.; Formal, Y.M. and K.Y.; Resources, K.Y.; Writing—original draft preparation, R.D.; Writing—review and editing, R.D., T.M., Y.M. and K.Y.; Visualization, R.D. and K.Y.; supervision, T.M., Y.M. and K.Y.; funding acquisition, K.Y. All authors have read and agreed to the published version of the manuscript.

Funding: This research was partly supported by JSPS KAKENHI Grant Nos. JP21H03431, JP19KT0020, JP19K11924, and JP20H04177.

Institutional Review Board Statement: The study was conducted according to the guidelines of the Declaration of Helsinki, and approved by Ethical Review Committee of Nara Institute of Science and Technology (2020-I-16).

Informed Consent Statement: Informed consent was waived because collected data does not include private information.

Data Availability Statement: The data presented in this study are available on request from the corresponding author.

Conflicts of Interest: The authors declare no conflict of interest. The funders had no role in the design of the study; in the collection, analyses, or interpretation of data; in the writing of the manuscript, or in the decision to publish the results.

References

1. Veronese, F.; Mangano, S.; Comai, S.; Matteucci, M.; Salice, F. Indoor Activity Monitoring for Mutual Reassurance. In Proceedings of the International Conference on Smart Objects and Technologies for Social Good, Venice, Italy, 30 November–1 December 2016; Springer: Berlin/Heidelberg, Germany, 2016; pp. 1–10.
2. *Statistical Handbook of Japan 2021*; Technical Report; Statistics Bureau Ministry of Internal Affairs and Communications: Tokyo, Japan, 2021.
3. Chan, M.; Campo, E.; Bourennane, W.; Bettahar, F.; Charlon, Y. Mobility behavior assessment using a smart-monitoring system to care for the elderly in a hospital environment. In Proceedings of the 7th International Conference on PErvasive Technologies Related to Assistive Environments, Island of Rhodes, Greece, 27–30 May 2014; ACM: New York, NY, USA, 2014; p. 51.
4. Morita, T.; Taki, K.; Fujimoto, M.; Suwa, H.; Arakawa, Y.; Suwa, H.; Yasumoto, K. BLE Beacon-based Activity Monitoring System toward Automatic Generation of Daily Report. In Proceedings of the 2018 IEEE International Conference on Pervasive Computing and Communications Workshops (PerCom Workshops), Athens, Greece, 19–23 March 2018; IEEE: New York, NY, USA, 2018.
5. Majumder, S.; Aghayi, E.; Noferesti, M.; Memarzadeh-Tehran, H.; Mondal, T.; Pang, Z.; Deen, M.J. Smart homes for elderly healthcare—Recent advances and research challenges. *Sensors* **2017**, *17*, 2496. [CrossRef] [PubMed]
6. Almeida, A.; Mulero, R.; Rametta, P.; Urošević, V.; Andrić, M.; Patrono, L. A critical analysis of an IoT—Aware AAL system for elderly monitoring. *Future Gener. Comput. Syst.* **2019**, *97*, 598–619. [CrossRef]
7. Vaiyapuri, T.; Lydia, E.L.; Sikkandar, M.Y.; Diaz, V.G.; Pustokhina, I.V.; Pustokhin, D.A. Internet of Things and Deep Learning Enabled Elderly Fall Detection Model for Smart Homecare. *IEEE Access* **2021**, *9*, 113879–113888. [CrossRef]
8. Moriya, K.; Nakagawa, E.; Fujimoto, M.; Suwa, H.; Arakawa, Y.; Kimura, A.; Miki, S.; Yasumoto, K. Daily living activity recognition with ECHONET Lite appliances and motion sensors. In Proceedings of the Pervasive Computing and Communications Workshops (PerCom Workshops), 2017 IEEE International Conference, Kona, HI, USA, 13–17 March 2017; IEEE: New York, NY, USA, 2017; pp. 437–442.
9. Stratogiannis, G.; Vlachostergiou, A.; Siolas, G.; Caridakis, G.; Mylonas, P.; Stafylopatis, A.; Kollias, S. User and home appliances pervasive interaction in a sensor driven smart home environment: The SandS approach. In Proceedings of the Semantic and Social Media Adaptation and Personalization (SMAP), 2015 10th International Workshop, Trento, Italy, 5–6 November 2015; IEEE: New York, NY, USA, 2015; pp. 1–6.
10. Peek, S.T.M.; Wouters, E.J.M.; van Hoof, J.; Luijkx, K.G.; Boeije, H.R.; Vrijhoef, H.J.M. Factors influencing acceptance of technology for aging in place: A systematic review. *Int. J. Med. Inform.* **2014**, *83*, 235–248. [CrossRef]
11. Levine, D.M.; Ouchi, K.; Blanchfield, B.; Diamond, K.; Licurse, A.; Pu, C.T.; Schnipper, J.L. Hospital-level care at home for acutely ill adults: A pilot randomized controlled trial. *J. Gen. Intern. Med.* **2018**, *33*, 729–736. [CrossRef]
12. Pouke, M.; Häkkilä, J. Elderly healthcare monitoring using an avatar-based 3D virtual environment. *Int. J. Environ. Res. Public Health* **2013**, *10*, 7283–7298. [CrossRef]

13. Sahami Shirazi, A.; Henze, N.; Dingler, T.; Pielot, M.; Weber, D.; Schmidt, A. Large-scale assessment of mobile notifications. In Proceedings of the SIGCHI Conference on Human Factors in Computing Systems, Toronto, ON, Canada, 26 Apirl–1 May 2014; ACM: New York, NY, USA, 2014; pp. 3055–3064.
14. Yan, Z.; Liu, C.; Niemi, V.; Yu, G. Exploring the impact of trust information visualization on mobile application usage. *Pers. Ubiquitous Comput.* **2013**, *17*, 1295–1313. [CrossRef]
15. Avenell, S.A. Facilitating spontaneity: The state and independent volunteering in contemporary Japan. *Soc. Sci. Jpn. J.* **2010**, *13*, 69–93. [CrossRef]
16. Matsui, T.; Onishi, K.; Misaki, S.; Fujimoto, M.; Suwa, H.; Yasumoto, K. Easy-to-Deploy Living Activity Sensing System and Data Collection in General Homes. In Proceedings of the 2020 IEEE International Conference on Pervasive Computing and Communications Workshops (PerCom Workshops), Austin, TX, USA, 23–27 March 2020; IEEE: New York, NY, USA, 2020; pp. 1–6.
17. Kashimoto, Y.; Morita, T.; Fujimoto, M.; Arakawa, Y.; Suwa, H.; Yasumoto, K. Sensing Activities and Locations of Senior Citizens toward Automatic Daycare Report Generation. In Proceedings of the Advanced Information Networking and Applications (AINA), 2017 IEEE 31st International Conference, Taipei, Taiwan, 27–29 March 2017; IEEE: New York, NY, USA, 2017; pp. 174–181.
18. Dawadi, R.; Asghar, Z.; Pulli, P. Internet of Things Controlled Home Objects for the Elderly. In Proceedings of the International Conference on Health Informatics, HEALTHINF, Porto, Portugal, 21–23 February 2017; Scitepress: Setubal, Portugal, 2017; pp. 244–251.
19. Kashimoto, Y.; Hata, K.; Suwa, H.; Fujimoto, M.; Arakawa, Y.; Shigezumi, T.; Komiya, K.; Konishi, K.; Yasumoto, K. Low-cost and device-free activity recognition system with energy harvesting PIR and door sensors. In Proceedings of the Adjunct Proceedings of the 13th International Conference on Mobile and Ubiquitous Systems: Computing Networking and Services, Hiroshima, Japan, 28 November–1 December 2016; ACM: New York, NY, USA, 2016; pp. 6–11.
20. Kashimoto, Y.; Fujiwara, M.; Fujimoto, M.; Suwa, H.; Arakawa, Y.; Yasumoto, K. ALPAS: Analog-PIR-sensor-based activity recognition system in smarthome. In Proceedings of the 2017 IEEE 31st International Conference on Advanced Information Networking and Applications (AINA), Taipei, Taiwan, 27–29 March 2017; IEEE: New York, NY, USA, 2017; pp. 880–885.
21. Civitarese, G.; Bettini, C. Monitoring objects manipulations to detect abnormal behaviors. In Proceedings of the Pervasive Computing and Communications Workshops (PerCom Workshops), 2017 IEEE International Conference, Kona, HI, USA, 13–17 March 2017; IEEE: New York, NY, USA, 2017; pp. 388–393.
22. Morita, T.; Fujiwara, M.; Arakawa, Y.; Suwa, H.; Yasumoto, K. Energy Harvesting Sensor Node Toward Zero Energy In-Network Sensor Data Processing. In Proceedings of the International Conference on Mobile Computing, Applications, and Services, Osaka, Japan, 28 February–2 March 2018; Springer: Berlin/Heidelberg, Germany, 2018; pp. 210–215.
23. Kim, H.H.; Ha, K.N.; Lee, S.; Lee, K.C. Resident location-recognition algorithm using a Bayesian classifier in the PIR sensor-based indoor location-aware system. *IEEE Trans. Syst. Man Cybern. Part C (Appl. Rev.)* **2009**, *39*, 240–245.
24. Shivakumar, N.S.; Sasikala, M. Design of vital sign monitor based on wireless sensor networks and telemedicine technology. In Proceedings of the 2014 International Conference on Green Computing Communication and Electrical Engineering (ICGCCEE), Coimbatore, India, 6–8 March 2014; IEEE: New York, NY, USA, 2014; pp. 1–5.
25. Bottazzi, D.; Corradi, A.; Montanari, R. Context-aware middleware solutions for anytime and anywhere emergency assistance to elderly people. *IEEE Commun. Mag.* **2006**, *44*, 82–90. [CrossRef]
26. Choudhury, B.; Choudhury, T.S.; Pramanik, A.; Arif, W.; Mehedi, J. Design and implementation of an SMS based home security system. In Proceedings of the 2015 IEEE International Conference on Electrical, Computer and Communication Technologies (ICECCT), Coimbatore, India, 5–7 March 2015; IEEE: New York, NY, USA, 2015; pp. 1–7.
27. Freitas, D.J.; Marcondes, T.B.; Nakamura, L.H.; Ueyama, J.; Gomes, P.H.; Meneguette, R.I. Combining cell phones and WSNs for preventing accidents in smart-homes with disabled people. In Proceedings of the 2015 7th International Conference on New Technologies, Mobility and Security (NTMS), Paris, France, 27–29 July 2015; IEEE: New York, NY, USA, 2015; pp. 1–5.
28. Zhang, Q.; Ren, L.; Shi, W. HONEY: A multimodality fall detection and telecare system. *Telemed. E-Health* **2013**, *19*, 415–429. [CrossRef]
29. Popescu, M.; Li, Y.; Skubic, M.; Rantz, M. An acoustic fall detector system that uses sound height information to reduce the false alarm rate. In Proceedings of the 2008 30th Annual International Conference of the IEEE Engineering in Medicine and Biology Society, Vancouver, BC, Canada, 20–25 August 2008; IEEE: New York, NY, USA, 2008; pp. 4628–4631.
30. Wang, J.; Zhang, Z.; Li, B.; Lee, S.; Sherratt, R.S. An enhanced fall detection system for elderly person monitoring using consumer home networks. *IEEE Trans. Consum. Electron.* **2014**, *60*, 23–29. [CrossRef]
31. Abbate, S.; Avvenuti, M.; Bonatesta, F.; Cola, G.; Corsini, P.; Vecchio, A. A smartphone-based fall detection system. *Pervasive Mob. Comput.* **2012**, *8*, 883–899. [CrossRef]
32. Mimamori. Long Distance Care. Available online: https://www.aivs.co.jp/watch_pri_en/ (accessed on 10 September 2021).
33. Canary. Elderly Monitoring System. Available online:https://www.techsilver.co.uk/product/elderly-monitoring-system/ (accessed on 10 September 2021).
34. Lively. Health Apps, Cell Phones for Seniors, Medical Alert. Available online: https://www.greatcall.com/ (accessed on 10 September 2021).
35. Grandcare. GrandCare—Technology For Seniors Made EASY. Available online: https://www.grandcare.com/ (accessed on 10 September 2021).
36. Alarm.com. A Smarter Solution for Independent Living. Available online: https://alarm.com/ (accessed on 10 September 2021).

37. Doxy Me:The Simple, Free & Secure Telemedicine Solution. Telemedicine Solution—Simple, Free, and Secure Available online: https://doxy.me/en/ (accessed on 10 September 2021).
38. VideoHealth. Telemedicine Software Platform for Providers. Available online: https://www.vidyohealth.com/ (accessed on 10 September 2021).
39. Petermans, J.; Piau, A. Gerontechnology: Don't miss the train, but which is the right carriage? *Eur. Geriatr. Med.* **2017**, *2017*, 281–283. [CrossRef]
40. Kaye, J.A.; Maxwell, S.A.; Mattek, N.; Hayes, T.L.; Dodge, H.; Pavel, M.; Jimison, H.B.; Wild, K.; Boise, L.; Zitzelberger, T.A. Intelligent systems for assessing aging changes: Home-based, unobtrusive, and continuous assessment of aging. *J. Gerontol. Ser. B Psychol. Sci. Soc. Sci.* **2011**, *66*, i180–i190. [CrossRef] [PubMed]
41. Fahim, M.; Fatima, I.; Lee, S.; Lee, Y.K. Daily life activity tracking application for smart homes using android smartphone. In Proceedings of the 2012 14th International conference on advanced communication technology (ICACT), PyeongChang, Korea, 19–22 February 2012; IEEE: New York, NY, USA, 2012; pp. 241–245.
42. Salman, H.M.; Wan Ahmad, W.F.; Sulaiman, S. Usability Evaluation of the Smartphone User Interface in Supporting Elderly Users from Experts' Perspective. *IEEE Access* **2018**, *6*, 22578–22591. [CrossRef]
43. Visuri, A.; van Berkel, N.; Okoshi, T.; Goncalves, J.; Kostakos, V. Understanding smartphone notifications' user interactions and content importance. *Int. J. Hum. Comput. Stud.* **2019**, *128*, 72–85. [CrossRef]
44. Ward, A.F.; Duke, K.; Gneezy, A.; Bos, M.W. Brain Drain: The Mere Presence of One's Own Smartphone Reduces Available Cognitive Capacity. *J. Assoc. Consum. Res.* **2017**, 2. [CrossRef]
45. Weber, D. Towards smart notification management in multi-device environments. In Proceedings of the 19th International Conference on Human-Computer Interaction with Mobile Devices and Services, Vienna, Austria, 4–7 September 2017; ACM: New York, NY, USA, 2017; p. 68.
46. Saikia, P.; Cheung, M.; She, J.; Park, S. Effectiveness of Mobile Notification Delivery. In Proceedings of the Mobile Data Management (MDM), 2017 18th IEEE International Conference, Daejeon, Korea, 29 May–1 June 2017; IEEE: New York, NY, USA, 2017; pp. 21–29.
47. Mehrotra, A.; Pejovic, V.; Vermeulen, J.; Hendley, R.; Musolesi, M. My phone and me: Understanding people's receptivity to mobile notifications. In Proceedings of the 2016 CHI Conference on Human Factors in Computing Systems, San Jose, CA, USA, 7–12 May 2016; ACM: New York, NY, USA, 2016; pp. 1021–1032.
48. Pielot, M.; Church, K.; De Oliveira, R. An in-situ study of mobile phone notifications. In Proceedings of the 16th international Conference on Human-Computer Interaction with Mobile Devices & Services, Toronto, ON, Canada, 23–26 September 2014; ACM: New York, NY, USA, 2014; pp. 233–242.
49. Schulze, F.; Groh, G. Conversational context helps improve mobile notification management. In Proceedings of the 18th International Conference on Human-Computer Interaction with Mobile Devices and Services, Florence, Italy, 6–9 September 2016; ACM: New York, NY, USA, 2016; pp. 518–528.
50. Künzler, F.; Kramer, J.N.; Kowatsch, T. Efficacy of mobile context-aware notification management systems: A systematic literature review and meta-analysis. In Proceedings of the 2017 IEEE 13th International Conference on Wireless and Mobile Computing, Networking and Communications (WiMob), Rome, Italy, 9–11 October 2017; IEEE: New York, NY, USA, 2017; pp. 131–138.
51. Horgas, A.L.; Wilms, H.U.; Baltes, M.M. Daily life in very old age: Everyday activities as expression of successful living. *Gerontologist* **1998**, *38*, 556–568. [CrossRef]
52. Shevchenko, Y.; Kuhlmann, T.; Reips, U.D. Samply: A user-friendly smartphone app and web-based means of scheduling and sending mobile notifications for experience-sampling research. *Behav. Res. Methods* **2021**. [CrossRef]
53. Marty-Dugas, J.; Smilek, D. The relations between smartphone use, mood, and flow experience. *Personal. Individ. Differ.* **2020**, *164*, 109966. [CrossRef]
54. Williams, D.; Ahamed, S.I.; Chu, W. Designing interpersonal communication software for the abilities of elderly users. In Proceedings of the Computer Software and Applications Conference Workshops (COMPSACW), 2014 IEEE 38th International, Vasteras, Sweden, 21–25 July 2014; IEEE: New York, NY, USA, 2014; pp. 282–287.
55. Gordon, M.L.; Garys, L.; Guestrin, C.; Bigham, J.P.; Trister, A.; Patel, K. App usage predicts cognitive ability in older adults. In Proceedings of the Conference on Human Factors in Computing Systems–Proceedings, Glasgow, UK, 4–9 May 2019; Association for Computing Machinery: New York, NY, USA, 2019.
56. Trudel, R.; Murray, K.B.; Kim, S.; Chen, S. The impact of traffic light color-coding on food health perceptions and choice. *J. Exp. Psychol. Appl.* **2015**, *21*, 255. [CrossRef] [PubMed]
57. Koenigstorfer, J.; Groeppel-Klein, A.; Kamm, F.; Rohr, M.; Wentura, D. The traffic light colors red and green in the context of healthy food decision-making. *ACR N. Am. Adv.* **2012**, 40, 945–946.
58. Nomali, M.; Mohammadrezaei, R.; Keshtkar, A.A.; Roshandel, G.; Ghiyasvandian, S.; Alipasandi, K.; Zakerimoghadam, M. Self-Monitoring by Traffic Light Color Coding Versus Usual Care on Outcomes of Patients With Heart Failure Reduced Ejection Fraction: Protocol for a Randomized Controlled Trial. *JMIR Res. Protoc.* **2018**, *7*, e9209. [CrossRef]
59. Dawadi, R.; Mizumoto, T.; Yasumoto, K. MutualMonitor: A Tool for Elderly People to Anonymously Monitor Each Other. In Proceedings of the 2018 Eleventh International Conference on Mobile Computing and Ubiquitous Network (ICMU), Auckland, New Zealand, 5–8 October 2018; IEEE: New York, NY, USA, 2018; pp. 1–6.

sensors

MDPI

Article

Design and Development of a Heterogeneous Active Assisted Living Solution for Monitoring and Following Up with Chronic Heart Failure Patients in Spain

Francisco José Melero-Muñoz [1,2,*], María Victoria Bueno-Delgado [2,3], Ramón Martínez-Carreras [2], Rafael Maestre-Ferriz [1], Miguel Ángel Beteta-Medina [1], Tomás Puebla-Martínez [1], Andrés Lorenzo Bleda-Tomás [1], Gorka Sánchez-Nanclares [4], Ricardo Pérez-de-Zabala [5] and Mónica Álvarez-Leon [6]

[1] Centro Tecnológico del Mueble y la Madera de Murcia, C/Perales s/n, 30510 Yecla, Spain
[2] Department of Information and Communication Technologies, Universidad Politécnica de Cartagena, Plaza Hospital 1, 30202 Cartagena, Spain
[3] E-Lighthouse Network Solutions S.L, C/Angel s/n, 30202 Cartagena, Spain
[4] Servicio Murciano de Salud, Edif. Habitamia 5º, 30100 Murcia, Spain
[5] MIWEnergia, Parque Científico de Murcia, 30100 Espinardo, Spain
[6] Minsait–Av. Bruselas 35, 28108 Alcobendas, Spain
* Correspondence: fj.melero@cetem.es

Citation: Melero-Muñoz, F.J.; Bueno-Delgado, M.V.; Martínez-Carreras, R.; Maestre-Ferriz, R.; Beteta-Medina, M.Á.; Puebla-Martínez, T.; Bleda-Tomás, A.L.; Sánchez-Nanclares, G.; Pérez-de-Zabala, R.; Álvarez-Leon, M. Design and Development of a Heterogeneous Active Assisted Living Solution for Monitoring and Following Up with Chronic Heart Failure Patients in Spain. *Sensors* **2022**, 22, 8961. https://doi.org/10.3390/s22228961

Academic Editors: Bijan Najafi, Enrico G. Caiani and Gérald Thouand

Received: 30 September 2022
Accepted: 17 November 2022
Published: 19 November 2022

Abstract: Heart failure is the most common disease among elderly people, and the risk increases with age. The use of smart Internet of Things (IoT) systems for monitoring patients with chronic heart failure (CHF) in a non-intrusive manner can result in better control of the disease, improving proactive healthcare through real-time and historical patient's data, promoting self-care in patients, reducing unneeded interaction between patients and doctors, reducing the number of hospitalizations and saving healthcare costs. This work presents an active assisted living (AAL) solution based on the IoT to provide a tele-assistance platform for CHF patients from the public health service of the region of Murcia in Spain, with formal and informal caregivers and health professionals also as key actors. In this article, we have detailed the methodology, results, and conclusions of the prevalidation phase for the set of IoT technologies to be integrated in the AAL platform, the first mandatory step before the deployment of a large-scale pilot that will lead to improving the innovation of the system from its current technology readiness level to the market. The work presented, in the framework of the H2020 Pharaon project, aims to serve as inspiration to the R&D community for the design, development, and deployment of AAL solutions based on heterogeneous IoT technologies, or similar approaches, for smart healthcare solutions in real healthcare institutions.

Keywords: AAL; IoT; healthcare; prevalidation; deployment; chronic heart failure; large-scale pilot; H2020

1. Introduction

In a rapidly ageing European society, there is a growing need for implementing information and communication technologies (ICT) and digital tools that improve the quality of life, independence, and overall health of older adults. In this context, the concept of ambient intelligence (AmI) appears to achieve a future where technology surrounds the users and helps them in their daily lives [1]. AmI leads to cutting-edge platforms referred to as active assisted living platforms [2,3]. The Internet of Things (IoT) [4] has also emerged as a set of technologies, systems, and design principles [5,6] that enable automation in many fields, such as remote and smart healthcare systems, playing an important role in AAL platforms and healthcare. A typical IoT environment consists of communication interfaces, sensors, advanced algorithms, and cloud interfaces [7]. Sensors are responsible for collecting data from various devices. Additionally, different communication technologies (wired and/or wireless), such as wireless sensor networks (WSN), provide network and communication infrastructure [8], while advanced algorithms are used to analyze and process data [9]. Numerous client/server requests can be exchanged in the cloud environment and allow the users to have access to various types of services simultaneously [10,11]. Due to cloud computing challenges and high requisites of emerging 5G, such as latency, reliability, resource constraints, etc. Fog computing is used to overcome these limitations and run the same applications anywhere close to users with real-time analysis and efficient decision-making features [12,13].

Considerable effort is being made in the R&D community to integrate IoT technologies, communications [14], databases, and computing [15], and to develop standardized and integrated AAL platforms [16], but the diversity of these types of systems has created a very fragmented market and a lack of standardized architectures and protocols [17].

In the healthcare industry, sanitary systems are being revolutionized by the IoT paradigm [18]. This plays an important role in telemonitoring in hospitals, especially at homes for elderly people with chronic diseases [19]. By using this technology, healthcare systems can experience major effects such as a reduction in response time to detect anomalies, high-quality care, low hospitalization costs, and high life expectancy [18].

In the specific case of heart failure (HF), the most common disease in elderly people and one that increases in prevalence with age [20], the use of an IoT system for monitoring patients with chronic heart failure (CHF) in a non-intrusive manner can result in better control of the disease, improving the proactive healthcare and reducing unneeded interaction between patients and doctors, thereby reducing the number of hospitalizations and saving healthcare costs.

HF is a leading cause of hospitalization, representing 1–2% of all hospital admissions [21,22]. In Spain, the prevalence of HF is around 5% (higher than in other EU countries and USA); the rate rises with age to 8% between the ages of 65 and 74, and to 16% in persons aged 75 years and over [23]. These rates mean enormous health care resources are required, e.g., in Spain, during the 2015–2019 period, costs of HF patients were EUR 15,373 per patient, with HF hospitalizations being the most important determinant (51.0%). Medication costs represented only a small proportion of total costs [24]. The latest data published by Eurostat revealed that in 2018, HF caused 19,142 deaths in Spain, constituting 4.5% of all deaths, 3.4% in men and 5.6% in women [25]. Within this frame of reference, cardiologists and family doctors believe that the key to a better control of patients with CHF is creating a proactive healthcare solution through a non-intrusive and integrated monitoring system that unifies patients' medical history and makes it available to all relevant healthcare professionals. This will improve the safety and efficiency of healthcare by increasing the availability of patients' real-time information. In addition, the health and care community are convinced that it is crucial to promote self-care in patients by providing them and their caregivers with health education and training and other social and health resources.

This work presents an AAL solution based on IoT to provide a tele-assistance platform for CHF patients from the public health service of the region of Murcia in Spain. Initially focused on CHF patients aged 55 and over, the study also included formal and informal caregivers and healthcare professionals. For the three user groups, two scenarios are considered to provide a comprehensive healthcare solution:

- Angel of Health, aiming at improving health and care services and follow-up for CHF patients and involving them in the health and care process from the data perspective;
- Care@Home, aiming at reducing the dependency of CHF patients and to detect emergency situations early.

Figure 1 summarizes the goals, roles and description of the two scenarios considered.

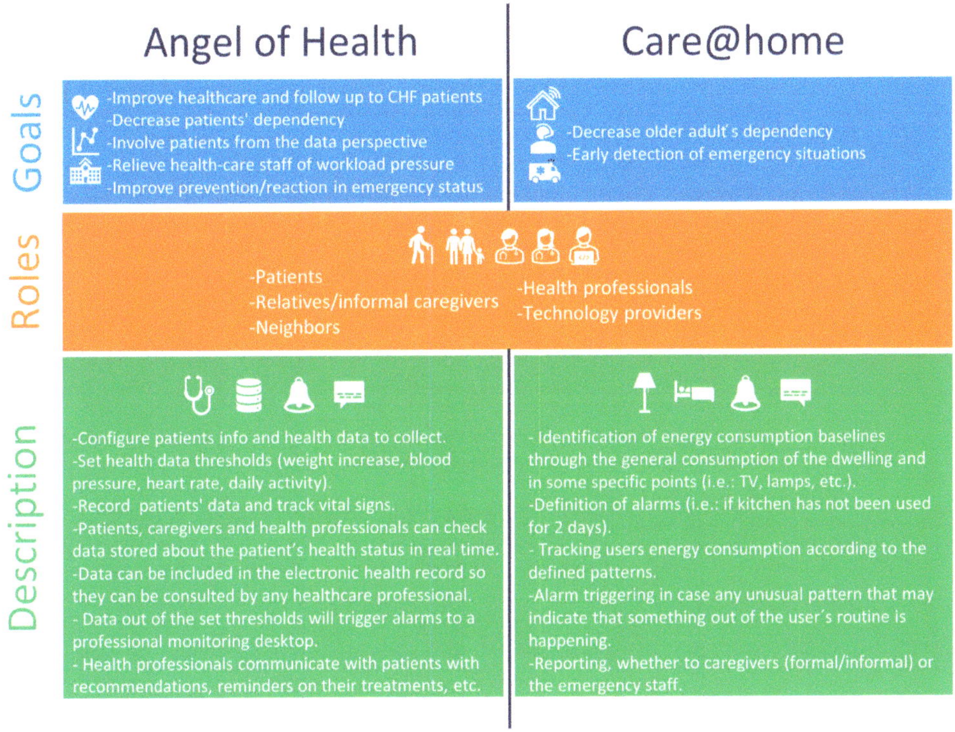

Figure 1. Scenarios of the Pharaon Murcia pilot.

This work is part of the overall research in progress regarding the H2020 Pharaon Project (Pilots for Healthy and Active Ageing) [26], which aims at providing a smart and active lifestyle for Europe's ageing population by creating a set of integrated and highly customizable interoperable open platforms with advanced services, devices, and tools in AAL, including IoT, artificial intelligence, robotics, cloud computing, smart wearables, big data, and intelligent analytics. Built upon mature existing state-of-the-art open platforms and technologies/tools, a user-centric approach is being followed for the deployment and the two-stages validation (prevalidation and large-scale pilots) in six different pilot sites: Murcia and Andalusia (Spain), Portugal, the Netherlands, Slovenia, and Italy.

The Pharaon ecosystem integrates a big set of services and functionalities which requires the involvement of a huge number of resources in order to achieve the necessary validation and trials in real-world scenarios. The validation within Pharaon is being performed through the six large-scale pilots proposed with different types of users, requirements, and chosen functionalities. Pharaon aims to carry out the unprecedented validation of different platforms simultaneously, each supporting a wide variety of advanced and customized assistive services and tools in six different large-scale pilots with the necessary resources. For example, the Murcia pilot will validate heart failure and non-intrusive home monitoring with alarm triggering, whereas the pilot in The Netherlands focuses on community building and providing tailored advice towards user empowerment in terms of health literacy.

This approach allows a sufficiently high number of users and healthcare professionals to use the system at each pilot site for a long-term period, and an unprecedented number of use cases, services, and technologies will be tested across all pilots.

In this work, the authors present the methodology, results, and conclusions of the prevalidation phase for the set of IoT solutions and their respective AAL platforms, which will introduce the tele-assistance platform for CHF patients from the public health service of the region of Murcia. The prevalidation stage includes the participation of all actors involved: patients, formal and informal caregivers, and healthcare professionals. Their early feedback regarding the functionality and usefulness of the tested technologies and platforms is the first mandatory step before the deployment of a large-scale pilot that will lead to improving the innovation of the system from its current technology readiness level (TRL) (6) to the market (9). The goal is to provide the R&D community with inspiration in the design and deployment of AAL solutions based on heterogeneous IoT technologies, or similar approaches, for smart healthcare solutions in real healthcare institutions.

The paper is organized as follows: Section 2 describes in detail the Pharaon Murcia pilot and the workplan followed for implementing the smart healthcare solution. Section 3 explains the engineering of user requirement tasks that were performed to identify the scenarios that define the use cases of the AAL healthcare solution and their requisites. Section 4 looks in depth at the system architecture that must perform the use cases and technical features of the hardware/software needed. Section 5 describes the roadmap implemented for the testing phase. Section 6 presents and discusses the results of the testing. Section 7 shows the conclusions and the future work recommendations. Finally, Section 8 presents a short discussion about the research limitations.

2. Pharaon Murcia Pilot: Overview and Workplan for the Healthcare Solution Deployment

The reduction in birth rate and the increase in life expectancy will, in the long term, lead to a progressive ageing of the population in the region of Murcia, which will manifest itself in an uninterrupted decline in the working-age population and a continued increase in the proportion of the population over 65 years of age [27]. One of the priorities of the research and innovation strategy for smart specialization in the region of Murcia [28] is related to health, biomedicine, and welfare, addressing, among other fields, housing care and ITC-supported social services, specialized care, access to services, and remote assistance.

The services and use cases to be deployed under Pharaon in the region of Murcia are aimed at building the foundations of a new telecare line in the region that will transcend the current model of health and care services that rely on the patients to notice when they need help. This new telecare model will allow patients to stay in their preferred environment and provide a more intense, effective, proactive, and less intrusive care and observation service. To do that, this work has been organized into a sequence of main steps, summarized in Figure 2. The first step consists of the elicitation and representation of the user requirements. This leads to the definition of the use case scenarios of the pilot and detailed technical requirements referring to the technologies to be used.

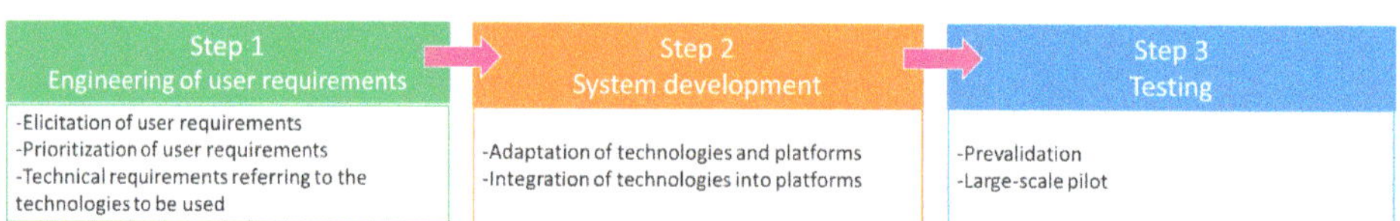

Figure 2. Work sequence within the Pharaon Murcia pilot.

The second step, informed by the results of step 1, has two main goals: to address the adaptations required for the compliance of all requirements, and the development and integration of the technologies and services in the platform.

Finally, step 3 consists of the testing phase, organized into two stages: (1) small-scale testing with a limited number of participants representing the different target users of the Murcia–Pharaon system. They provide early feedback regarding the functionality and usefulness of the system, enabling rapid and iterative improvements to address shortcomings discovered by actual system users while its implications are still manageable. The second stage is (2) running the large-scale pilot, which involves all the pilot users in real-world scenarios for an extended length of time.

3. Engineering of User Requirements in the Pharaon Murcia Pilot

In [29], the authors summarized the co-design and user requirement engineering work carried out in the pilot study of the region of Murcia. During the co-design phase, the methodology for the co-design and representation of user requirements was defined as goal models with a set of components: functional goals, quality goals, and emotional goals, following work in [30]. Corresponding use case scenarios and user stories were also defined.

The methodology entailed several up-to-date co-design methods for user requirements' elicitation. The ISO 9241-210 standard on ergonomics of human–system interaction [31] was followed. The original plan for eliciting and representing user requirements was modified due to the COVID-19 outbreak, following three phases:

The first phase is initial desk research on co-design workshops, data, and results from previous initiatives in which the public health service provider of the region of Murcia participated: ProEmpower [32], ReadiForHealth [33], INC3A [34], and CARPRIMUR [35]. They helped to identify an initial set of requirements from the stakeholder's perspective in the form of functional, quality, and emotional goals.

The second phase is the design and launch of a questionnaire addressing target users of the Pharaon system that helped to define a map of barriers and opportunities in the region regarding the assistance of patients suffering from CHF. The goal was to enrich the results in the previous phase with the opinion of other representatives that were also target groups (older adults, informal caregivers, and health and care professionals). The participants were reached through an online questionnaire that was duly promoted at a regional level, and they participated in a set of virtual co-design sessions in different focus groups. In total, 250 responses (56% were patients, relatives, or caretakers, and 44% were health and care professionals) were gathered, and they helped to complete the initial goal framework.

Next was the virtual co-design phase, consisting of the arrangement of a virtual workshop and the creation of different focus groups where representatives from the different target users of the Pharaon system were involved: health and care providers, older adults, and informal caregivers. In these workshops, different questions were posed regarding the CHF, target users, use cases, scenarios, and technologies. The discussions helped to confirm the goals and requirements identified and new ones appeared.

The outcomes of the three phases resulted in the definitive set of quality, functional, and emotional goals, and this was the basis for defining the three goal models of the Pharaon Murcia Pilot (Figure 3):

- Become involved in the health and care process;
- Improve patient care;
- Detect emergency situations.

The results also helped in identifying, clarifying, and organizing the system requirements in the Murcia pilot through the definition of seven use cases and the associated technology requirements [29]:

- Become involved in the health and care process;
- Assess personal situations and risks;
- Strengthen knowledge of healthy lifestyles and behaviors;
- Improve patient care;
- Boost disease follow-up;
- Upgrade interventions;
- Detect emergency situations.

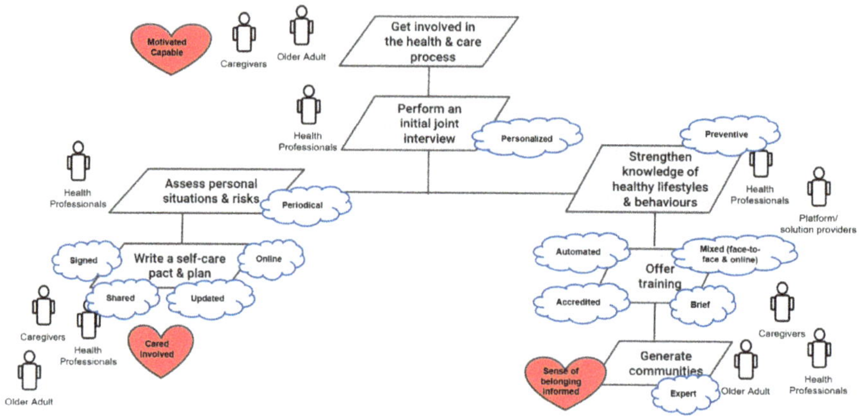

Figure 3. Representation of the Pharaon Murcia Pilot goal model "Get involved in the health and care process" [29].

4. Pharaon Murcia Pilot: Architecture and System Development

The technical requirements identified in the use cases found in step 1 of the workplan led to the selection of the technologies to be implemented (see Table 1) to cover the two scenarios, the goal models and the use cases with the software applications and platforms required (see Figures 4 and 5).

Table 1. Technologies of the Pharaon Murcia pilot.

Technology Name	Technology Classification	Description	Murcia Pilot Scenario	Hardware/Software Components	Technical Pre-Requirements
Amicare (Technical Research Centre of Furniture and Wood of the Region of Murcia)	Indoor non-intrusive tracking of daily habits with configurable triggering alarms on caregivers' smartphone	Non-invasive system that monitors older adults' daily habits and contributes to the peace of mind of relatives and caregivers thanks to the power of IoT. Out-of-sight textile, movement, and ambient sensors are safely connected through the cloud to any smartphone app with a preregistered user. The app can be configured and personalized by its user to trigger alarms if certain actions happen (or do not happen) within specified time windows, such as: "if not in bed at any time between 22:00–0:00", "if on the coach for more than 3 consecutive hours", "if away from bed for more than 45 consecutive minutes between 0:00 and 6:00".	Care@Home	Hardware components: -Textile sensor pad; -HW box including processor, wireless communications, on-board sensors (movement, humidity, luminosity, temperature), and textile sensor connectivity; -App for Android device (smartphone or tablet); -Web-based user interface for group of users (e.g., nursing homes).	Wi-Fi connection is required
uGRID (MIWenergía)	Energy Management Platform	The uGRID software aims to digitize energy consumption, providing the final consumer with more information about their demand of electrical energy. The purpose is to achieve the maximum possible energy efficiency and to control the electricity consumption by setting alerts and generating reports.	Care@Home	Hardware Components: -Dedicated database server; -Dedicated platform web server; -Power metering devices. Software Components: -Web platform based on PHP/Javascript; -MariaDB database (MySQL).	Wi-Fi connection is required

Table 1. *Cont.*

Technology Name	Technology Classification	Description	Murcia Pilot Scenario	Hardware/Software Components	Technical Pre-Requirements
Smartband Solution (based on Mi band 5 of Xiaomi) (Universidad Politécnica de Cartagena)	Wearable	Through a commercial smartband wirelessly connected to the patient's smartphone, the smartband solution allows recording, in a non-invasive way, real-time (and historical) data of patients regarding their heart rate and daily activity in a number of steps. The novelty of this solution is that it also provides the patient data to his/her caregiver and to his/her healthcare professional. Moreover, the healthcare professional can configure alarms if heart rate is lower or higher than a certain threshold.	Care@Home Angel of Health	Hardware Components: -Smartband; -Charger. Software Components: -App for Android devices (smartphone or tablet) to track user's heart rate and daily activity, reporting real-time and historical data to patients, caregivers, and healthcare professionals.	-Connection to a smartphone through Bluetooth is required -Internet connection is required for the smartphone
Onesait Healthcare Data (Indra Minsait)	Telemedicine Health Platform	It allows the treatment and follow-up of chronic patients at home. Through its two user interfaces, the system integrates the above technologies and offers tools for the bidirectional communication among healthcare professionals in the clinical setting and patients at home, so that patients are provided with personalized treatments according to their clinical conditions and progress. Onesait Healthcare currently allows the remote monitoring of patients suffering from specific chronic diseases, such as CHF, diabetes, or hypertension, while enabling mobility-ubiquity of users, personalization, and interoperability between systems (seamless systems).	Care@Home Angel of Health	Software Components: -MDM: includes managers for population information, catalogues and resources, along with a single sign on and an Audit and Log Server; -Onesait Healthcare Global Repository: health data storage standardized under HL7 FHIR® focused on interoperability; -Onesait Healthcare Professional Desktop: for professionals, citizens, managers, researchers, etc. that allows access to the different tools or applications; -Onesait Healthcare HomeCare: for health professionals to monitor patients; -Form Builder: tool for the configuration and parameterization of health questionnaires; -Alert: in charge of managing alarm triggers and its visualization by the operators to whom they are addressed; -MyHealth App: supports the functionalities to be used by Older Adults and their Caregivers.	-Access to smartphone is required for Patients and Caregivers. -Access to personal computer is required for health professional.

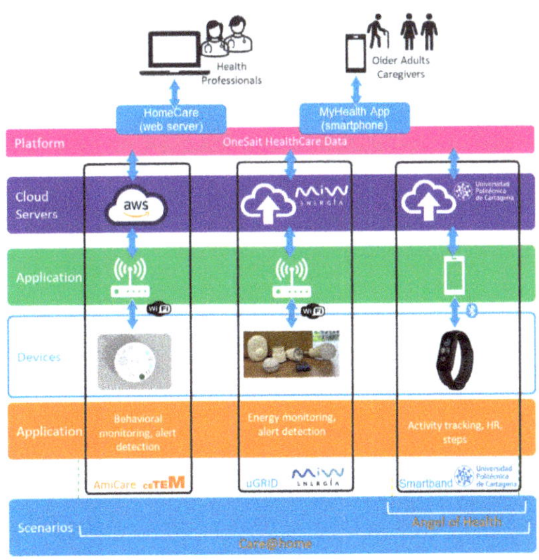

Figure 4. Overview of scenarios, technologies, and platform of the Pharaon Murcia pilot.

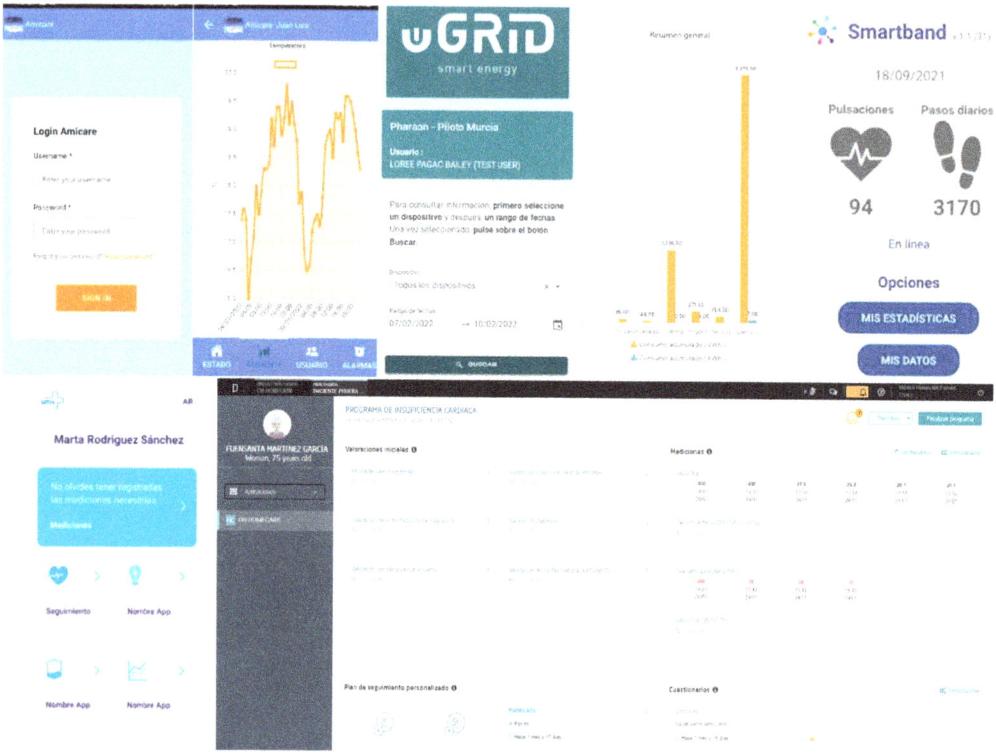

Figure 5. User interfaces of the technologies used in the Pharaon Murcia Pilot.

During step 2, technology providers worked on the adaptations of the technologies for the compliance of all users' needs identified in step 1 and for their integration into the Onesait Healthcare Data Platform, following the architecture agreed (see Figure 6).

For the six pilots, the architecture description was jointly carried out as a modelling exercise, having the defined use case scenarios and requirements for each pilot as the main input. Starting from the high-level abstraction and working towards adding more details, the explicit representation of the concept and the Pharaon ecosystem becomes clearer to everyone involved. It was decided that a technology-agnostic reference architecture model should be followed, focused on standard-based, non-AAL-specific, and somewhat recognized models, along with an adaptation of the 4 + 1 view model of architecture [36] with some additional views to ensure a high degree of coherence in the process of documenting the Pharaon architecture (see Figure 7).

Based on the experience and best practice of previous similar projects and initiatives [37–42], the Pharaon Reference Architecture is built on the common approach to "horizontal" functional layering that reflects the IoT implementations across various domains, and is expanded with two additional dimensions, cross-cutting functions, and properties.

Regarding the functional layers of the Murcia pilot architecture:

1. The device and network layer that represents the functional entities used for sensing, collecting data and actuating, and enabling network connectivity and transmission of data to other higher-level functional entities, includes Amicare, Miband 5, and the power metering devices that send energy consumption data to uGRID.
2. The platform layer, responsible for integration and interoperability of basic functional components for facilitating communication amongst them and exposing the functionality of these components and databases to provide basic data storage and

processing services. It comprises the storage and rule engines from Amicare, uGRID, and Smartband, and the Onesait Healthcare Data platform.

3. The service layer compiles all the services provided by Amicare, uGRID, and Smartband. It represents all functional components that support and ease application development, support mixture of different data streams, analytics and service components, and allow insights from data to be extracted and more complex data processing to be performed. Such services are executed in data centers (Smartband) or in cloud environments (Amicare and uGRID). They uniformly handle the underlying devices and networks, thus hiding the complexities of layers 1 and 2. Among these services, remote device management is included that can perform remote software upgrades, remote diagnostics or recovery, and dynamically reconfigure application processing such as setting event filters. Communication-related functions include publish/subscribe and message queue mechanisms. In general, data storage for anything from raw data to knowledge representations, and processing capabilities, such as data and event capture, filtering, and stream processing, are core services implemented by Amicare, uGrid, and Smartband.
4. The two user interfaces of the Onesait Healthcare data, the Homecare and MyHealth apps, along with the technologies' individual dashboard integrated in MyHealth app, are included in the Application Layer, responsible for representing data in rich visuals and/or interactive dashboards and providing direct functionality of the system from a user perspective. Besides visualization, it represents other functional components that are directly responsible for enabling application-specific visualization and user interaction (such as application-specific backend services).

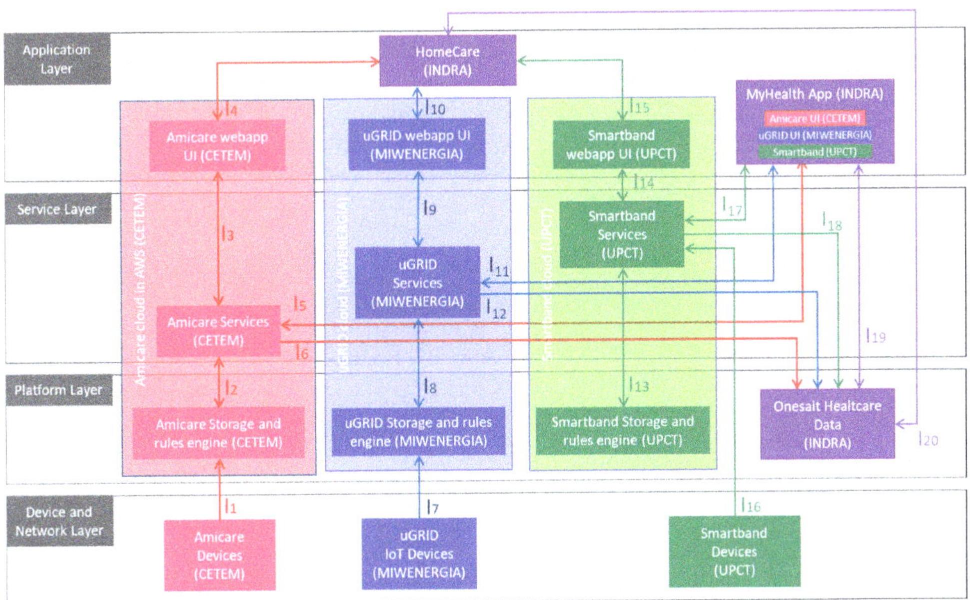

Figure 6. Murcia pilot high-level architecture.

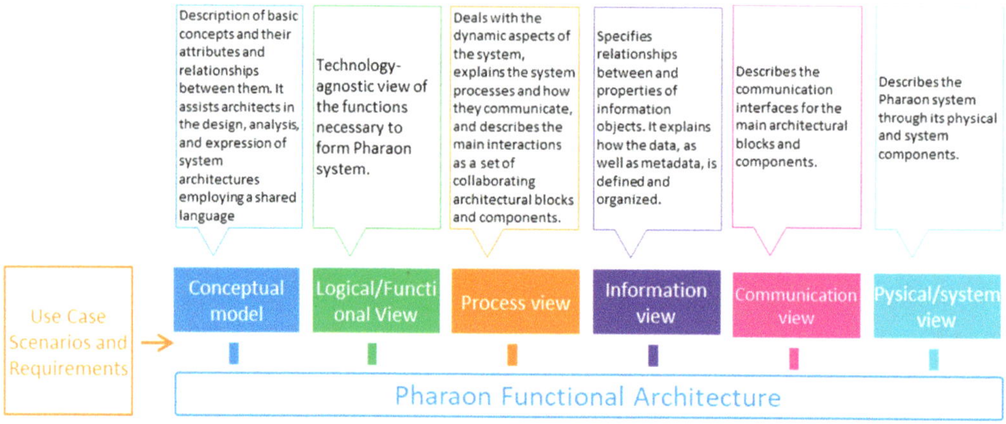

Figure 7. Pharaon architecture view model.

Figure 6 represents the architecture of the Murcia pilot at a high level, showing the elements of each of the horizontal functional layers and their cross-cutting functions, which address additional functionalities that are not linked to a single layer but whose provision requires spanning across several layers. This includes security, privacy, reliability, etc. Table 2 lists the technical description of each technology, including the interactions, technologies, protocols, and security implemented.

Table 2. Interactions between the elements of the Murcia pilot architecture and their technologies, protocols, and security.

	Interaction	Technologies and Protocols	Security
Amicare 1–6	I1	MQTT over Wi-Fi	Authentication
	I2	VNP	Authentication
	I3	REST API	TLS, AWS STS (Security Token Service)
	I4	HTTPS	JWT, SSL
	I5	REST API	TLS, AWS STS (Security Token Service)
	I6	REST API	Authentication in Rest services is through a JWT token signed with its private key
uGRID 7–12	I7	MQTT over Wi-Fi	Login
	I8	REST API	Uses SSL, API Key
	I9	REST API	Uses SSL, API Key
	I10	.NET Core (Blazor Server)	Uses SSL, Login
	I11	Ionic + Angular + Cordova	Uses SSL, Login
	I12	API REST/REST FHIR	Authentication in Rest services is through a JWT token signed with its private key
Smartband 13–18	I13	Local host socket	Authentication
	I14	HTTPS	Authentication (SSL)
	I15	HTTPS	JWT, SSL
	I16	Bluetooth	Authentication
	I17	HTTPS REST API	JWT, SSL
	I18	API REST/REST FHIR	Authentication in Rest services is through a JWT token signed with its private key
Onesait Healthcare Data 19–20	I19	API REST/REST FHIR	Oauth 2
	I20	API REST/REST FHIR	Oauth 2

5. Pharaon Murcia Pilot: Testing the Smart Healthcare Solution

A two-stage approach was defined for testing the Pharaon system with target users in all pilots:

1. The prevalidation stage, consisting of some initial real-world validation at a small scale, where a reduced number of participants representing the different target users (older adults, informal caregivers, and health professionals) test the technologies described above. The goal is to collect the opinion of the users regarding the use of these technologies in specific situations and to analyze the key performance indicators

(KPIs) focused on the users' level of autonomy, confidence, technology experience, usability, etc. These tests also help to detect bugs/problems/improvements and solve them before large-scale deployment.

2. The deployment of large-scale pilots with a significant number of target users testing the Pharaon system for a period of at least 12 months, where the impact is assessed at different levels according to a set of indicators agreed.

The present work includes the methods employed and results obtained during the prevalidation stage because, at the time of writing, the large-scale deployment had just started, and no significant data have been collected to evaluate the impact.

The six Pharaon pilots followed a common methodology described in a prevalidation protocol for all pilot sites to assess difficulties and willingness to use and for bug collection. For the Murcia pilot, this protocol comprised the following steps (see Figure 8):

- The initial approval of the study by the ethics authority where the prevalidation takes place. For the case of the Murcia pilot, the ethics committee of the Murcia Institute of Biomedical investigation, as the main ethics authority in the region in clinical research studies, issued a favorable opinion after the assessment of the proposed study, considering the relevance of its implementation, compliance with all basic principles in terms of ethics, and the suitability requirements of the protocol in relation to the research goals, along with the capacity of the researchers involved and the appropriateness of the available resources.
- The selection of the participants according to a set of inclusion and exclusion criteria determined by the nature and objectives of each pilot's requirements and the living conditions, digital skills, and health status of the volunteers (see Table 3).
- The definition of the key scenarios for users to perform and assess the technologies during the test sessions (see Table 4) and the definition of the key performance indicators (KPIs) (see Table 5).
- The pseudonymization of each participant and the compilation of socio-demographic data needed for correlation calculation under impact assessment.
- Technology set-up and configuration following a predefined protocol including all safety measures, explanations to users for each device that was installed, explanation and signature of the informed consent.
- The arrangement of test sessions where participants assessed a set of predefined scenarios and KPIs in two steps as well: 6 (1) where participants assessed the technologies individually, and 6 (2) where participants assessed the platform with the technologies integrated within it.
- The identification of bugs, their registration in a specific space created in Gitlab, and further improvements.

Table 3. Inclusion and exclusion criteria defined for users involved in the Pharaon Murcia pilot.

User Type	Inclusion Criteria	Exclusion Criteria
Older Adults	- CHF with any level of left ventricular ejection fraction (LVEF); - Frailty score preferably between 4 (vulnerable) and 6 (moderately frail) [1]; - Signed consent form; - Digital skills; - Availability of certain electronic devices and Wi-Fi Connection; - Digital skills, knowledge in e-health devices or higher knowledge in his/her pathologies.	- Planned heart transplantation, cardiac surgery, or left ventricular assisted device (LVAD) implant; - Chronic renal replacement therapy (haemodialysis, peritoneal dialysis, or transplant); - Evidence of active or suspected cancer or a history of malignancy in the last five years; - Inability or unwillingness to provide informed consent or to comply with study requirements; - Life expectancy below 1 year (due to another cause excluding HF) or a frailty scale of 7 and above; - Serious psychiatric illness; - Participation in another clinical trial.

Table 3. *Cont.*

User Type	Inclusion Criteria	Exclusion Criteria
Caregivers	- Older than 18 years old; - Digital skills; - Availability of electronic devices.	None
Health Professionals	- Digital skills; - Proactivity; - Motivation.	None

[1] According to the clinical frailty scale of the Canadian Study on Health and Ageing (CSHA) living at home [43].

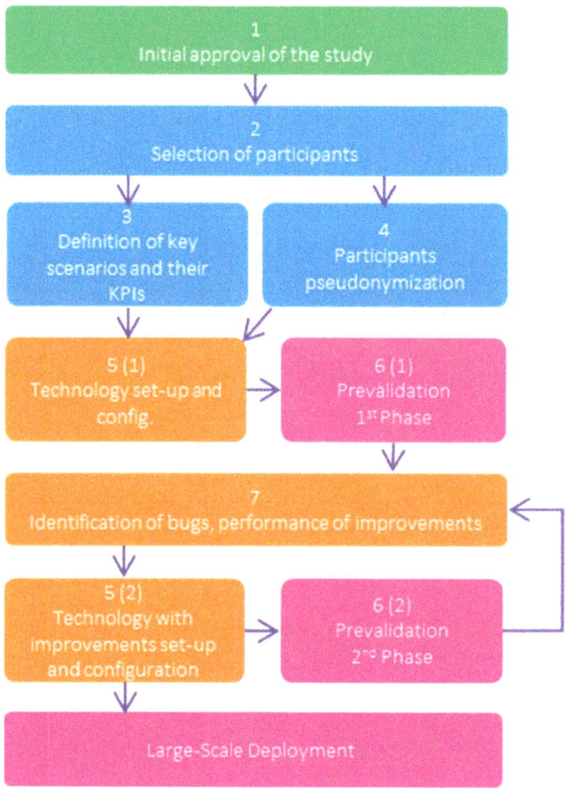

Figure 8. Prevalidation protocol followed by the six Pharaon pilots.

Table 4. Scenarios defined for each technology of the Pharaon Murcia pilot.

Technology	User Type	Scenario
Amicare	Patient	1. You want to find out the time you went to bed on a specific date according to Amicare. 2. You want to know the humidity in your home today.
	Caregiver	1. You want to check that Mrs. X's home temperature is between 19 °C and 25 °C. 2. You want to see if Mrs. X's movements in her flat were regular over the last 3 days. 3. You want to check that you will receive an alert when the alarm configuration is fulfilled.

Table 4. *Cont.*

Technology	User Type	Scenario
uGRID	Patient	1. You want to know your real-time electricity consumption. 2. You want to know if the kitchen has been used for cooking. 3. You want to know which electrical device consumes the most electricity in your home.
	Caregiver	1. You want to know which electrical device consumes the most electricity at the patient's home. 2. You want to know if Mrs. X has used the washing machine this week. 3. You want to see if you have received any alarms or events about changes in the power consumption patterns of Mrs. X.
Smartband	Patient	1. You want to see your heart rate. 2. You want to know how many steps you have taken today.
	Caregiver	1. You want to find Mr. X´s heart rate. 2. You want to know how many steps Mr. X has taken today.
Onesait Healthcare Data	Patient	1. You want to manually register a weight value/blood pressure. 2. You want to access your CHF program to fill in a health questionnaire. 3. You want to access your CHF follow-up program and consult your personal care plan.
	Caregiver	1. You want to manually register a weight/blood pressure value of Mrs. Z. 2. You want to access to the follow-up details of Mrs. Z. 3. You want to access Mrs. X´s CHF program to fill in a health questionnaire. 4. You want to access Mrs. X´s CHF follow-up program and consult her personal care plan.
	Health Professional	1. You want to access Mrs. X's CHF follow-up program. 2. You want to perform an initial clinical assessment of Mrs. X. 3. You want to create and publish the personal care plan (PCP) for Mrs. X through a template so she can consult it through the MyHealth App. 4. You want to customize what measures (vital signs) Mrs. X must include in the CHF follow-up. 5. You want to include a weight range between 50 and 62 kg, outside of which you will receive an alert for Mrs. X. 6. You want to customize what health questionnaires Mrs. X must fill in from MyHealth App and how she will have them available: only once, or any time to be filled whenever she wants. 7. You want to consult the most recent records of Mrs. X's heart rate and her heart rate history (Smartband). 8. You want to consult the history of personalized follow-up plans of Mrs. X (the details of the plan, which professional published it and when). 9. You want to know if Mrs. X has used the washing machine (or any other device that is being monitored) this week (MIW+). 10. You want to check that you will receive an alert when the alarm configuration is fulfilled (Amicare). 11. You want to check what patients have alerts from the list of patients of the CHF follow-up program. 12. You want to finalize Mrs. X´s CHF follow-up program.

Table 5. Key performance indicators used during the prevalidation stage.

KPI	Evaluation Area	Data Source	Kind of Data Source	Target
Identification of potential bugs	Number of bugs detected during test session	Gitlab Platform	Test of functionalities and usage scenarios	Table of bugs number
Usage difficulties	User acceptance	Questionnaire	After-Scenario Questionnaire (ASQ)	Medium score ≥ 5
Willingness to use	User acceptance	Questionnaire	System Usability Scale (SUS)	Medium score ≥ 68

Regarding the KPIs agreed upon (see Table 5), the one focused on usage difficulties found by the users has been evaluated through the ASQ questionnaire, with three questions (see Table 6) and a seven-level scale for each one. For the analysis of the results, the rates equal to or higher than 5 were interpreted as no difficulty. The questionnaire also comprised a "Comments" section for registering any difficulties or positive feedback about the scenario evaluated. The KPI focused on willingness to use was assessed by the System Usability Scale (SUS) questionnaire (see Table 7), a method that allows us to obtain a general view of subjective assessments of usability for a wide variety of products and services, including hardware, software, mobile devices, websites, and applications. Due to the heterogeneity of ICT solutions that comprise the Pharaon pilots, SUS was the user-testing tool agreed upon by all partners in the Pharaon project, even though there is other research focused on healthcare solutions that use the popular Technology Acquisition Model (TAM) questionnaire as the tool to measure the technology acceptability [44].

Table 6. Sample questions in the After Scenario Questionnaire (ASQ).

ASQ1: Overall, I Am Satisfied with the Ease of Completing the Tasks in This Scenario.								
Disagree	1	2	3	4	5	6	7	Agree
Comment: specification of agreement or disagreement.								
ASQ2: Overall, I Am Satisfied with the Amount of Time It Took to Complete the Tasks in This Scenario.								
Disagree	1	2	3	4	5	6	7	Agree
Comment: specification of agreement or disagreement.								
ASQ3: Overall, I Am Satisfied with the Support Information (Online-Help, Messages, Documentation) When Completing the Tasks.								
Disagree	1	2	3	4	5	6	7	Agree
Comment: specification of agreement or disagreement.								

Table 7. System Usability Scale (SUS).

SUS1. I Think That I Would Like to Use This System Frequently.						
Strongly disagree	1	2	3	4	5	Strongly agree
SUS2. I Found the System Unnecessarily Complex.						
Strongly disagree	1	2	3	4	5	Strongly agree
SUS3. I Thought the System Was Easy to Use.						
Strongly disagree	1	2	3	4	5	Strongly agree
SUS4. I Think I Would Need the Support of a Technical Person to Be Able to Use This System.						
Strongly disagree	1	2	3	4	5	Strongly agree
SUS5. I Found the Various Functions in This System Were Well Integrated.						
Strongly disagree	1	2	3	4	5	Strongly agree
SUS6. I Thought There Was Too Much Inconsistency in This System.						
Strongly disagree	1	2	3	4	5	Strongly agree
SUS7. I Would Imagine That Most People Would Learn to Use This System Very Quickly.						
Strongly disagree	1	2	3	4	5	Strongly agree
SUS8. I Found the System Very Cumbersome to Use.						
Strongly disagree	1	2	3	4	5	Strongly agree
SUS9. I Felt Very Confident Using the System.						
Strongly disagree	1	2	3	4	5	Strongly agree
SUS10. I Needed to Learn a Lot of Things Before I Could Get Going with This System.						
Strongly disagree	1	2	3	4	5	Strongly agree

The SUS scale comprises 10 items that participants had to score from 1 to 5 once the scenarios of each technology were finalized; the calculation of the results based on the participants' scores gives a general score between 0 and 100.

$$\text{SUS Score} = 2.5 \times [(\text{SUS1} + \text{SUS3} + \text{SUS5} + \text{SUS7} + \text{SUS9-5}) + (25 - \text{SUS2} - \text{SUS4} - \text{SUS6} - \text{SUS8} - \text{SUS10})]$$

Table 8 shows the general guideline on SUS Score interpretation provided by some authors [45]. The prevalidation protocol considered those solutions as accepted if the average score of SUS was graded as A (excellent) or B (Good).

Table 8. General guideline of SUS Score interpretation [45].

SUS Score	Grade	Adjective Rating
>80.3	A	Excellent
68.0–80.3	B	Good
68	C	Okay
51–68	D	Poor
<51	E	Awful

Finally, it is necessary to provide a deeper explanation of the way the steps 5–7 were performed during the prevalidation (see Figure 8):

- Steps 5 (1) and 6 (1) were carried out together for each participant (patient and his/her caregiver), that is, on a fixed date, the technologies were installed and configured at the participant's home. In that visit, the patient and his/her caregiver took time testing the technologies individually, performing the predefined scenarios for each technology shown in Table 4 and filling out the ASQ and SUS questionnaires.
- Step 7 was carried out, analyzing the results gathered in the steps before, identifying and solving bugs and problems, and improving the technologies accordingly. In some cases, the technologies or their improvements could involve new set-ups or configurations; then, step 5 (2) was included.
- After the first 5-6-7 iteration, step 6 (2) was performed in different sessions, organized by type of participant. On the one hand, healthcare professionals participated in the testing sessions, performing the scenarios defined with the Onesait Homecare Healthcare software tool, where all technologies were integrated (see Table 2). After that, they completed the ASQ and SUS. On the other hand, patients and caregivers performed the second testing round with the improved technology result of step 7 and 5 (2), and all of them already integrated in the MyHealth App. As in the previous testing session, they completed the ASQ and SUS.

It is important to remark that the second prevalidation phase was affected by COVID-19 restrictions, and some testing sessions were performed in hybrid mode following the premises of the Murcia health service.

6. Results

During the prevalidation stages, 44 people participated in testing sessions 6 (1) and 6 (2), namely:

- Prevalidation phase 1. Testing sessions in 6 (1): nine sessions were performed with a total of fifteen participants—nine older adults and six informal caregivers—from four municipalities of the region (Murcia, Cartagena, Alcantarilla and Yecla).
- Prevalidation phase 2. Testing session in 6 (2): twelve sessions were performed with 14 health professionals, ranging from cardiologists, to nurses, family doctors, neurologists, and pharmacists, and two sessions with eight older adults (two of them online), and seven informal caregivers.

Table 9 shows data from participants related to the following five variables: type of user, gender, year of birth, digital skills, and education level.

Table 9. Socio-demographic information of the recruited participants.

Type of User	Gender	Year of Birth	Digital Skills (*)	Education Level (**)	Pre-Validation Phases Involved
Patients	Female	1964	Some Experience	EQF Level 6	Phase1, Phase 2
	Female	1946	Some Experience	EQF Level 7	Phase1, Phase 2
	Male	1959	Some Experience	EQF Level 6	Phase1, Phase 2
	Male	1936	Some Experience	EQF Level 6	Phase1, Phase 2
	Male	1968	Experienced/Proficient	EQF Level 7	Phase1, Phase 2
	Female	1946	Some Experience	EQF Level 6	Phase1, Phase 2
	Male	1955	Some Experience	EQF Level 4	Phase1, Phase 2
	Female	1957	Some Experience	EQF Level 6	Phase1, Phase 2
	Female	1962	Some experience	EQF Level 5	Phase1, Phase 2
Caregivers	Female	1962	Some experience	EQF Level 5	Phase1, Phase 2
	Male	1995	Experienced/Proficient	EQF Level 7	Phase1, Phase 2
	Female	1972	Some experience with autonomy	EQF Level 5	Phase 2
	Male	1945	Some experience	EQF Level 6	Phase1, Phase 2
	Female	1956	Experienced/Proficient	EQF Level 3	Phase1, Phase 2
	Female	1955	Some experience	EQF Level 5	Phase1, Phase 2
	Male	1955	Some experience with autonomy	EQF Level 7	Phase1, Phase 2
Health Professionals	Male	1970	Some experience with autonomy	EQF Level 8	Phase 2
	Male	1977	Some experience with autonomy	EQF Level 8	Phase 2
	Female	1962	Some experience with autonomy	EQF Level 6	Phase 2
	Female	1982	Some experience with autonomy	EQF Level 6	Phase 2
	Male	1965	Some experience with autonomy	EQF Level 8	Phase 2
	Female	1974	Some experience with autonomy	EQF Level 6	Phase 2
	Male	1969	Experienced/Proficient	EQF Level 7	Phase 2
	Male	1974	Some experience with autonomy	EQF Level 8	Phase 2
	Female	1981	Some experience with autonomy	EQF Level 8	Phase 2
	Female	1973	Some experience with autonomy	EQF Level 8	Phase 2

Table 9. *Cont.*

Type of User	Gender	Year of Birth	Digital Skills (*)	Education Level (**)	Pre-Validation Phases Involved
	Female	1959	Some experience with autonomy	EQF Level 6	Phase 2
	Male	1973	Some experience with autonomy	EQF Level 8	Phase 2
	Female	1960	Some experience with autonomy	EQF Level 8	Phase 2
	Male	1979	Some experience with autonomy	EQF Level 8	Phase 2

(*) No experience, Some experience, Some experience with autonomy, Experienced/Proficient. (**) EQF Level 1: Primary Education; EQF Level 2: Academic Secondary School Lower Cycle, New Secondary School, and Lower Secondary School; EQF Level 3: Academic Secondary School Upper Cycle, Intermediate and Higher VET (up to 3rd grade); EQF Level 4: Post-secondary non-tertiary education; EQF Level 5: Short-cycle tertiary education; EQF Level 6: Bachelor's Degree, Higher Apprenticeship; EQF Level 7: Master's Degree, postgraduate certificate and diplomas; EQF Level 8: Doctorate or Equivalent.

Note that, although the number of participants (samples) in the prevalidation phase may seem small, many practitioners in the industry have adopted the five-users rule as standard practice for user-testing, which points out that five participants are enough for getting a useful result for testing usability.

All participants filled in the ASQ and SUS. The statistical methodology used consisted of a set of sequential procedures for handling the qualitative and quantitative research data: collection, counting, presentation, synthesis, and analysis. The analysis was performed using descriptive and exploratory analysis. The descriptive analysis served to describe the set of data, thus obtaining the parameters that distinguish the characteristics of a set of data. The reasons for carrying out this analysis are that it allowed us to know, in detail, the information we had and to know the way in which the information was structured. It helps to make deductions directly from the data and parameters obtained. The exploratory analysis consisted of a set of statistical techniques whose purpose was to obtain a basic understanding of the data, allowing the detection of salient features, such as unexpected and outliers. In this work, the mean was the main statistical technique used. The mean scores and global mean for each questionnaire and question are summarized in Tables 10–13.

Table 10. ASQ mean scores for the individual technologies during the pre-validation phase 1.

Technology	Type of User	Scenario	Mean Scores for Individual Questions			Global Mean
			ASQ1	ASQ2	ASQ3	
Amicare	Patients	1. You want to find out the time you went to bed on a specific date according to Amicare.	6.00	5.89	6.56	6.15
		2. You want to know the humidity in your home today.	6.89	6.78	6.67	6.78
	Caregivers	1. You want to check that Mrs. X's home temperature is between 19 °C and 25 °C.	6.67	6.83	7.00	6.83
		2. You want to see if Mrs. X's movements in her flat were regular over the last 3 days.	6.00	6.17	6.50	6.22
		3. You want to check that you receive an alert when the alarm configuration is fulfilled.	6.33	6.50	6.50	6.44

Table 10. Cont.

Technology	Type of User	Scenario	Mean Scores for Individual Questions			Global Mean
			ASQ1	ASQ2	ASQ3	
uGRID	Patients	1. You want to know your real-time electricity consumption.	5.75	5.38	6.25	5.79
		2. You want to know if the kitchen has been used for cooking.	5.88	5.88	6.00	5.92
		3. You want to know which electrical device consumes the most electricity in your home.	6.25	6.25	6.38	6.29
	Caregivers	1. You want to know which electrical device consumes the most electricity at the patient´s home.	6.67	6.83	7.00	6.83
		2. You want to know if Mrs. X has used the washing machine this week.	6.00	6.17	6.50	6.22
		3. You want to see if you have received any alarms or events about changes in power consumption patterns of Mrs. X.	6.33	6.50	6.50	6.44
Smartband Solution	Patients	1. You want to consult your heart rate.	6.86	6.86	6.71	6.81
		2. You want to know how many steps you have taken today.	7.00	7.00	6.86	6.95
	Caregivers	1. You want to consult Mr. X´s heart rate.	5.33	5.50	5.67	5.50
		2. You want to know how many steps Mr. X has taken today.	5.67	5.50	5.83	5.67

Table 11. SUS mean scores for the individual technologies during the pre-validation phase 1.

Technology	Type of User	Mean Scores for Individual Questions										SUS Mean Score
		SUS1	SUS2	SUS3	SUS4	SUS5	SUS6	SUS7	SUS8	SUS9	SUS10	
Amicare	Patient	4.44	1.44	4.78	1.78	4.67	1.33	4.44	1.44	4.56	1.44	88.61
	Caregiver	4.67	1.83	4.50	1.33	4.67	1.83	3.33	1.33	4.50	1.50	84.58
uGRID	Patient	4.00	1.50	4.38	1.75	4.38	1.63	3.88	1.50	4.50	1.75	82.50
	Caregiver	4.50	1.33	4.83	1.33	4.17	1.33	4.17	1.17	4.83	1.00	90.83
Smartband Solution	Patient	4.86	1.14	4.86	1.43	4.71	1.14	4.86	1.14	4.29	1.29	93.57
	Caregiver	4.60	2.00	4.80	1.20	5.00	1.00	3.80	1.20	4.80	1.20	91.00

In the first prevalidation phase, the mean score of the ASQ questions for each technology/scenario resulted in a value higher than 5 for all questions (see Table 10). Only one of nine patients (11.11%) scored questions ASQ1 and ASQ2 below 5 for Amicare scenario 1, considering the scenario somewhat difficult and not so fast to complete. The global mean for all questions reveals that almost all participants performed the requested scenarios with no difficulties, in an acceptable time, and felt supported throughout the process.

Regarding usability, the mean SUS scores obtained from all technologies (Table 10) were above 80.3, achieving an "A" grade (Excellent). Only two participants rated the smartband solution with 80 points.

From the results of the prevalidation phase 1, we concluded that the KPIs were reached for all technologies/scenarios, fulfilling the requisites of the first testing phase of prevalidation.

In the second prevalidation phase, the mean scores (Tables 12 and 13) were better than in the previous sessions in almost all technologies/scenarios, which means that participants are more satisfied with the solution in terms of difficulty and usability. From the eight patients that tested the MyHealth App module from Onesait Healthcare Data, the rates given to the SUS questions of only one patient (12.5%) achieved a score below 80.5. The six caregivers scored all individual ASQ questions with 5 or above, and for the Homecare module from Onesait Healthcare Data, at least one ASQ question from scenarios 5, 6, 9, 10, and 12 were scored as 4 by four professionals (28.5%). The rates given to the SUS questions of two health professionals (12.28%) received a score below 80.5.

Finally, it is important to mention that, during the first prevalidation phase, only two bugs (minor issues) were reported in Gitlab related to the smartband solution that were solved and the software was updated for step 6 (2). No bugs were reported during the second phase of pre-validation.

Table 12. ASQ mean scores for Onesait Healthcare Data with the integrated technologies during the pre-validation phase 2.

Technology	Type of User	Scenario	Mean Scores for Individual Questions			Total Mean
			ASQ1	ASQ2	ASQ3	
Onesait Healthcare Data	Patients	1. You want to manually register a weight value/blood pressure.	6.50	6.50	7.00	6.67
		2. You want to access your CHF program to fill in a health questionnaire.	7.00	6.50	7.00	6.67
		3. You want to access your CHF follow-up program and consult your personal care plan.	7.00	6.50	7.00	6.67
	Caregivers	1. You want to manually register a weight/blood pressure value of Mrs. Z.	7.00	7.00	7.00	7.00
		2. You want to access to the follow-up details of Mrs. Z.	6.67	6.67	6.83	6.72
		3. You want to access Mrs. X´s CHF program to fill in a health questionnaire.	6.67	6.67	7.00	6.78
		4. You want to access Mrs. X´s CHF follow-up program and consult her personal care plan.	6.83	7.00	7.00	6.94
	Health Professionals	1. You want to access Mrs. X CHF follow-up program.	7.00	6.67	6.83	6.83
		2. You want to perform the initial clinical assessment to Mrs. X.	6.75	6.58	6.58	6.64
		3. You want to create and publish the personal care plan (PCP) for Mrs. X through a template so she can consult it through the MyHealth App.	6.58	6.50	6.50	6.53
		4. You want to customize what measures (vital signs) Mrs. X must include in the CHF follow-up.	6.92	6.83	6.92	6.89
		5. You want to include a weight range between 50 and 62 kg, outside of which you will receive an alert for Mrs. X.	6.83	6.58	6.58	6.67
		6. You want to customize what health questionnaires Mrs. X must fill in from MyHealth App and how she will have them available: only once, or anytime to be filled in whenever she wants.	6.42	6.83	6.42	6.56
		7. You want to consult the last records of Mrs. X's heart rate and her heart rate history (Smartband).	5.92	6.25	6.08	6.08
		8. You want to consult the history of personalized follow-up plans of Mrs. X (the details of the plan, which professional published it and when).	6.92	6.92	6.92	6.92
		9. You want to know if Mrs. X has used the washing machine (or any other device that is being monitored) this week (MIW+).	6.60	6.50	6.40	6.50
		10. You want to check that you receive an alert when the alarm configuration is fulfilled (Amicare).	6.33	6.25	6.25	6.28
		11. You want to check what patients have alerts from the list of patients of the CHF follow-up program.	6.58	6.67	6.50	6.58
		12. You want to finalize the Mrs. X´s CHF follow-up program.	6.67	6.67	6.67	6.67

Table 13. SUS mean scores for Onesait Healthcare Data with the integrated technologies during the pre-validation phase 2.

Technology	Type of User	Mean Scores for Individual Questions										SUS Mean Score
		SUS1	SUS2	SUS3	SUS4	SUS5	SUS6	SUS7	SUS8	SUS9	SUS10	
Onesait Healthcare Data	Patients	4.38	1.50	4.75	1.75	4.75	1.50	3.88	1.38	4.38	2.00	85.00
	Caregivers	4.67	1.17	5.00	1.00	4.83	1.16	4.16	1.16	4.83	1.16	92.50
	Health Professionals	4.33	1.67	4.75	2.00	4.33	1.25	3.92	1.33	4.58	1.25	86.04

7. Conclusions and Future work

This work presented an innovative AAL solution built upon novel and mature existing IoT technologies with the aim of providing a tele-assistance platform for CHF patients from the public health service of the region of Murcia in Spain. The solution has been designed with a user-centric approach, but also involves formal and informal caregivers and health professionals. From the authors' point of view, this work has been used as a useful guideline to provide the R&D community with inspiration for the design and deployment of AAL solutions based on heterogeneous IoT technologies, or similar approaches, for smart healthcare solutions in real healthcare institutions.

It is important to note some key points of the work presented, as follows. The starting point must be a set of needs identified. Normally, healthcare professionals or institutions are the ones who detect the need. In this work, the needs were detected by the healthcare institution of the Murcia region for CHF patients. After that, a workplan of well-defined steps must be agreed upon by all parties. It must include at least engineering of user requirements, design of the system architecture and development, and testing of the solution for a final step of large-scale deployment. For each step, a clear methodology must be defined and executed with a set of expected outputs identified, which will serve as input for the next step(s). The results obtained from each step must be analyzed, identifying the improvements to implement and lessons learnt.

Regarding the steps defined for this work in depth, it is important to highlight some important issues regarding timing, participants, key points of the execution, and lessons learnt. In the engineering of user requirements, at least six months of implementation were required, with the participation of representatives of all target users involved in the final AAL solution. In a healthcare solution such as the one presented, patients, formal and informal caregivers, and healthcare professionals are expected to be involved at least. Moreover, a high number of participants help to better define the co-design and representation of user requirements as goal models, use case scenarios, and user stories. In this work, up to 250 people participated in the different initiatives, workshops, and meetings organized to define them.

The step focused on architecture definition, development, and integration is a technical approach with a high load of software development. The duration will depend on the maturity of the technologies involved, the functional requirements, and the final healthcare product expected. In this work, this step was planned for one year of work and was focused on two sequential goals. First, the definition of a functional architecture with a set of functional layers and others spanned across several layers, such as security, privacy and reliability, defining the interactions and protocols to use. Second, the development and adaptations of three innovative IoT technologies—Amicare, Smartband, uGRID—and the Homecare and MyHealth App as modules of the Onesait Healthcare Data platform. The main lesson learnt was the importance of using open and interoperable interfaces. Although some IoT solutions are shown as plug and play, minor or major adoptions must almost always be carried out. Data privacy and security are also a must.

Regarding the prevalidation step of the AAL solution, the mandatory assessment of the functionality, usability, and acceptance of the solution by the final users is noteworthy. It entails the participation of all actors involved in this work: patients, formal and informal caregivers, and health professionals. Moreover, all technologies must be monitored during the prevalidation to ensure good performance. In this work, the prevalidation involved the testing of each stand-alone IoT technology and the testing of all IoT technologies integrated in the final healthcare platform solution. Over two months, the authors collected feedback on the tested solution in both phases with 44 participants involved. From the results, the authors concluded that, on average, the rate of acceptance and usability of all technologies and software solutions was higher than expected, and no major bugs or unexpected issues were detected. Then, the product will be ready to be launched in a large-scale pilot.

As a final conclusion, the steps executed, explained in detail in this paper, have been essential in utilizing the future actions that will improve the innovation of the healthcare platform solution presented at its current TRL of 6, to launch a final development in view of creating a large-scale pilot, where around 450 participants are expected in the Murcia region, and to design and develop the impact, exploitation, and business plan.

8. Discussion

Although the work presented has been performed to a high professional quality, following guidelines agreed under a H2020 European project framework, and the results of the work performed have been very positive and useful, it is also important to mention that the authors have identified in this contribution a set of limitations without undermining the quality and integrity of the research. In this section, we summarize them and discuss how they could impact the work and results, and we provide countermeasures and alternatives for future work if this applies.

This work has been performed thanks to the collaborative work of six different organizations and professionals of different disciplines. In addition, more than 300 participants have been reached to participate during the different stages of the research work, which involved difficult organization and coordination. The COVID-19 outbreak added more difficulties due to the impossibility of performing face-to-face meetings and other organizational problems, which resulted in all tasks being redesigned and rescheduled. Time constraints also affected the research negatively, setting as future work some tasks that could have been finished or that at least had sufficient results which could have been included at the time this contribution was written, e.g., the experience of the large-scale pilot under execution.

Some research/development limitations have been also identified during the development of the integrated ICT solution, due to the architecture complexity and high-quality requirements and/or the lack of knowledge of some specific programming languages mandatory for the final integration with the platforms provided by Indra-Minsait.

Data and statistics have also been a limiting factor of the research performed. During the prevalidation phase, 44 participants were reached. Although it is a sufficient sample size for analyzing usability using SUS, the use of other evaluation questionnaires may need more samples, or some researchers may even consider 44 a low number of participants in a testing phase of a large healthcare ICT solution.

The lack of similar research works in the scientific literature has also limited the comparison with similar approaches and the enrichment that a state of art offer to the research contributions.

Finally, we remark that the work developed has been an ad hoc solution for specific target users and healthcare providers, and may show a strong regional focus, limiting its impact. However, some regional, national, and international stakeholders have already shown interest in the work for replicating the solution, or part of it, for other healthcare systems or types of patients (e.g., palliative healthcare).

Author Contributions: Conceptualization, F.J.M.-M., M.V.B.-D., R.M.-C., R.M.-F., M.Á.B.-M., T.P.-M., A.L.B.-T., G.S.-N., R.P.-d.-Z. and M.Á.-L.; Data curation, F.J.M.-M., R.M.-C. and T.P.-M.; Formal analysis, F.J.M.-M., M.V.B.-D., R.M.-C., R.M.-F., M.Á.B.-M., T.P.-M., A.L.B.-T., G.S.-N., R.P.-d.-Z. and M.Á.-L.; Funding acquisition, F.J.M.-M. and M.V.B.-D.; Investigation, F.J.M.-M., R.M.-C. and M.Á.B.-M.; Methodology, F.J.M.-M., M.V.B.-D. and R.M.-C.; Project administration, F.J.M.-M. and M.V.B.-D.; Resources, F.J.M.-M., M.V.B.-D., R.M.-C., R.M.-F., M.Á.B.-M., T.P.-M., A.L.B.-T., G.S.-N., R.P.-d.-Z. and M.Á.-L.; Software, F.J.M.-M., M.V.B.-D., R.M.-C., R.M.-F., M.Á.B.-M., T.P.-M., A.L.B.-T., G.S.-N., R.P.-d.-Z. and M.Á.-L.; Supervision, F.J.M.-M., M.V.B.-D. and R.M.-F.; Validation, F.J.M.-M., M.V.B.-D., R.M.-C. and G.S.-N.; Visualization, F.J.M.-M., M.V.B.-D. and R.M.-F.; Writing—original draft, F.J.M.-M., M.V.B.-D. and R.M.-C.; Writing—review and editing, F.J.M.-M., M.V.B.-D. and R.M.-F. All authors have read and agreed to the published version of the manuscript.

Funding: Funding for this research is provided by the European Commission under the EU Horizon 2020 Pharaon Project 'Pilots for Healthy and Active Ageing', Grant Agreement no. 857188, Grant PID2020-112675RB-C41 funded by MCIN/AEI/10.13039/501100011033, GO2EDGE (Ref. RED2018-02585-T) and Onofre-3 Grant PID2020-112675RB-C41 funded by MCIN/AEI/10.13039/501100011033.

This publication is based upon work from COST Action CA16226 Indoor living Space Improvement: Smart Habitat for the Elderly supported by COST (European Cooperation in Science and Technology). COST is a funding agency for research and innovation networks. Our Actions help connect research initiatives across Europe and enable scientists to grow their ideas by sharing them with their peers. This boosts their research, career and innovation. www.cost.eu (accessed on 29 September 2022).

Institutional Review Board Statement: The study was conducted in accordance with the Declaration of Helsinki and approved by the Permanent Commission of the Internal Scientific Committee of the Biomedical Research Institute of Murcia (IMIB).

Informed Consent Statement: Informed consent was obtained from all the participants involved in the study.

Data Availability Statement: All the data reported in this work are private, including the software developments, available in the private repository GitLab https://gitlab.com/pharaongroup (accessed on 29 September 2022).

Acknowledgments: The authors acknowledge the support provided by the partners involved in the Murcian Pilot and the full consortium of the Pharaon project.

Conflicts of Interest: The authors declare no conflict of interest.

References

1. Costa, A.; Julián, V.; Novais, P. Advances and trends for the development of ambient-assisted living platforms. *Expert Syst.* **2017**, *34*, 2. [CrossRef]
2. Ahmed, S.; Ilyas, M.; Raja, M.Y.A. Smart living: Ubiquitous services powered by ambient intelligence (AmI). In Proceedings of the 2019 IEEE 16th International Conference on Smart Cities: Improving Quality of Life Using ICT & IoT and AI (HONET-ICT), Charlotte, NC, USA, 6–9 October 2019; pp. 043–048.
3. Dunne, R.; Morris, T.; Harper, S. A survey of ambient intelligence. *ACM Comput. Surv.* **2022**, *54*, 1–27. [CrossRef]
4. Ashton, K. That 'Internet of Things' Thing. *RFID J.* **2009**, *22*, 97–114.
5. Bahga, A.; Madisetti, V. Internet of Things: A Hands-On Approach. 2014. Available online: www.internet-of-things-book.com (accessed on 1 November 2021).
6. Zhou, H. *The Internet of Things in the Cloud. A Middleware Perspective*; CRC Press: Boca Raton, FL, USA, 2012; Available online: https://www.crcpress.com/The-Internet-of-Things-in-the-Cloud-A-Middleware-Perspective/Zhou/p/book/9781439892992 (accessed on 1 November 2021).
7. Höller, J.; Tsiatsis, V.; Mulligan, C.; Karnouskos, S.; Avensand, S.; Boyle, D. *From Machine-to-Machine to the Internet of Things: Introduction to a New Age of Intelligence*; Academic Press: Cambridge, MA, USA, 2014; Elsevier: Amsterdam, The Netherlands, 2014; ISBN 978-0-12-407684-6.
8. Fosso Wamba, S.; Anand, A.; Carter, L. A literature review of RFID-enabled healthcare applications and issues. *Int. J. Inf. Manag.* **2013**, *33*, 875–891. [CrossRef]
9. Park, J.H.; Yen, N.Y. Advanced algorithms and applications based on IoT for the smart devices. *J. Ambient. Intell. Humaniz. Comput.* **2018**, *9*, 1085–1087. [CrossRef]
10. Asghari, P.; Rahmani, A.M.; Javadi, H.H.S. Internet of Things applications: A systematic review. *Comput. Netw.* **2019**, *148*, 241–261. [CrossRef]
11. Sultan, N. Making use of cloud computing for healthcare provision: Opportunities and challenges. *Int. J. Inf. Manag.* **2014**, *34*, 177–184. [CrossRef]
12. Farahani, B.; Firouzi, F.; Chang, V.; Badaroglu, M.; Constant, N.; Mankodiya, K. Towards fog-driven IoT eHealth: Promises and challenges of IoT in medicine and healthcare. *Future Generat. Comput. Syst.* **2018**, *78*, 659–676. [CrossRef]
13. Sheikh Sofla, M.; Haghi Kashani, M.; Mahdipour, E.; Faghih Mirzaee, R. Towards effective offloading mechanisms in fog computing: A systematic survey. *Multimed. Tools Appl.* **2022**, *81*, 1997–2042. [CrossRef]
14. Perera, C.; Zaslavsky, A.B.; Christen, P.; Georgakopoulos, D. Context Aware computing for the Internet of Things: A survey. *IEEE Commum. Surv. Tuts.* **2013**, *16*, 414–454. [CrossRef]
15. Cirillo, F.; Solmaz, G.; Berz, E.L.; Bauer, M.; Cheng, B.; Kovacs, E. A standard-based open source IoT platform: FIWARE. *IEEE Internet Things Mag.* **2019**, *2*, 12–18. [CrossRef]
16. Choi, D.; Choi, H.; Shon, D. Future changes to smart home based on AAL healthcare service. *J. Asian Archit. Build. Eng.* **2019**, *18*, 190–199. [CrossRef]
17. Abtoy, A.; Abdellah, T.; Abderahim, T. Ambient Assisted living system's models and architectures: A survey of the state of the art. *J. King Saud Univ.-Comput. Inf. Sci.* **2020**, *32*, 1–10.
18. Kashani, M.; Madanipour, M.; Nikravan, M.; Asghari, P.; Mahdipour, E. A systematic review of IoT in healthcare: Applications, techniques and trends. *J. Netw. Comput. Appl.* **2021**, *192*, 103164. [CrossRef]

19. Kyriazakos, S.; Prasad, R.; Mihovska, A.; Pnevmatikakis, A.; op den Akker, H.; Hermens, H.; Barone, P.; Mamelli, A.; de Domenico, S.; Pocs, M.; et al. eWALL: An open-source cloud-based eHealth platform for creating home caring environments for older adults living with chronic diseases or frailty. *Wirel. Pers. Commun.* **2017**, *97*, 1835–1875. [CrossRef]
20. Seferović, P.M.; Polovina, M.; Bauersachs, J.; Arad, M.; Ben Gal, T.; Lund, L.H.; Felix, S.B.; Arbustini, E.; Caforio, A.L.P.; Farmakis, D.; et al. Heart failure in cardiomyopathies: A position paper from the Heart Failure Association of the European Society of Cardiology. *Eur. J. Heart Fail.* **2019**, *21*, 553–576. [CrossRef]
21. Roger, V.L. Epidemiology of heart failure. *Circ. Res.* **2013**, *113*, 646–659. [CrossRef]
22. François, A.; Zannad, F.; Filippatos, G. Epidemiology of acute heart failure syndromes. *Heart Fail. Rev.* **2007**, *12*, 91–95.
23. Sayago-Silva, I.; García-López, F.; Segovia-Cubero, J. Epidemiology of heart failure in Spain over the last 20 years. *Rev. Española Cardiol.* **2013**, *66*, 649–656. [CrossRef]
24. Escobar, C.; Varela, L.; Palacios, B.; Capel, M.; Sicras, A.; Sicras, A.; Hormigo, A.; Alcázar, R.; Manito, N.; Botana, M. Costs and healthcare utilisation of patients with heart failure in Spain. *BMC Health Serv. Res.* **2020**, *20*, 964. [CrossRef]
25. According to Data from the National Statistics Institute. Available online: www.ine.es (accessed on 1 September 2020).
26. Pharaon EU H2020 Project (GA, No. 857188): Pilots for Healthy and Active Ageing. Available online: https://www.pharaon.eu/ (accessed on 14 August 2022).
27. National Statistics Institute, Population Projection 2020–2070. September 2020. Available online: https://www.ine.es/en/metodologia/t20/meto_propob_2020_2070_en.pdf (accessed on 14 August 2022).
28. RIS3 MUR. Available online: https://www.ris3mur.es/?lang=en (accessed on 13 September 2022).
29. Martínez, R.; Moreno-Muro, F.J.; Melero-Muñoz, F.J.; Bueno-Delgado, M.V.; Garrido-Lova, J.; Sánchez-Melero, M.; Tuveter, K. Co-design and Engineering of User Requirements for a Novel ICT Healthcare Solution in Murcia, Spain. In *Smart Objects and Technologies for Social Good*; Pires, I.M., Spinsante, S., Zdravevski, E., Lameski, P., Eds.; GOODTECHS 2021. Lecture Notes of the Institute for Computer Sciences, Social Informatics and Telecommunications Engineering; Springer: Cham, Switzerland, 2021; Volume 401. [CrossRef]
30. Taveter, K.; Sterling, L.; Pedell, S.; Burrows, R.; Taveter, E.M. A Method for Eliciting and Representing Emotional Requirements: Two Case Studies in e-Healthcare. In Proceedings of the 2019 IEEE 27th International Requirements Engineering Conference Workshops (REW), Jeju Island, Republic of Korea, 23–27 September 2019; pp. 100–105. [CrossRef]
31. ISO 9241-210:2010, Ergonomics of Human-System Interaction—Part 210: Human-Centred Design for Interactive Systems; ISO: Geneva, Switzerland. Available online: https://www.iso.org/standard/52075.html (accessed on 14 August 2022).
32. ProEmpower H2020 Project (GA No. 727409): Procuring Innovative ICT for Patient Empowerment and Self-Management for Type 2 Diabetes Mellitus. Available online: https://proempower-pcp.eu/ (accessed on 14 September 2022).
33. Readi for Health FP7 Project (GA No. 320021): Regional Digital Agendas for Europe. Available online: https://cordis.europa.eu/project/id/320021 (accessed on 14 September 2022).
34. IN3CA CIP Project (GA No. 621006): INclusive INtroduction of INtegrated CAre. Available online: https://cordis.europa.eu/project/id/621006 (accessed on 14 September 2022).
35. CARPRIMUR Project: Training and Assistance Initiative for Cardiologists, Physicians, Nurses and Patients. Available online: https://carprimur.com (accessed on 14 September 2022).
36. Kruchten, P.B. The 4+1 View Model of architecture. *IEEE Softw.* **1995**, *12*, 42–50. [CrossRef]
37. Synchronicity H2020 Project (GA No. 732240): Delivering an IoT enabled Digital Single Market for Europe and Beyond. Available online: https://cordis.europa.eu/project/id/732240 (accessed on 14 September 2022).
38. AUTOPILOT H2020 Project (GA No. 731993): AUTOmated Driving Progressed by Internet of Things. Available online: https://autopilot-project.eu/ (accessed on 14 September 2022).
39. ACTIVAGE H2020 Project (GA No. 732679): ACTivating InnoVative IoT Smart Living Environments for AGEing Well. Available online: http://activageproject.eu/ (accessed on 14 September 2022).
40. IoF 2020 H2020 Project (GA No. 731884): Internet of Food and Farm 2020. Available online: https://www.iof2020.eu/ (accessed on 14 September 2022).
41. Monica H2020 Project (GA No. 732350): Management of Networked IoT Wearables—Very Large Scale Demonstration of Cultural Societal Applications. Available online: https://www.monica-project.eu/ (accessed on 14 September 2022).
42. Create-IoT H2020 Project (GA No. 732929): Cross Fertilisation through Alignment Synchronisation and Exchanges for IoT. Available online: https://cordis.europa.eu/project/id/732929 (accessed on 14 September 2022).
43. McDowell, I.; Hill, G.; Lindsay, J. An Overview of the Canadian Study of Health and Aging. *Int. Psychogeriatr.* **2001**, *13*, 7–18. [CrossRef] [PubMed]
44. Mishra, R.K.; Park, C.; Momin, A.S.; Rafaei, N.E.; Kunik, M.; York, M.K.; Najafi, B. Care4AD: A Technology-Driven Platform for Care Coordination and Management: Acceptability Study in Dementia. *Gerontology* **2022**, 1–12. [CrossRef] [PubMed]
45. Alathas, H. How to Measure Product Usability with the System Usability Scale (SUS) Score. Available online: https://uxplanet.org/how-to-measure-product-usability-with-the-system-usability-scale-sus-score-69f3875b858f (accessed on 16 September 2022).

sensors

MDPI

Review

Can the Eight Hop Test Be Measured with Sensors? A Systematic Review

Luís Pimenta [1], Nuno M. Garcia [2], Eftim Zdravevski [3], Ivan Chorbev [3], Vladimir Trajkovik [3], Petre Lameski [3], Carlos Albuquerque [4,5,6] and Ivan Miguel Pires [1,2,*]

1 Escola de Ciências e Tecnologia, University of Trás-os-Montes e Alto Douro, Quinta de Prados, 5001-801 Vila Real, Portugal; al70827@alunos.utad.pt
2 Instituto de Telecomunicações, Universidade da Beira Interior, 6200-001 Covilhã, Portugal; ngarcia@di.ubi.pt
3 Faculty of Computer Science and Engineering, University Ss Cyril and Methodius, 1000 Skopje, North Macedonia; eftim.zdravevski@finki.ukim.mk (E.Z.); ivan.chorbev@finki.ukim.mk (I.C.); trvlado@finki.ukim.mk (V.T.); petre.lameski@finki.ukim.mk (P.L.)
4 Health Sciences Research Unit: Nursing (UICISA: E), Nursing School of Coimbra (ESEnfC), 3046-851 Coimbra, Portugal; calbuquerque@essv.ipv.pt
5 Higher School of Health, Polytechnic Institute of Viseu, 3504-510 Viseu, Portugal
6 Child Studies Research Center (CIEC), University of Minho, 4710-057 Braga, Portugal
* Correspondence: impires@it.ubi.pt; Tel.: +351-966-379-785

Abstract: Rehabilitation aims to increase the independence and physical function after injury, surgery, or other trauma, so that patients can recover to their previous ability as much as possible. To be able to measure the degree of recovery and impact of the treatment, various functional performance tests are used. The Eight Hop Test is a hop exercise that is directly linked to the rehabilitation of people suffering from tendon and ligament injuries on the lower limb. This paper presents a systematic review on the use of sensors for measuring functional movements during the execution of the Eight Hop Test, focusing primarily on the use of sensors, related diseases, and different methods implemented. Firstly, an automated search was performed on the publication databases: PubMed, Springer, ACM, IEEE Xplore, MDPI, and Elsevier. Secondly, the publications related to the Eight-Hop Test and sensors were filtered according to several search criteria and 15 papers were finally selected to be analyzed in detail. Our analysis found that the Eight Hop Test measurements can be performed with motion, force, and imaging sensors.

Keywords: Eight Hop Test; systematic review; measurement; sensors; diseases

Citation: Pimenta, L.; Garcia, N.M.; Zdravevski, E.; Chorbev, I.; Trajkovik, V.; Lameski, P.; Albuquerque, C.; Pires, I.M. Can the Eight Hop Test Be Measured with Sensors? A Systematic Review. *Sensors* **2022**, *22*, 3582. https://doi.org/10.3390/s22093582

Academic Editor: Andrea Cataldo

Received: 23 March 2022
Accepted: 6 May 2022
Published: 8 May 2022

1. Introduction

The Eight Hop Test is a hop exercise that consists of jumping with one leg in a circuit in the form of the number eight [1,2]. This test is helpful to evaluate the physical strength of individuals that suffered from some disease related to the lower limb [2,3].

Different kinds of sensors available in the market allow the measurement of patterns related to different movements [4–6]. It can handle the creation of automatic methods to the empowerment of the physical treatments [7–9]. These methods are important to give the same opportunities to rural environments in terms of treatments and monitoring of health conditions [10–12]. There are different types of sensors, but the sensors that are especially important for these measurements are the sensors for motion detection, which are available in commonly used mobile devices [13–15]. The positioning of these devices has different limitations, but it is relatively easy due to different support straps that allow these devices' statical position [16,17]. Therefore, the technology may help clinicians or scientists to study the detailed biomechanical parameters of jumping during rehabilitation programs as relevant variables for clinically significant scores and decide on the initiation of RTS with more confidence [18].

This review is included in a project related to the automation of the measurement of different results of the different physical functional tests, including the Heel-Rise Test [19,20], Timed-Up and Go Test [6,21], Ten Meter Walk Test [22], Six-Minute Walk Test [23], Functional Reach Test [24], 30-s Chair Stand Test [25,26], and Sit-to-Stand Test [27], where the development of different solutions involved in further studies. Furthermore, it is helpful and benefits the creation of Enhanced Living Environments [28,29]. The Eight Hop Test is mainly associated with different musculotendinous injuries, such as Cruciate Ligament of the Knee, Medial patellofemoral ligament, and Achilles tendon [18] and with injuries on the anterior cruciate ligament and gluteus medius [30,31].

The physical functional tests, such as bilateral or unilateral vertical and horizontal jump tests, require muscle strength and neuromuscular coordination for dynamic joint stability, which deteriorates with a knee injury [32,33]. In this regard, functional tests have been widely evaluated in laboratories using motion capture cameras, force platforms, and contact mattresses [32,34] to better understand biomechanical changes after knee injuries.

The monitoring of the biomechanics of the lower limbs during functional activities may help in the decision making related to the performance of sports or working activities after injury and prevention of osteoarthritis [35]. The analyzed test in this review allows the doctors to check the deficit of the significant underlying muscle deficits and ligament instability are still present throughout the post-operative rehabilitation period [35].

The study's purpose consists of a systematic review related to the measurement of the results of the Eight Hop Test with sensors, including motion force and imaging sensors. With the Eight Hop Test, the sensors can reach different results.

This paragraph ends the introductory section of this systematic review. Next, Section 2 describes research questions, inclusion criteria, search strategy, and analyzed study characteristics. Section 3 shows each study's results and summary. Third, in Section 4, we discuss and highlight the most critical points, and finally, Section 5 concludes the paper.

2. Methodology

2.1. Research Questions

The questions of this systematic review were focused on: (RQ1) Which devices can be used in the Eight Hop Test? (RQ2) Which data are related to the different types of diseases diagnosed by the Eight Hop Test? (RQ3) What are the benefits of implementing technological methods for measuring the results of the Eight Hop Tests?

2.2. Inclusion Criteria

The exercises and the sensors/equipment that had been used on the measurements of the Eight Hop Test results were based on the following inclusion criteria: (1) studies that measured different parameters related to the Eight Hop Test with sensors/equipment; (2) studies that used Eight Hop Test as the primary method; (3) studies that relate the Eight Hop Test with some diseases; (4) studies that clearly present results and population; (5) studies that were published between 2012 and 2021; (6) studies written in English.

2.3. Search Strategy

The Eight Hop Test consists of activities based on hopping used for rehabilitation purposes, and some studies only use a few of them. The search was performed with a Natural Language Processing (NLP)-based framework [36] in the following databases: IEEE Xplore; PMC; Pubmed Central; Springer; Association for Computing Machinery (ACM); Elsevier; and Multidisciplinary Digital Publishing Institute (MDPI). The keywords applied for this research were: "Eight hop test sensors"; "Eight hop test exercises"; and "hopping". Each study was filtered using the defined criteria presented in Section 2.2. The research was performed on 2 November 2021.

2.4. Extraction of Study Characteristics

There are specific parameters extracted from the studies. The information from the studies was divided and presented in Table 1 by the following terms: year of publication; population; the purpose of the studies; sensor/equipment; types of methods; and diseases. Some studies do not mention all the information required in Table 1, but they always gave precious data to help answer the questions in Section 2.1. The lack of information provided in the studies was compensated by contacting the respective authors of the analyzed studies. The information related to implementing the Eight Hop Test with technological equipment is limited, and this subject needs more research.

Table 1. Study analysis.

Paper	Year of Publication	Location	Population	Purpose of the Study	Sensors/Equiment	Diseases Studied
Baxter et al. [37]	2021	United States of America	10 healthy adults	The study analyses which Hop exercises makes a stronger and durable Achilles tendon	Reflective marker; 12 camera motion capture system; Force plate	Achilles tendon fracture
Ebert et al. [38]	2021	Australia	34 males and 16 females	The study focus on which Hop Tests syncs better with isokinetic knee extensor strength and the deficits after an injury in the anterior cruciate ligament reconstruction	Stopwatch; Accelerometer; Velcro strap; Isokinetic dynamometer	Anterior cruciate ligament
Ebert et al. [39]	2021	Australia	34 males and 16 females	This study aimed to find if the eight hop tests can identify limb asymmetry after anterior cruciate ligament reconstruction	Accelerometer; Stopwatch	Anterior cruciate ligament
Joschtel et al. [40]	2021	Australia	46 children	Comparison fundamental movement skill proficiency in children with bronchiectasis with measured Physical Ability	ActiGraph GT3X; Accelerometer	Bronchiectasis
Lawson et al. [41]	2021	United Kingdom	111 males and 108 females	Studying the fundamental movement skill levels in primary school children	SECA portable stadiometer; Nikon video camera;	Healthy
Biesert et al. [42]	2020	Sweden	24 patients	This study proposed an evaluation of a medial patellofemoral ligament using patient reported measures and functional testing	Goniometer	Patellofemoral ligament
Ergişi et al. [43]	2020	Turkey	15 males, 1 female and 8 healthy male controls	This study examines the functional outcomes, static-dynamic postural stability of patients with an associated gluteus medius treated injury	Wireless electromyography; Bipolar adhesive surface electrodes	Gluteus medius

Table 1. *Cont.*

Paper	Year of Publication	Location	Population	Purpose of the Study	Sensors/Equiment	Diseases Studied
Dingenen et al. [44]	2019	Belgium	16 non-injured participants & 28 anterior cruciate ligament reconstructed participants	The study had 2 purposes. The first one was to examine the test-retest reliability of single hop tests in the forward, medial and rotational direction. The second one was to detect limb asymmetries of the medial rotational hop tests in comparison with forward hop tests	Measuring tape	Anterior cruciate ligament
Sancho et al. [45]	2019	United Kingdom	15 male recreational runners	The study examines the best hopping exercises in runners with Achilles tendinopathy based on a self-reported pain	Metronome; Ultrasound scanner	Achilles tendinopathy
Owusu-Akyaw et al. [46]	2018	United States of America	8 male subjects	Comparison between anterior cruciate ligament deficient and intact knees in subjects due to patellofemoral joint and mechanics	Magnetic Resonance (MR) scanner	Anterior cruciate ligament; Patellofemoral joint osteoarthritis
Reuter et al. [47]	2017	Germany	8 professional athletes	This study proposed to show a relation of different dynamic postural control tests in healthy professional athletes and their measures	Measuring tape	Healthy
Lidstone et al. [48]	2016	United States of America	8 college-aged males	This study investigates changes in plantar flexor contractile component length, changes in plantar flexor muscle activity and tendon length and how it could reduce mechanical efficiency during exhaustive stretch-shortening cycle exercises	Wireless electrode; Ultrasound scanner; Athletic Tape; Retro-reflective markers; MX03 + NIR Cameras	Healthy
Wibawa et al. [49]	2016	Indonesia	10 healthy subjects	Analyses muscle activities like normal walking, one-legged forward and side jumping with a Musculoskeletal Modeling System	9 m long walkway; force plates; Vicon Motion System; Ten cameras; Reflective markers; Electrodes	Healthy
Furlong et al. [50]	2014	Ireland	7 healthy active adults	Analyses the center of pressure locations during two-legged hopping	Cameras; Force Plates; metronome; Retro-reflective markers	Healthy
Waldhelm et al. [51]	2012	United States of America	15 college-age males	This study determines which exercises related to strength, endurance, flexibility, motor control and function are more reliable in clinical measurements	Biodex System 3 pro; Biodex Balance System SD	Healthy

3. Results

3.1. Summary of the Search Process Results

As presented in Figure 1, the review has 9608 articles, 2770 of which are duplicates, and 6747 are marked as ineligible. For these reasons, the articles were removed correctly. The remaining 91 were filtered as well. In the filtering process (including the complete text evaluation), we found that 10 were Review/Survey, 61 were not related, one presented a quiz, two were not written in the English language, and two were not available. The remaining 15 papers were included in the qualitative synthesis and quantitative synthesis. In summary, we examined 15 scientific articles.

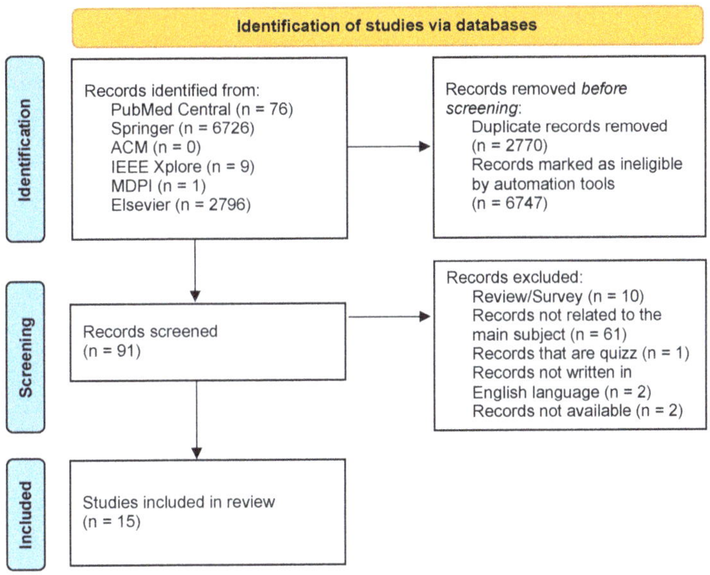

Figure 1. Flow diagram of identification and inclusion of papers.

Based on the results presented in Table 1, the analyzed studies were published between 2012 and 2021, reporting that the major part of the studies was published in 2021 (five studies), and, before that, only 10 studies are scarcely distributed between 2012 and 2020. By analyzing the location where the studies have been performed, a major part of the studies were performed in the United States of America (four studies), Australia (three studies), and United Kingdom (two studies), where the remaining studies are distributed by the globe. Regarding the sensors/equipment used, the most relevant used are cameras (five studies), reflective markers (four studies), force plates (four studies), electrodes (three studies), accelerometer (three studies), stopwatches (two studies), ultrasound scanner (two studies), metronome (two studies), and measuring tape (two studies), where the remaining sensors/equipment are used in only one study. Regarding the different types of diseases, only nine studies included people with specific diseases, including injury in an anterior cruciate ligament (four studies), Achilles tendon injury (two studies), bronchiectasis (one study), gluteus medius injury (one study), and Patellofemoral ligament injury (one study). In addition, all the studies used statistical and mathematical methods to prove the test's veracity.

3.2. Main Results, Benefits and Limitations of the Selected Studies

In Table 2, we summarize the main results, benefits and limitations of the selected studies relevant to measurement of the results of the Eight Hop Test.

Table 2. Study results and benefits.

Study	Results and Benefits	Limitations
Baxter et al. [37]	The results gave enough data to develop a method to measure exercise progression that helps increase the Achilles tendon's strength based on the magnitude duration and rate of tendon loading	Only eight healthy adults were included in the study, the population is limited. Test were made only on healthy people that contradicts the purpose of the study (rehabilitation)
Ebert et al. [38]	The results show that specific hop tests such as single medial and single countermovement jump correlated the most with isokinetic knee extensor when the more sophisticated testing equipment is missing. The hop measurements of study can inform the clinician of the possible existence of significant underlying quadriceps deficits are still present even after the operative rehabilitation period.	N/D
Ebert et al. [39]	The final results showed that single lateral hop, single medial hop, timed speedy hop, and single countermovement jump were the best physical exercises to demonstrate the functional limb asymmetry among the patients.	N/D
Joschtel et al. [40]	Results showed that children who suffer from bronchiectasis are more likely to not achieve age equivalency for locomotor skills and for object control skills. However, there were no differences for sedentary activities, light-intensity activities and games, waling, and running.	The children who met their age equivalency for fundamental skill had more time spent in daily physical activity than the other who did not.
Lawson et al. [41]	The results find that any child could master all the fundamental skills mentioned. However, the study gave precious knowledge, it was found that to improve essential skills in all children, the effort should focus on stability skills and force/power production.	N/D
Biesert et al. [42]	The results showed that patients had worse results than the control group in all tests, which led the study to conclude that patients with a medial patellofemoral ligament reconstruction do not regain normal knee function.	N/D
Ergişi et al. [43]	The study results showed that patients with an antegrade trochanteric are more likely to have a good balance but poor functional performance.	The results cannot be explained by the study, further studies are needed.
Dingenen et al. [44]	Results showed that medial and rotational hope tests have the probability of showing limb asymmetries in a person with anterior cruciate ligament reconstructed compared to the forward hope test.	There are no sensors included in the study, for the tests, a rolling tape was used. The uninjured were only tested twice and the ACL-reconstructed participants once.
Sancho et al. [45]	The results showed that education and training with pain-guided hopping has positive impacts in recreational runners with Achilles Tendinopathy	Parts of the results were justified by the participants. Three participants did not follow up the advised activities.

Table 2. *Cont.*

Study	Results and Benefits	Limitations
Owusu-Akyaw et al. [46]	The results found that the anterior cruciate ligament was associated with decreased patellar cartilage thickness by noticing that exercise would induce cartilage strain compared to the uninjured knees	The first limitation is the fact that only eight subjects were used for the study. Second, they were all male, that excludes a comparison with female subjects.
Reuter et al. [47]	Results demonstrated a correlation between the single-leg hop test and the star excursion balance test in terms of performance. These two exercises are the most efficient to determine overall postural control in athletes	The population of the study was formed by male athletes only, that excludes a possible comparison with female athletes.
Lidstone et al. [48]	The results found that the mechanical efficiency of hopping did not change and remained the same.	The population of the study was formed by male participants only, which excluded a possible comparison with female participants.
Wibawa et al. [49]	Results showed that the study can be used as baseline for scientific work, to get more reliable and robust musculoskeletal models, as it contributes to an uncertainty reduction.	The first limitation is that six subjects had to be excluded due to abnormal walking, marker trajectory errors, and errors in marker data. That leads to a small population. Second, the Modeling software can possibly miscalculate the knee net moment, absence of co-contraction, and simplified knee joint.
Furlong et al. [50]	The results showed that using retro-reflective markers in specific joints can determine the center of pressure during quiet standing and two-legged hopping at a particular frequency.	The results are limited to quiet standing and two-legged hopping in healthy adults. For that reason, more investigation is required to assure the accuracy of the method in walking and running or with clinical populations.
Waldhelm et al. [51]	Results showed that endurance tests are the most reliable for clinical measurements, followed by flexibility, strength, motor control, and functional.	The population of the study was formed by male participants only, which excludes a possible comparison with female participants.

3.3. *Qualitative Synthesis of the Most Relevant Works*

Baxter et al. [37] used four different types of sensors/equipment to implement several methods. The study used reflective markers, a 12-camera motion capture system, three-embedded force plates, and open-source musculoskeletal modeling software to perform the analysis. The study analyzes which rehabilitation exercises, such as single-leg hop, make a more robust and durable Achilles tendon. For that reason, eight young adults were put to the test. During the tests, enough data was collected to develop an exercise progression that helps increase the Achilles tendon's strength based on the magnitude duration and rate of tendon loading. In conclusion, peak Achilles tendon loads varied more than 12-fold, from 0.5 bodyweights during a seated hell raise to 7.3 bodyweights during a forward single leg hop.

In [38], the authors used motion and force sensors, such as accelerometer and dynamometer, to study which Hop Tests syncs better with isokinetic knee extensor strength, the deficits after an injury in the anterior cruciate ligament reconstruction, and which hop test correlates more with isokinetic knee extensor strength. Thirty-four males and sixteen females with surgery in the past 9–12 months (16–50 years of age) were assessed for the study. The hop tests presented in the study are single (SHD), triple (THD), and triple crossover (TCHD) hop for distance, six minute timed hop (6 MTH), single medial (MHD), and single lateral (LHD) hop for distance, single countermovement jump (SLCMJ) and timed speedy hop (SHT). The results show that specific hop tests such as single medial and single countermovement jump correlated the most with isokinetic knee extensor when the more sophisticated testing equipment is lacking.

Ebert et al. [39] used an accelerometer and a stopwatch to find if the eight hop tests can identify limb asymmetry after anterior cruciate ligament reconstruction. For this study, fifty patients (34 males and 16 females) were assessed 9–12 months following anterior cruciate ligament reconstruction. The test was made in both limbs in a randomized order. These included single (SHD), triple (THD), and triple crossover (TCHD) hop for distance, six minute timed hop (6 MTH), single medial (MHD). Single lateral (LHD) hop for distance, single countermovement jump (SLCMJ), and timed speedy hop (TSHT). The results showed that single lateral hop, single medial hop, timed speedy hop, and single countermovement jump was the best physical exercises to demonstrate the functional limb asymmetry among the patients. In conclusion, if the purpose is to detect lingering functional deficits, it is recommended to incorporate the previous-hop test mentioned.

The authors of [40] tested the fundamental skill and physical activity of children with bronchiectasis using ActiGraph GT3x and an accelerometer to show if the performance is affected by the disease. Forty-six children with bronchiectasis (mean age 7.5 $\pm$ 2.6 years, 63% Male) were recruited from the Queensland Children's Hospital, Brisbane. The children were measured by normal quotidian activities like sedentary, light-intensity, games, walking, running, and moderate-to-vigorous activities. The results showed that children with bronchiectasis tend to delay their fundamental skills development. Fewer than 5% of children demonstrated mastery in the run, gallop, hop, and leap, while fewer than 10% demonstrated the ability to perform the two-handed strike, overarm throw, and underarm throw. Only eight of the 46 children (17.4%) achieved their age equivalency for locomotor skills, while just four (8.7%) completed their object control skills. It is important to note that children in their age equivalency had significantly more time in daily physical activity during the tests.

The authors of [41] used SECA portable stadiometer, Nikon video camera, and Windows Media Player 2013 to examine primary school children's fundamental movement skill proficiency levels. It recruited 219 participants (111 boys, 108 girls) aged between 7–10 years from three schools in central England to perform eight skills related to locomotor, object control, and stability skills. The eight fundamental skills involved run, jump, hop, skip, catch, overarm throw, underarm throw, and stability. The results find that any child could master all the fundamental skills mentioned. The conclusion says that to improve essential skills in all children, the effort should focus on stability skills (improving coordination) and force/power production.

In [42], the authors proposed an evaluation of a medial patellofemoral ligament using patient-reported measures and functional testing. For this study, 24 patients with a medical record between 2008 and 2013 were examined with a control group of uninjured persons of the same age and gender. The evaluation had two phases. In the first part, questionnaires evaluated the knee function based on the Tegner score, the knee injury and osteoarthritis outcome score (KOOS), the Lysholm score, SF-36, and EQ-5D-3L. The second part was the functional performance that involved square jump, steps down test, and the single-leg hop for distance. The results were: patients 11.5 sets for the square jump versus control 21 sets, 11.5 sets for the step-down test versus control 22 sets, and 77 cm for the single-leg hop for distance versus control 126 cm. The patients showed worse results than the control group in all tests, which led the study to conclude that patients with a medial patellofemoral ligament reconstruction do not regain normal knee function.

In [43], the authors examine the functional outcomes, including static-dynamic postural stability of patients with an associated gluteus medius treated injury. For this study, 16 patients were chosen with the clinical record (treated with an antegrade trochanteric IMN) between January 2009 and July 2013 and eight healthy male controls. Some data was gathered before the physical activity, including muscle strength, static and dynamic postural stability, and fall risk. The measurements included the participation of imaging sensors electromyography (EMG). The study results showed that patients with an antegrade trochanteric IMN are more likely to have a good balance but poor functional performance. Still, more studies are needed to find the reason behind the results.

In [44], the author did not use any sensor to perform the test, which was evaluated by an examinator. The study considered 16 non-injured participants and 32 anterior cruciate ligament reconstruction participants. It was intended to examine the test-retest reliability of single-hop tests in the forward, medial and rotational direction and then detect limb asymmetries of the medial rotational hop tests, compared to forward hop tests made for the participants with a reconstructed anterior cruciate ligament. For the tests, they used some hop exercises like the single hop for distance (SH), triple hop for distance (TH), medial side triple hop for distance (MSTH), and 90° medial rotation hop for distance (MRH). The non-injured participants were tested twice, and the anterior cruciate ligament participants once. To prove the methods, it was calculated the intraclass correlation coefficients (ICCs), the standard errors of measurement (SEM), and the most negligible detectable differences (SDD). In the end, these exercises are reliable for rehabilitation purposes. Medial and rotational hope tests have the probability of showing limb asymmetries in a person with a reconstructed anterior cruciate ligament compared to the forward hope test.

In [45], the study's authors examine the best exercises (including hopping) in runners with Achilles tendinopathy based on self-reported pain. Fifteen male runners with Achilles tendinopathy were tested by loading the Achilles tendon with running, sprinting, hopping, jumping, and morning stiffness. The pain was measured before and after the workout with a numeric rating scale where zero means "no pain" and ten is "the worst possible pain". In total, 100% of the participants were recruited, 87% retention, and 93% followed-up. Exercise adherence was 70%. However, fidelity was 50%. Three participants suffered adverse events due to not following the advised exercises. Still, five participants were satisfied, and eight were very satisfied. In conclusion, the recommended education and training with pain-guided hopping positively impacts recreational runners with Achilles Tendinopathy.

Owusu-Akyaw et al. [46] used a magnetic resonance scanner to extract images from the knees before and after each subject performed a series of 60 single-legged hops. Then the images were converted into three-dimensional surface models of cartilage and bone to assess the cartilage characteristics in terms of thickness distribution. Eight male subjects with unilateral anterior cruciate ligament consented to participate in this study. The results found that the anterior cruciate ligament was associated with decreased patellar cartilage thickness by noticing that exercise would induce cartilage strain compared to the uninjured knees.

Reuter et al. [47] took the top German decathlon team, a group of eight professional athletes, to perform some high-end exercises to study postural control while exercising. Star Excursion Balance Test (SEBT), single hop for distance (SLH), crossover hop for distance (COH), triple hop for distance (TH) were used to perform the studies. The results demonstrated a correlation between the single-leg hop test and the star excursion balance test in terms of performance. These two exercises are the most efficient to determine overall postural control in athletes. For measurements, a measuring tape was used.

The author of [48] used Wireless electrodes, ultrasound probe, athletic tape, retro-reflective markers, and MX03 + NIR Cameras to perform the studies in eight college-aged males with no musculoskeletal injury. The study's primary purpose was to investigate changes in plantar flexor contractile component length, plantar flexor muscle activity, and tendon length and how it could reduce mechanical efficiency during exhaustive hopping exercises. For the study, eight college-aged males with no musculoskeletal injury, neuromuscular disease, or functional limitation in their legs participated in a complete hopping exercise to the absolute limit. In that time, the data was collected and analyzed. The results found that the mechanical efficiency of hopping did not change and remained the same.

Wibawa et al. [49] used a Gait Laboratory (Dept of Rehabilitation Medicine, UMCG, The Netherlands) to perform the tests, including imaging and force sensors to analyze muscle activities like normal walking, one-legged forward, and side jumping with a Musculoskeletal Modeling System. A nine-meter-long walkway, force plates, Vicon Motion System, cameras, reflective markers, and electrodes were used to measure and analyze ten healthy subjects (six males and four females) during the exercises. Each subject was evaluated, and then the values obtained by doing muscle activity were compared with the Musculoskeletal Modeling System. Some individuals were excluded during the study due to abnormal walking, marker trajectory error, and errors in market data. The other included three trials contributing enough data to conclude the investigation. The electrodes measured the right leg of the subjects. The correlation between sensors and the Eight Hop Test can be a game-changing move when the problem is finding how the injuries comport during the exercises. In conclusion, the study showed differences between the data and the model extracted from the Musculoskeletal Modeling System.

In [50], the authors used force plates, cameras, retro-reflective markers, and a digital metronome to analyze the center of pressure locations during two-legged hopping. By following the university ethics committee's approval, eight healthy and active adults (five females; three males) consented to participate in the study, doing at least ten jumps in a specific frequency. The attachment made the measurements of retro-reflective markers to a particular joint (metatarsophalangeal joint). The results showed that using retro-reflective markers in specific joints can determine the center of pressure during quiet standing and two-legged hopping at a particular frequency. Still, the results are limited to quiet standing and two-legged hopping in healthy adults. For that reason, more investigation is required to assure the accuracy of the method in walking and running or with clinical populations.

The authors of [51] used a population of fifteen college-age males, with right lower extremity dominance, to determine which exercises related to strength, endurance, flexibility, motor control, and function are more reliable in clinical measurements. It used Biodex System 3 pro and Biodex Balance System SD to collect some data before their studies began. For each test, there was a different exercise. Strength: eight isometric tests and a sit-up test. Endurance tests: the trunk flexor test, trunk extensor test, and bilateral side bridge tests. Flexibility tests: the sit-and-reach test and active range of the trunk and hip joint motions. Motor control: limb balance test and proprioception via passive reposition tests of the hips. Functional: squat test and single-leg hop test for time and distance. The results showed that endurance tests are the most reliable for clinical measurements, followed by flexibility, strength, motor control, and functionality.

3.4. Relationship between Studies, Sensors and Diseases

For this section, Table 3 represents a relation between sensors and diseases used to prove that studying different types of diseases directly related to the Eight Hop test is important. Still, the combination of the different methods and well-formed strategies has equal importance when monitoring people with various diseases using the Eight Hop test.

Table 3. Relation between diseases and sensors used.

Study	Sensors Categories						Diseases						
	Medical Sensors	**Motion Sensors**	**Time Counting Sensors**	**Imaging Sensors**	**Force Sensors**	**Support Equipment/Consumables**	**Anterior cruciate ligament**	**Healthy**	**Bronchiectasis**	**Achilles Tendon**	**Gluteus Medius**	**Patellofemoral ligament**
Baxter et al. [37]				X	X	X				X		
Ebert et al. [38]		X	X		X		X					
Ebert et al. [39]		X	X				X					
Joschtel et al. [40]		X							X			
Lawson et al. [41]				X		X		X				
Biesert et al. [42]						X						X
Ergişi et al. [43]	X										X	
Dingenen et al. [44]						X	X					
Sancho et al. [45]			X	X						X		
Owusu-Akyaw et al. [46]				X			X					X
Reuter et al. [47]						X		X				
Lidstone et al. [48]				X		X		X				
Wibawa et al. [49]				X	X	X		X				
Furlong et al. [50]			X	X		X		X				
Waldhelm et al. [51]					X			X				

4. Discussion

4.1. Summary of Relationship between Sensors and Diseases

The Eight Hop Test, specifically, was not present in any studies. However, the data collected from each study can help us understand which sensors are used in Hop Tests since the Eight Hop Test is a part of Hop Tests.

Some studies were based on problems related to a specific disease, and Figure 2 demonstrates various diseases that conducted the studies, where six studies were made with healthy people, four studies were performed with people with anterior cruciate ligament reconstruction, and the other studies included people with Achilles tendon (two studies) and patellofemoral ligament (two studies) as the leading injury. Gluteus Medius and Borchiectasis were mentioned in one publication each.

As presented in Table 4, the sensors available in the different studies were distributed in various categories, such as medical sensors, motion sensors, time counting equipment, imaging sensors, force sensors, and support equipment/consumables. Regarding the sensors used in the various studies, force sensors (four studies), such as force plates, dynamometer, Biodex System 3 pro, and Biodex Balance System SD had the most variety with four different sensors, followed by imaging sensors (seven studies), such as cameras, magnetic resonance scanner, and ultrasound scanner, and motion sensors (three studies), such as accelerometer, Vicon motion system, and Actigraph GT3x that had a variety of three different sensors.

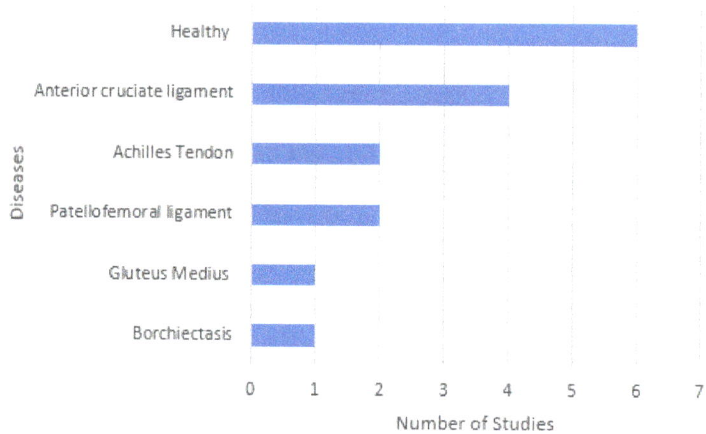

Figure 2. Distribution of the various diseases by the studies.

Table 4. Relation between sensors used and its categories.

Sensors Categories	Sensors
Medical sensors	Electromyography
Motion Sensors	Accelerometer Vicon Motion System Actigraph GT3x
Time counting equipment	Stopwatches Metronome
Imaging sensors	Cameras Magnetic Resonance scanner Ultrasound scanner
Force sensors	Force Plates Dynamometer Biodex System 3 pro Biodex Balance System SD
Support equipment/consumables	Reflective markers Electrodes 9-m long walkway Athletic tape Measuring tape Velcro strap Stadiometer Goniometer

4.2. *Relationship between Ages of Participants and Studies*

Regarding people's gender, the studies averaged 20.6 males and 12.8 females, including children and adults. Figure 3 presents the distribution of the ages of the different participants in the analyzed studies, where more than seven studies included individuals aged between 20 and 34 years old.

All studies used statistical and mathematical methods to study the results. The most used feature was the distance followed by time and the number of repetitions. Not all studies used sensors as the primary source of collecting data, where some of them were based on measuring distances and examining motion captures.

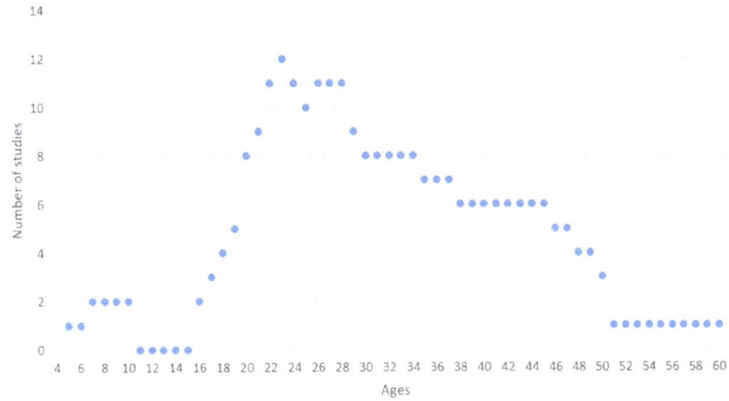

Figure 3. Relation of ages and number of studies.

4.3. Final Remarks

We can conclude that more studies are needed to develop a global solution for precise measurements. There is no proven evidence that the use of sensors in the Eight Hop test is essential, but according to the studies, it helps contribute to the fidelity and viability of the measurements.

After a deep analysis of the fifteen studies presented in this systematic review, we can find answers to our main questions. Regarding the RQ1, "Which devices can be used to perform studies in the Eight Hop Test?", we verified that the most common sensors used were the imaging sensors such as cameras, magnetic resonance, and ultrasound scanners. Furthermore, we have force sensors, time counting sensors, and motion sensors. More than half of the studies mention the need for support equipment/consumables, helping in the measuring, and completing the purpose of the sensors in these studies.

Concerning RQ2, "Which data are related to the different types of diseases diagnosed by the Eight Hop Test?", the analyzed studies show that different diseases require different sets of sensors and sensor data. For the same disease, for example Anterior cruciate ligament, different studies use different sensors. Additional research is needed to find out which sensor gives the best results, since no comparative analysis could be performed due to the varying experimental setups of the analyzed studies. Table 3 shows the relation between the sensors and the various diseases.

Finally, regarding RQ3, "What are the benefits of implementing technological methods for the measurements of the results of the Eight Hop Tests?", we verified that in addition to the limitations presented in Table 2 the studies showed some benefits related to rehabilitation and empowerment on the clinical information.

More studies are needed on this topic, but one thing is sure: we can use sensors to measure possible results prevenient from the Eight Hop Test.

5. Conclusions

In this review, a total of 15 studies were selected based on the inclusion criteria and thoroughly analyzed. The review identified which sensors are used in the Eight Hop Test, which are the most used sensors, the relevance of sensors in measurements, which diseases are related to the Eight Hop Test, and which methods can be used to perform the Eight Hop Test. It is important to mention that there is a lack of studies to develop a method for analyzing the Eight Hop Test with sensors. However, the sensors increase the viability of the measurements and help clinical teams to perform better diagnostics in health.

As future work, a mobile application will be developed to create a new method for the commodity measurement of the results of the Eight Hop Test that will be integrated with other ongoing studies related to the construction of a Personal Digital Life Coach.

Author Contributions: Conceptualization, L.P., N.M.G. and I.M.P.; methodology, L.P.; validation, N.M.G. and I.M.P.; formal analysis, L.P.; investigation, L.P.; data curation, L.P.; writing—original draft preparation, L.P., N.M.G. and I.M.P.; writing—review and editing, L.P., N.M.G., E.Z., I.C., P.L., V.T., C.A. and I.M.P.; supervision, N.M.G. and I.M.P.; funding acquisition, C.A., N.M.G. and I.M.P. All authors have read and agreed to the published version of the manuscript.

Funding: This work is funded by FCT/MEC through national funds and, when applicable, co-funded by the FEDER-PT2020 partnership agreement under the project UIDB/50008/2020. This work is also funded by National Funds through the FCT—Foundation for Science and Technology, I.P., within the scope of the project UIDB/00742/2020.

Institutional Review Board Statement: Not applicable.

Informed Consent Statement: Not applicable.

Data Availability Statement: Not applicable.

Acknowledgments: This article is based upon work from COST Action IC1303-AAPELE—Architectures, Algorithms, and Protocols for Enhanced Living Environments and COST Action CA16226–SHELD-ON—Indoor living space improvement: Smart Habitat for the Elderly, supported by COST (European Cooperation in Science and Technology). COST is a funding agency for research and innovation networks. Our Actions help connect research initiatives across Europe and enable scientists to grow their ideas by sharing them with their peers. It boosts their research, career, and innovation. More information is available at www.cost.eu. Furthermore, we would like to thank the Politécnico de Viseu for their support.

Conflicts of Interest: The authors declare no conflict of interest.

References

1. Shigaki, L.; Rabello, L.M.; Camargo, M.Z.; da Costa Santos , V.B.; de Oliveira Gil , A.W.; de Oliveira, M.R.; da Silva Junior, R.A.; de Souza Guerino Macedo , C. Comparative analysis of one-foot balance in rhythmic gymnastics athletes. *Rev. Bras. Med. Esporte* **2013**, *19*, 104–107. [CrossRef]
2. Malik, S.; Anand, P.; Bhati, P.; Hussain, M.E. Effects of dry cupping therapy on pain, dynamic balance and functional performance in young female with recreational runners chronic plantar fasciitis. *Sport. Orthop. Traumatol.* **2022**, *38*, 159–170. [CrossRef]
3. Popovski, G.; Ponciano, V.; Marques, G.; Pires, I.M.; Zdravevski, E.; Garcia, N.M. Personal Digital Life Coach for Physical Therapy. In Proceedings of the 2020 IEEE International Conference on Big Data, Atlanta, GA, USA, 10–13 December 2020; pp. 3797–3802.
4. Aminian, K.; Najafi, B. Capturing human motion using body-fixed sensors: Outdoor measurement and clinical applications. *Comput. Animat. Virtual Worlds* **2004**, *15*, 79–94. [CrossRef]
5. Aroganam, G.; Manivannan, N.; Harrison, D. Review on wearable technology sensors used in consumer sport applications. *Sensors* **2019**, *19*, 1983. [CrossRef]
6. Ponciano, V.; Pires, I.M.; Ribeiro, F.R.; Villasana, M.V.; Crisóstomo, R.; Canavarro Teixeira, M.; Zdravevski, E. Mobile Computing Technologies for Health and Mobility Assessment: Research Design and Results of the Timed Up and Go Test in Older Adults. *Sensors* **2020**, *20*, 3481. [CrossRef]
7. Van Hees, V.T.; Fang, Z.; Langford, J.; Assah, F.; Mohammad, A.; da Silva, I.C.; Trenell, M.I.; White, T.; Wareham, N.J.; Brage, S. Autocalibration of accelerometer data for free-living physical activity assessment using local gravity and temperature: An evaluation on four continents. *J. Appl. Physiol.* **2014**, *117*, 738–744. [CrossRef]
8. Kärkkäinen, M.; Rikkonen, T.; Kröger, H.; Sirola, J.; Tuppurainen, M.; Salovaara, K.; Arokoski, J.; Jurvelin, J.; Honkanen, R.; Alhava, E. Physical tests for patient selection for bone mineral density measurements in postmenopausal women. *Bone* **2009**, *44*, 660–665. [CrossRef]
9. Madsen, L.P.; Hall, E.A.; Docherty, C.L. Assessing Outcomes in People With Chronic Ankle Instability: The Ability of Functional Performance Tests to Measure Deficits in Physical Function and Perceived Instability. *J. Orthop. Sports Phys. Ther.* **2018**, *48*, 372–380. [CrossRef]
10. Hardman, R.; Begg, S.; Spelten, E. Healthcare professionals' perspective on treatment burden and patient capacity in low-income rural populations: Challenges and opportunities. *BMC Fam. Pract.* **2021**, *22*, 1–15. [CrossRef]
11. Theobald, S.; Brandes, N.; Gyapong, M.; El-Saharty, S.; Proctor, E.; Diaz, T.; Wanji, S.; Elloker, S.; Raven, J.; Elsey, H. Implementation research: New imperatives and opportunities in global health. *Lancet* **2018**, *392*, 2214–2228. [CrossRef]
12. Badawy, S.M.; Radovic, A. Digital approaches to remote pediatric health care delivery during the COVID-19 pandemic: Existing evidence and a call for further research. *JMIR Pediatr. Parent.* **2020**, *3*, e20049. [CrossRef] [PubMed]
13. Voicu, R.-A.; Dobre, C.; Bajenaru, L.; Ciobanu, R.-I. Human physical activity recognition using smartphone sensors. *Sensors* **2019**, *19*, 458. [CrossRef]

14. Pires, I.M.S. Aplicação Móvel e Plataforma Web para Suporte à Estimação de Gasto Energético em Actividade Física. 2012. Available online: https://www.semanticscholar.org/paper/Aplica%C3%A7%C3%A3o-m%C3%B3vel-e-plataforma-Web-para-suporte-%C3%A0-do-Engenharia/59009c3b133519f852985effe98a5fe4a5131952 (accessed on 5 May 2022).
15. Pires, I.M.; Hussain, F.; Garcia, N.M.M.; Lameski, P.; Zdravevski, E. Homogeneous Data Normalization and Deep Learning: A Case Study in Human Activity Classification. *Future Internet* **2020**, *12*, 194. [CrossRef]
16. Pires, I.; Felizardo, V.; Pombo, N.; Garcia, N.M. Limitations of energy expenditure calculation based on a mobile phone accelerometer. In Proceedings of the 2017 International Conference on High Performance Computing & Simulation (HPCS), Genoa, Italy, 17–21 July 2017; pp. 124–127.
17. Pires, I.M.; Garcia, N.M.; Pombo, N.; Flórez-Revuelta, F. Limitations of the Use of Mobile Devices and Smart Environments for the Monitoring of Ageing People. In Proceedings of the 4th International Conference on Information and Communication Technologies for Ageing Well and e-Health HSP, Madeira, Portugal, 22–23 March 2018.
18. Ahmadian, N.; Nazarahari, M.; Whittaker, J.L.; Rouhani, H. Quantification of Triple Single-Leg Hop Test Temporospatial Parameters: A Validated Method Using Body-Worn Sensors for Functional Evaluation after Knee Injury. *Sensors* **2020**, *20*, 3464. [CrossRef] [PubMed]
19. Pires, I.M.; Andrade, M.; Garcia, N.M.; Crisóstomo, R.; Florez-Revuelta, F. Measurement of heel-rise test results using a mobile device. In Proceedings of the Doctoral Consortium—DCPhyCS, (PhyCS 2015), Angers, France, 1–13 February 2015; SciTePress (INSTICC): Setúbal, Portugal, 2015; pp. 9–18.
20. Pires, I.M.; Ponciano, V.; Garcia, N.M.; Zdravevski, E. Analysis of the Results of Heel-Rise Test with Sensors: A Systematic Review. *Electronics* **2020**, *9*, 1154. [CrossRef]
21. Ponciano, V.; Pires, I.M.; Ribeiro, F.R.; Marques, G.; Garcia, N.M.; Pombo, N.; Spinsante, S.; Zdravevski, E. Is The Timed-Up and Go Test Feasible in Mobile Devices? A Systematic Review. *Electronics* **2020**, *9*, 528. [CrossRef]
22. Pires, I.M.; Lopes, E.; Villasana, M.V.; Garcia, N.M.; Zdravevski, E.; Ponciano, V. A Brief Review on the Sensor Measurement Solutions for the Ten-Meter Walk Test. *Computers* **2021**, *10*, 49. [CrossRef]
23. Pires, I.M.; Denysyuk, H.V.; Villasana, M.V.; Sá, J.; Marques, D.L.; Morgado, J.F.; Albuquerque, C.; Zdravevski, E. Development Technologies for the Monitoring of Six-Minute Walk Test: A Systematic Review. *Sensors* **2022**, *22*, 581. [CrossRef]
24. Pires, I.M.; Garcia, N.M.; Zdravevski, E. Measurement of Results of Functional Reach Test with Sensors: A Systematic Review. *Electronics* **2020**, *9*, 1078. [CrossRef]
25. Adusumilli, G.; Joseph, S.E.; Samaan, M.A.; Schultz, B.; Popovic, T.; Souza, R.B.; Majumdar, S. iPhone Sensors in Tracking Outcome Variables of the 30-Second Chair Stand Test and Stair Climb Test to Evaluate Disability: Cross-Sectional Pilot Study. *JMIR MHealth UHealth* **2017**, *5*, e166. [CrossRef]
26. Pires, I.M.; Marques, D.; Pombo, N.; Garcia, N.M.; Marques, M.C.; Flórez-Revuelta, F. Measurement of the Reaction Time in the 30-S Chair Stand Test using the Accelerometer Sensor Available in off-the-Shelf Mobile Devices. In Proceedings of the 4th International Conference on Information and Communication Technologies for Ageing Well and e-Health HSP, Madeira, Portugal, 22–23 March 2018.
27. Marques, D.L.; Neiva, H.P.; Pires, I.M.; Zdravevski, E.; Mihajlov, M.; Garcia, N.M.; Ruiz-Cárdenas, J.D.; Marinho, D.A.; Marques, M.C. An Experimental Study on the Validity and Reliability of a Smartphone Application to Acquire Temporal Variables during the Single Sit-to-Stand Test with Older Adults. *Sensors* **2021**, *21*, 2050. [CrossRef] [PubMed]
28. Autexier, S.; Goleva, R.; Garcia, N.M.; Stainov, R.; Ganchev, I.; Mavromoustakis, C.X.; Dobre, C.; Chorbev, I.; Trajkovik, V.; Zdravevski, E. End-users' AAL and ELE service scenarios in smart personal environments. In *Enhanced Living Environments: From Models to Technologies*; Goleva, R.I., Ganchev, I., Dobre, C., Garcia, N., Valderrama, C., Eds.; Institution of Engineering and Technology: London, UK, 2017; pp. 101–131. ISBN 978-1-78561-211-4.
29. Sousa, P.S.; Sabugueiro, D.; Felizardo, V.; Couto, R.; Pires, I.; Garcia, N.M. mHealth Sensors and Applications for Personal Aid. In *Mobile Health*; Adibi, S., Ed.; Springer Series in Bio-/Neuroinformatics; Springer: Cham, Switzerland, 2015; Volume 5, pp. 265–281. ISBN 978-3-319-12816-0.
30. Jeanfavre, M.; Humphrey, A.; Klein, M. CPQ Orthopaedics (2021) 5: 3 Systematic Review. *CPQ Orthop.* **2021**, *5*, 1–105.
31. Bendixen, S. *The Utilization of Lower Quarter Testing to Assess Return to Sport Readiness in an Athlete following Achilles Tendon Repair: A Case Report*; University of Iowa: Iowa City, IA, USA, 2018.
32. Xergia, S.A.; Pappas, E.; Georgoulis, A.D. Association of the Single-Limb Hop Test With Isokinetic, Kinematic, and Kinetic Asymmetries in Patients After Anterior Cruciate Ligament Reconstruction. *Sports Health Multidiscip. Approach* **2015**, *7*, 217–223. [CrossRef] [PubMed]
33. Logerstedt, D.; Grindem, H.; Lynch, A.; Eitzen, I.; Engebretsen, L.; Risberg, M.A.; Axe, M.J.; Snyder-Mackler, L. Single-Legged Hop Tests as Predictors of Self-Reported Knee Function After Anterior Cruciate Ligament Reconstruction: The Delaware-Oslo ACL Cohort Study. *Am. J. Sports Med.* **2012**, *40*, 2348–2356. [CrossRef] [PubMed]
34. Wren, T.A.L.; Mueske, N.M.; Brophy, C.H.; Pace, J.L.; Katzel, M.J.; Edison, B.R.; Vandenberg, C.D.; Zaslow, T.L. Hop Distance Symmetry Does Not Indicate Normal Landing Biomechanics in Adolescent Athletes With Recent Anterior Cruciate Ligament Reconstruction. *J. Orthop. Sports Phys. Ther.* **2018**, *48*, 622–629. [CrossRef] [PubMed]
35. Grindem, H.; Snyder-Mackler, L.; Moksnes, H.; Engebretsen, L.; Risberg, M.A. Simple decision rules can reduce reinjury risk by 84% after ACL reconstruction: The Delaware-Oslo ACL cohort study. *Br. J. Sports Med.* **2016**, *50*, 804–808. [CrossRef] [PubMed]

36. Zdravevski, E.; Lameski, P.; Trajkovik, V.; Chorbev, I.; Goleva, R.; Pombo, N.; Garcia, N.M. Automation in Systematic, Scoping and Rapid Reviews by an NLP Toolkit: A Case Study in Enhanced Living Environments. In *Enhanced Living Environments*; Ganchev, I., Garcia, N.M., Dobre, C., Mavromoustakis, C.X., Goleva, R., Eds.; Lecture Notes in Computer Science; Springer International Publishing: Cham, Switzeland, 2019; Volume 11369, pp. 1–18. ISBN 978-3-030-10751-2.
37. Baxter, J.R.; Corrigan, P.; Hullfish, T.J.; O'Rourke, P.; Silbernagel, K.G. Exercise Progression to Incrementally Load the Achilles Tendon. *Med. Sci. Sports Exerc.* **2021**, *53*, 124–130. [CrossRef]
38. Ebert, J.R.; Edwards, P.; Preez, L.D.; Furzer, B.; Joss, B. Knee extensor strength, hop performance, patient-reported outcome and inter-test correlation in patients 9–12 months after anterior cruciate ligament reconstruction. *Knee* **2021**, *30*, 176–184. [CrossRef]
39. Ebert, J.R.; Du Preez, L.; Furzer, B.; Edwards, P.; Joss, B. Which Hop Tests Can Best Identify Functional Limb Asymmetry in Patients 9-12 Months After Anterior Cruciate Ligament Reconstruction Employing a Hamstrings Tendon Autograft? *Int. J. Sports Phys. Ther.* **2021**, *16*, 393. [CrossRef]
40. Joschtel, B.; Gomersall, S.R.; Tweedy, S.; Petsky, H.; Chang, A.B.; Trost, S.G. Fundamental movement skill proficiency and objectively measured physical activity in children with bronchiectasis: A cross-sectional study. *BMC Pulm. Med.* **2021**, *21*, 269. [CrossRef]
41. Lawson, C.; Eyre, E.L.J.; Tallis, J.; Duncan, M.J. Fundamental Movement Skill Proficiency Among British Primary School Children: Analysis at a Behavioral Component Level. *Percept. Mot. Skills* **2021**, *128*, 625–648. [CrossRef] [PubMed]
42. Biesert, M.; Johansson, A.; Kostogiannis, I.; Roberts, D. Self-reported and performance-based outcomes following medial patellofemoral ligament reconstruction indicate successful improvements in knee stability after surgery despite remaining limitations in knee function. *Knee Surg. Sports Traumatol. Arthrosc.* **2020**, *28*, 934–940. [CrossRef]
43. Ergişi, Y.; Kafa, N.; Tokgöz, M.A.; Demir, E.; Kanık, Z.H.; Sezgin, E.A.; Ataoğlu, M.B. Is gluteus medius injured in patients treated with a trochanter tip entry intramedullary nail? Clinical, electrophysiological and functional outcomes. *Jt. Dis. Relat. Surg.* **2020**, *31*, 312–319. [CrossRef] [PubMed]
44. Dingenen, B.; Truijen, J.; Bellemans, J.; Gokeler, A. Test–retest reliability and discriminative ability of forward, medial and rotational single-leg hop tests. *Knee* **2019**, *26*, 978–987. [CrossRef] [PubMed]
45. Sancho, I.; Morrissey, D.; Willy, R.W.; Barton, C.; Malliaras, P. Education and exercise supplemented by a pain-guided hopping intervention for male recreational runners with midportion Achilles tendinopathy: A single cohort feasibility study. *Phys. Ther. Sport* **2019**, *40*, 107–116. [CrossRef]
46. Owusu-Akyaw, K.A.; Heckelman, L.N.; Cutcliffe, H.C.; Sutter, E.G.; Englander, Z.A.; Spritzer, C.E.; Garrett, W.E.; DeFrate, L.E. A comparison of patellofemoral cartilage morphology and deformation in anterior cruciate ligament deficient versus uninjured knees. *J. Biomech.* **2018**, *67*, 78–83. [CrossRef]
47. Reuter, S.; Forkel, P.; Imhoff, A.B.; Beitzel, K. Postural control in elite decathlon athletes: Are various modes of dynamic assessment needed? *J. Sports Med. Phys. Fit.* **2017**, *57*, 936–941. [CrossRef]
48. Lidstone, D.E.; van Werkhoven, H.; Stewart, J.A.; Gurchiek, R.; Burris, M.; Rice, P.; Feimster, G.; McBride, J.M. Medial gastrocnemius muscle-tendon interaction and architecture change during exhaustive hopping exercise. *J. Electromyogr. Kinesiol.* **2016**, *30*, 89–97. [CrossRef]
49. Wibawa, A.D.; Verdonschot, N.; Halbertsma, J.P.K.; Burgerhof, J.G.M.; Diercks, R.L.; Verkerke, G.J. Musculoskeletal modeling of human lower limb during normal walking, one-legged forward hopping and side jumping: Comparison of measured EMG and predicted muscle activity patterns. *J. Biomech.* **2016**, *49*, 3660–3666. [CrossRef]
50. Furlong, L.-A.M.; Harrison, A.J. A motion analysis marker-based method of determining centre of pressure during two-legged hopping. *J. Biomech.* **2014**, *47*, 1904–1908. [CrossRef]
51. Waldhelm, A.; Li, L. Endurance tests are the most reliable core stability related measurements. *J. Sport Health Sci.* **2012**, *1*, 121–128. [CrossRef]

MDPI
St. Alban-Anlage 66
4052 Basel
Switzerland
Tel. +41 61 683 77 34
Fax +41 61 302 89 18
www.mdpi.com

Sensors Editorial Office
E-mail: sensors@mdpi.com
www.mdpi.com/journal/sensors

www.ingramcontent.com/pod-product-compliance
Lightning Source LLC
Chambersburg PA
CBHW040826250926
51850CB00007BA/1272
9783036570266